Population, Migration and Settlement in Australia and the Asia-Pacific

The chapters in this book reflect on the work of seminal Australian geographer, the late Professor Graeme Hugo. Graeme Hugo was widely respected because of his impressive contributions to scholarship and policy in the fields of migration, population and development, which spanned several decades. This collection of works contains contributions from authors whose own research has been influenced by Hugo and includes numerous authors who worked closely with Hugo throughout his career. The collection provides an opportunity to reflect on Hugo's legacy, and also to foreground contemporary scholarship in his key areas of research focus. The chapters are organised into two thematic threads. Part 1 contains works relating to 'Population, Migration and Settlement in Australia', whereas Part 2 focuses on 'Labour and Environmental Migration in the Asia-Pacific'. Together, these two thematic threads provide broad coverage of Graeme Hugo's key areas of research focus. The chapters also serve as a reminder of Hugo's steadfast concern with producing careful scholarship for the public good, and seek to prompt continued work in this vein.

The chapters were originally published in special issues of *Australian Geographer*.

Natascha Klocker is a human geographer whose research focuses on equity, discrimination and inclusion/exclusion – most often explored through the lens of migration and ethnic diversity. Her current research explores the environmental and agricultural knowledges, capacities and contributions of migrants and refugees.

Olivia Dun is a human geographer with an environmental science, migration studies and international development background. Her research examines the links between environmental change, agriculture and migration.

Population, Migration and Settlement in Australia and the Asia-Pacific

In Memory of Graeme Hugo

Edited by
Natascha Klocker and Olivia Dun

LONDON AND NEW YORK

First published 2018 by Routledge

2 Park Square, Milton Park, Abingdon, Oxfordshire OX14 4RN
52 Vanderbilt Avenue, New York, NY 10017

Routledge is an imprint of the Taylor & Francis Group, an informa business

First issued in paperback 2019

British Library Cataloguing in Publication Data
A catalogue record for this book is available from the British Library

ISBN13: 978-1-138-55128-2 (hbk)
ISBN13: 978-0-367-89194-7 (pbk)

Typeset in Minion Pro
by diacriTech, Chennai

Publisher's Note
The publisher accepts responsibility for any inconsistencies that may have arisen during the conversion of this book from journal articles to book chapters, namely the possible inclusion of journal terminology.

Disclaimer
Every effort has been made to contact copyright holders for their permission to reprint material in this book. The publishers would be grateful to hear from any copyright holder who is not here acknowledged and will undertake to rectify any errors or omissions in future editions of this book.

Contents

CONTENTS

CONTENTS

Citation Information

The chapters in this book were originally published in *Australian Geographer*. When citing this material, please use the original volume, issue, date of publication and page numbering for each article, as follows:

Chapter 1
Editorial
Natascha Klocker and Olivia Dun
Australian Geographer, volume 47, issue 4 (December 2016) pp. 373–376

Chapter 2
Remembering Graeme Hugo, Population Geographer
Ruth Fincher
Australian Geographer, volume 47, issue 4 (December 2016) pp. 377–382

Chapter 3
Migration moderate, 'Master Weaver' and inspirational team leader: reflecting on the lasting legacy of Graeme Hugo in three spheres of migration policy
Marie McAuliffe
Australian Geographer, volume 47, issue 4 (December 2016) pp. 383–389

Chapter 4
Rethinking Australian migration
Stephen Castles
Australian Geographer, volume 47, issue 4 (December 2016) pp. 391–398

Chapter 5
Ageing research in Australia: reflecting on Graeme Hugo's four decades of contribution
Debbie Faulkner, Helen Barrie Feist and Judith Lewis
Australian Geographer, volume 47, issue 4 (December 2016) pp. 399–415

Chapter 6
Social Applications of Geographical Information Systems: technical tools for social innovation
Danielle Taylor and Jarrod Lange
Australian Geographer, volume 47, issue 4 (December 2016) pp. 417–433

For any permission-related enquiries please visit:
http://www.tandfonline.com/page/help/permissions

Notes on Contributors

Karen Agutter is a Researcher at the Department of History, School of Humanities, University of Adelaide, Australia.

Rachel A. Ankeny is a Professor at the Department of History, School of Humanities, University of Adelaide, Australia.

Neil Argent is a Professor at the Division of Geography and Planning, University of New England, Australia.

Emma Baker is an Associate Professor at the School of Architecture and Built Environment, University of Adelaide, Australia.

Charlotte Bedford is an Independent Research Consultant based in London, UK.

Richard Bedford is Professor of Population Geography at the University of Waikato and Auckland University of Technology, New Zealand.

Smita Bhutani is a Professor at the Centre of Advanced Study in Geography, Panjab University, India.

Fidelma Breen is a PhD candidate at the Discipline of Geography, Environment and Population, School of Social Sciences, University of Adelaide, Australia.

Stephen Castles is a Professor at the Department of Sociology and Social Policy, University of Sydney, Australia.

Neil T. Coffee is based at the Centre for Population Health Research, School of Health Science, University of South Australia, Australia.

John Connell is a Professor at the School of Geosciences, University of Sydney, Australia.

Olivia Dun is a Postdoctoral Fellow at the School of Geography, University of Melbourne, Australia.

Amani Elnasri is based at the School of Economics, University of New South Wales, Australia.

Carol Farbotko is a human geographer at the School of Land and Food—Geography and Environmental Studies, University of Tasmania, Australia.

Debbie Faulkner is based at the Centre for Housing, Urban and Regional Planning, School of Social Sciences, University of Adelaide, Australia.

Helen Barrie Feist is a University Research Fellow at the Hugo Centre for Migration and Population Research, University of Adelaide, Australia.

Ruth Fincher is a Professor at the School of Geography, University of Melbourne, Australia.

Kate Golebiowska is Senior Research Fellow at Northern Institute, Charles Darwin University, Australia.

Trevor Griffin is based at the Hugo Centre for Migration and Population Research, University of Adelaide, Australia.

Amandeep Kaur is based at the Centre of Advanced Study in Geography, Panjab University, India.

Natascha Klocker is a human geographer in the School of Geography and Sustainable Communities, University of Wollongong, Australia.

Jarrod Lange is Senior Research Consultant at the Hugo Centre for Migration and Population Research, University of Adelaide, Australia.

Judith Lewis is a Research Associate at the Discipline of Geography, Environment and Population, School of Social Sciences, University of Adelaide, Australia.

Nancy Lutkehaus is Professor of Anthropology and Political Science at the Department of Anthropology, University of Southern California, USA.

Jane McAdam is Scientia Professor of Law and Director of the Andrew & Renata Kaldor Centre for International Refugee Law at the University of New South Wales, Australia.

Marie McAuliffe is a PhD candidate at the School of Demography, College of Arts and Social Sciences, Australian National University, Australia.

Richard McGrath is a Lecturer in Health Sciences at the School of Health Sciences, University of South Australia, Australia.

Karen E. McNamara is a Senior Lecturer at the School of Earth and Environmental Sciences, The University of Queensland, Australia.

Lisel O'Dwyer is based at the School of Social and Policy Studies, Flinders University, Australia.

Peter Smailes is a Visiting Research Fellow in a research-only affiliate capacity at the Hugo Centre for Migration and Population Research, University of Adelaide, Australia.

Yan Tan is an Associate Professor at the Department of Geography, Environment and Population, University of Adelaide, Australia.

Danielle Taylor is a Research Associate at the Hugo Centre for Migration and Population Research, University of Adelaide, Australia.

Janet Wall is an Administrator at the School of Social Sciences, Faculty of Arts, University of Adelaide, Australia.

Margaret Walton-Roberts is an Associate Dean at the Department of Geography and Environmental Studies and the Balsillie School of International Affairs, Wilfrid Laurier University, Canada.

Glenn Withers is a Professor at the Research School of Economics, Australian National University, Australia.

Janette Young is Program Director for the Bachelor of Health Sciences at the School of Health Sciences, University of South Australia, Australia.

Margaret Young is based at the School of Social Sciences, Faculty of Arts, University of Adelaide, Australia.

In May 2016 refugees were, once again, brought to the forefront of an Australian federal election campaign. This has been a regular occurrence since 2001, when Prime Minister John Howard's allegations that asylum seekers had thrown their children overboard[1] helped to justify his government's increasingly restrictive border protection measures. In 2016, Immigration Minister Peter Dutton raised a different set of concerns about humanitarian arrivals to Australia in response to the Australian Greens' proposal that the annual refugee intake be increased to 50 000 people. Dutton's assertions were two-pronged. First, many refugees are not 'numerate or literate in their own language, let alone English', and would 'languish in unemployment queues and on Medicare … there's no sense in sugar-coating that, that's the scenario' (Bourke 2016). Second, Dutton warned, 'These people would be taking Australian jobs, there's no question about that' (Bourke 2016). The fact that the two prongs of Dutton's argument (unemployment and job stealing) directly contradicted each other is perhaps neither here nor there—both sought to emphasise (in the minister's own words) the 'huge cost' of resettling refugees (Bourke 2016). As the media seized on these comments, *The Guardian*'s Ben Doherty and Helen Davidson decided to run a fact check. They brought a voice of reason into the debate—it was the voice of geographer Graeme Hugo AO.[2]

Doherty and Davidson (2016) found the arguments needed to refute Dutton in a 2011 report authored by Graeme and his colleagues.[3] This report, *Economic, Social and Civic Contributions of First and Second Generation Humanitarian Entrants* (Hugo 2011), had been commissioned by the then Department of Immigration and Citizenship (DIAC). The over-arching message of the report, as summarised by the Department itself, provides a stark counterpoint to the typically negative assertions that have come to frame refugees and asylum seekers in public debate:

> The research found the overwhelming picture, when one takes the longer term perspective of changes over the working lifetime of Humanitarian Program entrants and their children, is one of *considerable achievement and contribution*. (Policy Innovation, Research and Evaluation Unit, DIAC 2011; emphasis added).

The report carefully demonstrated—using solid empirical evidence—the long-term convergence of refugee employment rates with those of Australian-born persons; high levels of labour force engagement and tertiary education amongst second-generation humanitarian arrivals; the contributions of refugees to regional development; their entrepreneurialism, civic engagement, volunteering and their connections with the communities in which they have settled (Hugo 2011).

Our purpose here is not to enter into a detailed discussion of that particular report, or to reflect on the latest political manoeuvres surrounding asylum seekers and refugees in Australia. Rather, we open with this example because it raises an important point about the legacy of Graeme Hugo, who passed away on 20 January 2015, only a few months after being diagnosed with cancer, at the age of 68. The example underscores the ongoing influence of Graeme's extraordinary and vast body of work. Hugo's scholarship was careful and evidence based; his voice offered a measured, reliable and insightful contribution to Australian population and immigration debates over many decades. Such contributions are recounted ably and poignantly in the three Thinking Space pieces that begin this special issue (by Ruth Fincher, Marie McAuliffe and Stephen Castles), so too in an obituary prepared by John Connell (2015) in this journal. A detailed overview of Hugo's publications has also been made available on the website of the Australian Population and Migration Research Centre.[4]

Graeme's legacy raises a valuable provocation for Australian geographers regarding the important role of the 'geographer' in public debate. An Editorial published in *The Guardian (UK)* in 2015 described geography as a 'subject for our times'—because 'it is inherently multidisciplinary in a world that increasingly values [and we would add *needs*] people who have the skills … to work across the physical and social sciences'. Hugo's research exemplified and made the most of the diverse and valuable skill set of the geographer. Moreover, he brought those skills to bear for the public good. There has perhaps never been a time when the skills *and voices* of geographers are more urgently needed —in response to a whole host of contemporary environmental and social challenges. Hugo's passing offers a prompt for all of us to reflect on the role that we would like to see academic geography play in public debate. How can we, as Australian geographers, enhance our discipline's position—such that politicians, journalists and other opinion-makers turn to geographers for evidence, insights and advice at times when measured and informed interjections are urgently needed?

When we issued our call for papers for a special issue of *Australian Geographer* in memory of Graeme Hugo we received an overwhelming response. There were many who wanted to pay tribute. In the interests of inclusivity (one of Graeme's often-expressed personal values), there are two issues of collected papers, organised into thematic threads. This first special issue, 'Population, Migration and Settlement in Australia', will be accompanied by a second one in early 2017, 'Labour and Environmental Migration in the Asia-Pacific'. Together, these two issues of *Australian Geographer* provide broad coverage of many of Graeme Hugo's key areas of research focus.

This first special issue commences with three Thinking Spaces: Ruth Fincher remembers Graeme Hugo 'the population geographer', Marie McAuliffe reflects on Hugo's contributions in the area of migration policy, and Stephen Castles prompts us to rethink Australian migration through a piece that was originally prepared, at Graeme's invitation, for an international symposium in 2014. In addition, this special issue contains eight papers, many of which have been prepared by Hugo's close colleagues, friends and students. The first two papers provide detailed overviews of his contributions to two key areas of research: population ageing in Australia (Debbie Faulkner, Helen Barrie Feist and Judith Lewis), and social applications of Geographical Information Systems (Danielle Taylor and Jarrod Lange). The authors of the remaining six papers opted to pay tribute to Hugo by presenting their own empirical work. Many of the research projects discussed were undertaken collaboratively with Graeme, or were supervised by him. Others were inspired by his research. Kate Golebiowska, Amani Elnasri and Glenn Withers present empirical evidence of the impacts of Australian immigration policy on regional population growth, and tackle the oft-made and nefarious contention that immigration contributes to unemployment amongst the Australian-born. Next, Karen Agutter and Rachel Ankeny consider the role of migrant hostels in the residential settlement patterns of immigrants through careful archival and interview-based data gathered as part of their 'Hostel Stories' research project. Janette Young, Lisel O'Dwyer and Richard McGrath's paper considers the question of what constitutes 'successful' migration, through a focus on British migration to Australia in the post-Second World War period. Linking their research findings to the present day, they argue that careful consideration ought to be given to the potentially negative outcomes of targeted migrant recruitment schemes. Continuing on this theme, Fidelma Breen outlines the experiences of Irish migrants who came to Australia under the Temporary Work (Skilled) (subclass 457) visa. For some of these migrants, misunderstandings about temporary skilled migration to Australia—and a lack of clarity around skill recognition and qualification transferability—had enormous personal and familial consequences.

The final two papers in this issue align with Hugo's research interest in population settlement patterns in Australia. Neil Coffee, Jarrod Lange and Emma Baker build on work that they conducted with Graeme in 2000. They present a novel spatial approach that enables comparison and visualisation of population density change across Australian cities, over time. Finally, Peter Smailes, Trevor Griffin and Neil Argent offer the results of a longitudinal study of population distribution in Hugo's home State: South Australia. Their paper considers how the concentration of population and economic activity into regional cities (in this case, Port Lincoln) impacts the demographic sustainability of their broader regions.

As we have read the contributions made to this special issue we have been reminded of the overwhelming range, scope and importance of Hugo's work. We have also felt reassured that there are others continuing this work—many of whom worked alongside Graeme Hugo, and made important contributions to his own scholarship over many years. Our sincere thanks go to Janet Wall for her generous assistance throughout this process.

Notes

1. The 'Children Overboard' incident, as it has become known, took place over 3 days between 6 and 8 October 2001, 1 month before the Australian federal election was held on 10 November 2001. The incident involved a group of asylum seekers who were heading towards Australian waters on board a boat labelled Suspected Illegal Entry Vehicle 4 (SIEV 4) by Australian authorities. SIEV 4 was intercepted by HMAS Adelaide. Photos presented to the media at the time appeared to show the SIEV 4 asylum seekers throwing their children overboard, and Prime Minister Howard stated: 'I don't want in Australia people who would throw their own children into the sea' (Bowden 2001). These assertions were proven false; the photos were in fact of asylum seekers abandoning the sinking vessel (SIEV 4) and being rescued by the crew of HMAS Adelaide (see Odgers 2002; Senate Select Committee on the Scrafton Evidence 2004 for further detail).
2. Officer of the Order of Australia: 'The Officer of the Order of Australia is awarded for distinguished service of a high degree to Australia or humanity at large' (https://www.itsanhonour.gov.au/honours/awards/medals/officer_order_australia.cfm).
3. The report was authored by Graeme Hugo with the assistance of: Sanjugta Vas Dev, Janet Wall, Margaret Young, Vigya Sharma and Kelly Parker.
4. http://www.adelaide.edu.au/apmrc/pubs/

References

Bourke, L. 2016. "Peter Dutton Says 'Illiterate and Innumerate' Refugees Would Take Australian Jobs." *The Sydney Morning Herald*. Accessed July 21, 2016. http://www.smh.com.au/federal-politics/federal-election-2016/peter-dutton-says-illiterate-and-innumerate-refugees-would-take-australian-jobs-20160517-goxhj1.html.

Bowden, T. 2001. "Navy Chief Enters Asylum Seekers Debate." *Australian Broadcasting Corporation, TV Program Transcript*. Accessed August 1, 2016. Broadcast November 8, 2001. http://www.abc.net.au/7.30/content/2001/s412083.htm.

Connell, J. 2015. "Obituary: Graeme John Hugo 1946–2015." *Australian Geographer* 46 (2): 271–279.

Doherty, B., and H. Davidson. 2016. "Fact Check: Was Peter Dutton Right About 'Illiterate' Refugees 'Taking Jobs'?" *The Guardian Australia*. Accessed July 21, 2016. http://www.theguardian.com/australia-news/2016/may/18/fact-check-was-peter-dutton-right-about-illiterate-refugees-taking-jobs.

Hugo, G. 2011. *Economic, Social and Civic Contributions of First and Second Generation Humanitarian Entrants*, Final Report to Department of Immigration and Citizenship (with the assistance of Vas Dev, S., Wall, J., Young, M., Sharma, V. and Parker, K.) Australian Population and Migration Research Centre (APMRC), University of Adelaide.

Odgers, S. J. 2002. *Report of Independent Assessor to Senate Select Committee on A Certain Maritime Incident*. Forbes Chambers, August 21, 2002. Accessed August 1, 2016. http://www.aph.gov.au/~/media/wopapub/senate/committee/maritime_incident_ctte/report/odgers_pdf.ashx.

Policy Innovation, Research and Evaluation Unit, Department of Immigration and Citizenship. 2011. "About the Research: *Economic, Social and Civic Contributions of First and Second Generation Humanitarian Entrants*." Accessed July 29, 2016. http://www.border.gov.au/ReportsandPublications/Documents/research/economic-social-civic-contributions-about-the-research2011.pdf.

Senate Select Committee on the Scrafton Evidence. 2004. *Report—Senate Select Committee on the Scrafton Evidence*. Parliament House, Commonwealth of Australia. Accessed July 29, 2016. www.aph.gov.au/~/media/wopapub/senate/committee/scrafton_ctte/report/report_pdf.ashx.

The Guardian (UK). 2015. "The Guardian View on Geography: It's the Must-have A-level." Accessed July 21, 2016. https://www.theguardian.com/commentisfree/2015/aug/13/the-guardian-view-on-geography-its-the-must-have-a-level.

Natascha Klocker

Australian Centre for Cultural Environmental Research, School of Geography and Sustainable Communities, University of Wollongong, Wollongong, NSW 2522, Australia

Olivia Dun

School of Geography, University of Melbourne, Melbourne, VIC 3010, Australia

Remembering Graeme Hugo, Population Geographer

Ruth Fincher

Graeme Hugo was a population geographer, his academic training having been both in geography and demography. He was widely published, internationally renowned and appreciated over many decades, as was amply testified in the glorious obituaries written by his close colleagues after his untimely death (see Bedford 2015; Connell 2015; Stimson et al. 2015). Through all his published work we see his passion as a population geographer to bring social geography's commitment to exploring people's attachment to place and their mobility between places into close relationship with demography's capacity to measure and analyse long-term changes in the nature of the population and to project futures from this. Graeme was utterly committed, as well (and this shows in his published work too), to working together with policymakers by providing them with the best possible evidence about the populations of concern to them and drawing out for them the implications of that evidence. In the paragraphs to follow I offer an appreciation of the areas of research in population geography to which, it seems to me, Graeme Hugo contributed particularly strongly. It is impossible to write about Graeme's research legacy, however, without remembering the generous and inclusive way in which he conducted his life as an academic geographer, in the actual doing of his research. So I conclude with a comment about the legacy of academic practice that he leaves for us.

In Graeme Hugo's remarkably long and varied list of publications, four (often inter-linked) research themes seem to me to characterise his contributions to population geography and to exhibit his fundamental intellectual priorities. First, he made pioneering conceptual and empirical contributions to understanding the spatial complexity and circularity of migration patterns, seeing these as contextually grounded in place. This interest endured through his career, with special reference to the Asia-Pacific region in which his PhD research in Indonesia had been deeply located. Second, Hugo documented over many years the settlement patterns of the Australian population, ensuring that the spatial nature of data from the population census was made very evident to a range of publics. Third, he brought his depth of academic knowledge to bear in discussions of Australian immigration trends and outcomes, in a manner that was measured and never shrill, and which made him an interlocutor respected by governments as well as academic colleagues from a range of disciplines. Fourth, he associated the issue of environmental change with the very social phenomenon of migration, writing recently on climate change as an initiator of migration in the Asia-Pacific region.

First, then, from the early 1980s Graeme Hugo made significant and original contributions in pointing out and documenting the circularity of migration patterns, first in internal migration and later in international migration. His work demonstrated the spatial and social complexity of people's mobility, and raised the question of the relationship between so-called 'temporary' migration and 'permanent' migration (with the influential idea of the latter that 'real' migration consists of moving from one place to another irrevocably, and staying put in the destination). The New Zealand population geographer Dick Bedford (2015, 62) notes that as early as 1973 Hugo was corresponding with colleagues to say that he was finding in his Indonesian research population mobility that differed dramatically from common understandings: '[i]nstead of moving into local towns and cities, people were moving back and forth between village homes, local communities and ancestral hearths; circulation within rural areas rather than long-term rural–urban migration was much more common and important'. In an early paper, laying out his pioneering perspective on circular migration in Indonesia, Hugo (1982) regretted that surveys of population mobility in that country focused only on permanent migration rather than also recognising non-permanent movement, thus making people in Java (particularly) seem immobile and fixed in their ways when the evidence he found contradicted that. By 1988, he was extending his data-rich analysis of Indonesian population mobility to demonstrate the presence of Indonesian labourers in international labour migration flows to and from the Middle East, Malaysia and Singapore. Peppering his writing with lively questions about the social changes that spatially broad circular migration might bring to Indonesian communities, Hugo (1988) again commented on the need for improvement in the data collected in Indonesia so that it could capture the kinds of mobility that were actually taking place and in so doing help development planning. Fast forward to the twenty-first century, when Hugo (2004) was explaining how circulation is now the dominant form of international migration, in part because moving to keep in touch with relatives in other parts of the world is cheaper—one does not have to leave them behind. But also a move away from permanent migration and to circulation is related to the emergence of highly dominant global cities whose workforces rely upon transient business experts settling there for a few years and then moving to work somewhere else. Developing this defining insight of the circularity of migration was an agenda-changing conceptual and empirical contribution to population geography and related disciplines. Hugo was very prominent in the group who first saw this insight and later developed it.

A second important contribution of Graeme's population geography has been its consistent and high-quality documentation of the spatial patterns of the Australian population. In any Australian university library there will be a set of *Atlases of the Australian People*, giving cartographic presentation to the data of the population censuses of the 1980s and 1990s, for many of which Graeme Hugo was the senior author. And Hugo worked the census data into useful forms, not just for presentation in atlases. Stimson et al. (2015) pay particular tribute to an index he developed of the accessibility to services experienced by the population in different Australian locations. They say: 'one of his lasting legacies will be the Accessibility/Remoteness Index for Australia (ARIA) created initially for the Department of Health and Ageing in 1997–98 and later updated for the Australian Bureau of Statistics as ARIA+ in 2001 and 2006 and ARIA++ in 2011' (Stimson et al. 2015, 117; see also Taylor and Lange, this issue). In making the effort to communicate the spatial dimensions of the population in Australia, Hugo clearly agreed with the point made recently by population geographer Martin Bell (2015, 302) in his essay on the need to retain consistent population censuses, that an 'understanding of spatial differentiation and spatial interaction remains a central focus, and an abiding contribution of the discipline [of geography]'. Not only was Hugo's documentation about the population as a whole, it was also about particular social dimensions of the population and their spatial manifestations. In particular, he was long interested in the ageing of the population and the consequences of that demographic fact (see Faulkner, Feist, and Lewis, this issue). The importance of population ageing, as he saw it, is conveyed in his discussion of the distinction between young-old and old-old groups within the Australian population, their different rates of growth and their different housing preferences and forms of mobility (Hugo 1987). This interest in drawing out the special place of ageing in population futures was extended internationally in later writing (e.g. Hugo 1997). I see this concern of Hugo's with the constant and visible documenting of the spatial characteristics of Australia's population as a major contribution to basic, national self-knowledge: few other geographers have made such an effort and it is an effort that has served many others well, in a range of professions and endeavours that require planning based on place-sensitive evidence.

As a third distinctive contribution of Graeme Hugo's, I note his writing and commentary on Australian immigration, particularly the numbers and characteristics of immigrants coming to Australia under different categories of national migration policy. (Hugo wrote on emigration from Australia, as well, but I refer to his larger body of work on immigration, here.) He drew on the wealth of statistics collected by the Commonwealth government's Immigration Departments, over the years, to form penetrating analyses of the national migration situation as it changed. This is evident, for example, in his 2006 discussion of temporary migration to Australia (Hugo 2006), in which he examined flows of Working Holiday Makers, Temporary Business Entrants, Overseas Students and temporary migrants from New Zealand, demonstrating their significant presence since the mid-1990s in different Australian locations and labour markets. In a parallel with his early work on circular migration in Indonesia and its absence from the official statistics, Graeme pointed out strongly that temporary migration is highly significant in Australia and that at the time of writing this was not recognised, as research and official data collections continued to focus on so-called 'permanent' migration flows. Hugo often combined with scholars from other disciplines to share perspectives on Australian immigration, and one has the sense that he was vitally interested in generating cross-disciplinary discussion and knowledge about this highly significant aspect of Australia's population growth and composition. He was one of four editors, for example, with economists and sociologists, of the widely read book *Australian Immigration: A Survey of the Issues* (Wooden et al. 1994), organised by the federal government's then Bureau of Immigration Research (disbanded in 1996) as it sought to build the knowledge developed by Australia's university researchers about immigration and its outcomes, and to communicate this to government and the Australian public. In 2002, he was engaging with legal researchers and lawyers about skilled migration policies and their contribution to the national interest (Hugo 2002). More recently, Hugo collaborated with sociologists Castles and Vasta to edit a special issue of the *Journal of Intercultural Studies*, seeking to update outdated assumptions about Australian migration (see Castles, Hugo, and Vasta 2013). To this special issue, Graeme's written contribution was to clarify the way in which remittances, diasporas and return migrations from Australia can contribute to economic development in different parts of Asia, with Australia linked in to such development. Graeme Hugo's input as a population geographer to informed discussions about Australian immigration over the last three decades, discussions that occurred both within the scholarly community and beyond it, were of lasting benefit. When the Bureau of Immigration Research was closed by the incoming federal government in 1996, and its funding of research on immigration ceased, Graeme's voice as a highly skilled and reasoned researcher making assessments of the impacts and form of Australia's immigration continued (thankfully) to be prominent.

I select as a fourth contribution made by Graeme Hugo as a population geographer his association of migration with environments. In most countries, discussions of international and internal migration and research about it do not dwell on matters environmental. But Hugo, the geographer and student of Australia's population and the history of debates about it, knew well that in Australia discussions about population size and its distribution were rooted in questions of carrying capacity (earlier on) and sustainability (more recently) (see Hugo 2011). At a time when national leaders were focusing once again on questions of population size and whether we should create a 'Big Australia', Graeme Hugo advocated strongly that geographers should be involved in these national debates, bringing forward their capacity to ask environmental questions that were equally social, and social questions that were equally environmental. He had connected international migration and environmental questions previously, developing a major interest in whether and how environmental change can be a cause of migration flows (Hugo 1996). Recently, his work relating environmental change and migration focused on climate change and its likely effect on migration in different regions of Asia (Bardsley and Hugo 2010; Hugo 2013). At the start of his own review essay, in the edited book in which he drew together many existing papers about environmental change as a possible initiator of international and internal migration, Hugo (2013) stressed the complexity of this relationship, and its situatedness. Not only is the precise form and temporality of environmental change in a place of significance, but so too are the resources and circumstances of the communities living in the affected locations. In no sense, he says, can it be assumed that there will be a linear relationship between climate change and one-way migration flows. With this recent scholarly work, Graeme Hugo exhibited his capacity to draw in to discussions of population mobility the long-standing concerns of geography with environmental change, and equally to demonstrate to his colleagues in geography the importance of understanding the complexities of migrants' decision making.

If Graeme Hugo's intellectual contributions to geography, as a population geographer, have been in the four areas I have noted, it must be said that he did far more than this for geography as a discipline. He practised his geography in an exemplary way; he made the discipline of geography visible nationally and internationally, and brought great credit to it. Readers who knew Graeme will have their own knowledge of how he did this. I mention just three of the ways that I saw him, in action, as a geographer benefiting us all. First, he mentored many PhD students, particularly students who came to work with him from Asia: I count in his CV a list of 64 PhD students supervised to completion, and more will have completed since then. One must not forget the even larger number of Masters and undergraduate Honours research projects he supervised. Many of those students will now be working in research or governmental settings, knowing and grateful that their training and skills are from population geography. Second, Hugo's links to the Australian Research Council (ARC), the source of nationally competitive research funding for Australian university academics outside the fields of medicine and health, were extraordinary. He won much funding from the ARC, earning a string of the most prestigious fellowships for himself and his research program, through his career. Also, he led the 'College of Experts' panel for the social sciences, the core group conducting peer review and evaluation of social science applications for ARC funding, for many years and with great distinction and fairness. His presence as a population geographer at the very centre of the nation's top funding program, both as a recipient of major grants and as a leader of its evaluations, made the discipline of geography visible on the national research stage. Third, Hugo had exemplary relationships with national policymakers concerned with migration, in Australia and in numerous countries in Asia and elsewhere. Not all geographers have as a major priority the translation of their research into forms of use to policymakers, and not all seek to engage in constant conversations with policymakers, but Graeme Hugo was committed to those tasks. He accomplished this work brilliantly and in an effective and low-key way that was appreciated. An overseas colleague said to me that Graeme Hugo convinced policymakers around the world that geography was a worthwhile discipline. He could always contribute to their conversations in ways that the policymakers found enlightening and helpful.

Graeme Hugo, the eminent population geographer, was a superb ambassador for the discipline of geography, and his intellectual and personal contributions will long be remembered and celebrated.

Disclosure statement

No potential conflict of interest was reported by the author.

References

Bardsley, D. K., and G. J. Hugo. 2010. "Migration and Climate Change: Examining Thresholds of Change to Guide Effective Adaptation Decision-making." *Population and Environment* 32 (2/3): 238–262.

Bedford, R. 2015. "Obituary: Graeme John Hugo AO, FASSA, 1946–2015." *New Zealand Geographer* 71: 61–2.

Bell, M. 2015. "W(h)ither the Census?." *Australian Geographer* 46 (3): 299–304.

Castles, S., G. Hugo, and E. Vasta, eds. 2013. "Rethinking Migration and Diversity in Australia: Introduction." *Journal of Intercultural Studies* 34 (2): 115–121.

Connell, J. 2015. "Obituary: Graeme John Hugo 1946–2015." *Australian Geographer* 46 (2): 271–279.

Hugo, G. J. 1982. "Circular Migration in Indonesia." *Population and Development Review* 8 (1): 59–83.

Hugo, G. 1987. "Ageing in Australia: The Spatial Implications." *Urban Policy and Research* 5 (1): 24–26.

Hugo, G. 1988. "Population Movement in Indonesia Since 1971." *Tijdschrift voor Economische en Sociale Geografie* 79 (4): 242–256.

Hugo, G. 1996. "Environmental Concerns and International Migration." *International Migration Review* 30 (1): 105–131.

Hugo, G. 1997. "Asia and the Pacific; A Much Older Future for a Vast, Varied Region." *Global Aging Report* 2 (4): 3.

Hugo, G. 2002. "Migrants and Demography: Global and Australian Trends and Issues for Policymakers, Business and Employers." *Georgetown Immigration Law Journal* 16 (3): 649–683.

Hugo, G. 2004. "A new Global Migration Regime." *Around the Globe* 1 (3): 18–23.

Hugo, G. 2006. "Temporary Migration and the Labour Market in Australia." *Australian Geographer* 37 (2): 211–231.

Hugo, G. 2011. "Geography and Population in Australia: A Historical Perspective." *Geographical Research* 49 (3): 242–260.

Hugo, G., ed. 2013. *Migration and Climate Change*. Cheltenham, UK: Edward Elgar.

Stimson, R., A. Maude, C. Forster, P. Smailes, D. Rudd, M. Bell, and D. Forbes. 2015. "Obituary: Professor Graeme John Hugo BA, MA, PhD, FASSA, AO, 5 December 1946–20 January 2015." *Geographical Research* 53 (2): 115–118.

Wooden, M., R. Holton, G. Hugo, and J. Sloan. 1994. *Australian Immigration: A Survey of the Issues*. Canberra: Australian Government Publishing Service (AGPS).

Migration moderate, 'Master Weaver' and inspirational team leader: reflecting on the lasting legacy of Graeme Hugo in three spheres of migration policy

Marie McAuliffe

Sometimes you don't fully understand a person's worth until they are gone. They can be inadvertently taken for granted, sometimes overlooked or undervalued until they are no longer there to provide advice, assistance or an expert point of view. That was not the case with Professor Graeme Hugo. In the Australian and international academic circles of geography, demography and migration, there are few names that have been so enduring, and even fewer people who have been so influential.

Much has been written on Graeme's achievements, particularly in academia. Graeme Hugo was a highly prodigious and prolific scholar, producing over 200 peer-reviewed articles, over 30 books, more than 260 book chapters, around 30 book reviews and over 1000 conference papers (or around 25 per year between 1975 and 2013) (Bedford and Nieuwenhuysen 2015; Connell 2015). In a 2015 blog by Norwegian Professor Jorgen Carling on the 'names worth knowing' in international migration, he noted that the academic with the most publications in migration journals was Graeme Hugo (Carling 2015). The co-founder of the Migration Policy Institute in Washington, DC, Dr Demetrios Papademetriou, referred to his passing as the 'loss of a giant of migration studies' (Papademetriou 2015). Professor John Connell named Graeme as the most cited Australian geographer ever (Connell 2015); he was also included in the list of geography's global 'master weavers', based on citations (Bodman 1992). Just reading Graeme's 109-page CV is an effort, particularly as it is so dense. There is no padding. It makes you feel tired, and then when you try and imagine the work involved, well it's nothing short of overwhelming.

It is also important to recall Graeme's achievements as a teacher and mentor. In addition to the many thousands of hours tutoring and lecturing students over several decades, Graeme supervised dozens of research students in their pursuit of a total of 32 Honours theses, more than 50 Masters theses and around 65 PhD theses. Many of his students have gone on to great careers and have become influential scholars in their own right. He also contributed as an examiner, marking 30 Masters and PhD theses. I benefited personally from Graeme's enthusiasm and knowledge as a supervisor for my doctoral research until his untimely passing. Amongst one of his last batches of work were comments he made on a paper I had co-authored on media and migration. As was typical, Graeme's feedback was positive, constructive and insightful. I was very grateful, although expressions of gratitude to Graeme were easily made, partly because he was humble but also because they were often reciprocated.

Fittingly, much has also been written on Graeme's generosity both academically and personally (Bedford and Nieuwenhuysen 2015; Connell 2015; Papademetriou 2015). Much less, however, has been written on his contributions to migration policy—in Australia, in our neighbouring region and on global governance of international migration—and there has been an absence of discussion of Graeme's highly effective and understated leadership style. It is in these two related areas that, almost 2 years since his passing, there is much to reflect on. Collective contemplation of how we can ensure greatness such as Graeme's is able to survive and flourish in the current academic system to the benefit of both academia and policy is important for current and future generations. It is perhaps also something we owe to him and his extraordinary legacy.

Contributions to Australian population, migration and settlement policy

Graeme Hugo thought 'big' and he thought 'long'. His horizon was different from most, and his way of communicating his vision was almost unique. He was one of the most passionate people on migration but also one of the best listeners. Every interaction seemed to take place from a position of intellectual enquiry rather than intellectual superiority (which could so easily have been the case). Graeme's enthusiasm for learning and (re) thinking about migration never diminished.

Graeme contributed to population, migration and settlement policy in many different ways and over many years. He was Chair of the federal government's Demographic Change and Liveability Advisory Panel set up to inform the then government's sustainable population strategy. Never one to shy away from sensitive, complex or weighty issues, Graeme would bring his intellectual rigour and political nous to even the most fraught areas of immigration policy. He had no hesitation in taking up the invitation to contribute as a member of the advisory body overseeing the Australian government's irregular migration research program, and was extremely well placed to contribute in an even-handed and thoughtful way. As the then director of the research program, I again had the benefit of Graeme's wisdom, expertise and deep knowledge. Graeme, along with a handful of other Australian and overseas academics and migration practitioners, helped shepherd the research program as we sought to expand the middle ground on a highly polarised topic by undertaking and commissioning non-partisan research and analysis that was as objective as possible. Graeme was committed to building the evidence base on irregular migration as well as regular migration, settlement, and population change, and it was here that he truly excelled. But he was no wallflower when it came to taking on poorly constructed or underdone policy, as one of Graeme's statements from 2013 indicates:

> Ageing and migration are two dominant issues in all Western countries, including Australia, but the two issues are not separate. While many people think of migration as a 'silver bullet' solution for an ageing workforce, the reality is that those migrants will naturally age too. In fact, we currently have millions of people ageing in our community who came to this country as migrants … I don't think policymakers have got a grasp of this situation, and what it will mean for today's migrants in another 30–40 years in a range of areas, such as housing and health care.[1]

Policymakers need and admire scholars who are able to think critically, not merely engage in criticism. The most influential critical thinkers outside of government are the ones with policy nous—who know how to engage with the policy challenges and understand that, just like many academics, many policymakers are inspired to contribute to public service, and principally to improve the lives of people both near and far. Graeme understood this. He was one of migration's best 'policy scholars'.[2] He worked collegially with policymakers across Australia, in various capacities, and remained keenly interested in ensuring research had policy utility. He was invited to chair or join, for example, many policy-related committees and advisory bodies at the national and State levels, including (in addition to those mentioned above) the National Sustainability Council, the Aged Care Finance Committee, the Australian Urban Research Infrastructure Network, the Federal Housing Supply Council, the South Australian Government Population Committee and the Social Inclusion Board of South Australia.

Graeme produced numerous reports for policymakers over the years. Perhaps one of his most influential was his study on the contributions of refugees in Australia. Prepared for the (then) Department of Immigration and Citizenship, Graeme's 2011 report titled the *Social, Economic and Civic Contributions of Humanitarian Migrants to Australia* was one of the key pieces of evidence used by the then Minister of Immigration to support the increase in Australia's refugee intake (ABC 2011).

Ultimately, Graeme challenged policymakers to apply different frames of reference to policy conundrums. He helped shift those frames to inspire Australian policymakers to think beyond the geographic and geopolitical limits of their own 'Australianness', and better understand their increasing connections with countries in their region as well as further beyond.

Building bridges with Indonesia and beyond

Just as he pushed the boundaries with Australian policymakers, Graeme Hugo pushed the boundaries with Australian researchers. In his own words, from a 2014 Australian Research Council grant application, he explained:

> Australian international migration research has tended to be very inwardly-focused on the impacts on Australia and its national interest. Yet migration is a key element linking Australia with the Asia-Pacific region and my work is seeking to better understand these linkages. I believe this understanding can be the basis of improving Australia's economic, political and cultural engagement in the region during the 'Asian Century.'[3]

Graeme led by example, forging strong links and generating productive collaborations between researchers, migration practitioners and policymakers in Australia and the rest of the Asia-Pacific region. To some extent, this came naturally as he was a passionate internationalist with a special fondness for Indonesia, having conducted fieldwork for his doctoral research in Indonesia in 1973.

As was the case in Australia, Graeme worked with the people working on what he saw as the important issues. His enthusiasm and talent for geography was evident as early as his high school years when he won the John Lewis Medal and Royal Geographical Society Prize for topping geography in the South Australian matriculation examinations, and yet he wasn't blinded by the discipline to all else. Disciplinary excellence in geography, and later demography, formed a strong foundation but he saw the value in applying multi-disciplinary approaches. He was a strong academic who saw the real benefit in engaging with people working on migration from all angles, including policymakers, practitioners and civil society. In Indonesia, for example, he worked for the Ministry of Population and Environment, the Department of Manpower and Transmigration, and the Indonesian Central Planning Agency. Such was the level of trust and confidence he inspired that a glance through his work is testament to his standing—so often the sources in his tables and figures read 'Immigrasi unpublished data' or 'departmental unpublished data'.

Indonesia may have been the starting point, but Graeme's interests and expertise took him much further afield. Graeme had worked closely with researchers and policymakers across Asia, and supervised many Asian PhD students including from Indonesia, Malaysia, Thailand, Vietnam, Pakistan, Sri Lanka and Bangladesh.

Global migration governance

Graeme was unashamedly a migration moderate, seeking to expand the middle ground for the betterment of migrants and societies. His commitment to building a strong evidence base and sharing his knowledge in order to make a positive difference was indefatigable, for he strongly believed in formulating ways to better manage migration globally. He wrote papers for the 2005 Global Commission on International Migration as an Asia-Pacific lead, contributed to and attended the United Nations High Level Dialogue on Migration and Development in 2013, and was a member of the Consultative Committee of the Nansen Initiative on Population Displacement due to Disasters. He worked closely with international organisations in numerous capacities on a multitude of topics, and was often the person to call if an Asia-Pacific perspective on an aspect of migration was needed.

Part of Graeme's interest in global governance of migration was drawn from his strong sense of equity and opportunity:

> There is criticism that migration research and policy has focused excessively on immigration settlement and destinations and I have argued that our better understanding of the process requires greater balance to initiate more work on emigration and origins. The work on diasporas is an important part of this and will shed important light not only on migrant identity but also on impacts on origins and destinations.[3]

An intellectual leader as well as a talented team leader

The pressure to achieve within an academic reward system that values publication in peer-reviewed journals is acknowledged as having some serious downsides (Smith 2006). The 'publish or perish' culture has been found to stifle research innovation (Foster, Rzhetsky, and Evans 2015), lower research publication standards (Colquhoun 2011), encourage peer-review fraud (Prosser Scully 2015) and negatively affect the ability of researchers to work on applied research tailored to policymakers (Cherney et al. 2012). At the same time there is also pressure to undertake innovative research, publish in the top journals and present evidence to policy audiences and ultimately influence policy (Cherney et al. 2012). Clearly Graeme Hugo was a master at being able to publish high-quality research and influence policy but many academics cannot, and for good reason. Graeme's unusual success provides an opportunity to reflect on aspects of his approach.

The social sciences and humanities have tended to privilege individualism, presupposing that intellectual pursuits are solitary endeavours, forged principally by scholars on their own (who nevertheless may collaborate but typically by bringing their own specific talents and experiences to the table). These days there is also pressure to build and maintain one's 'profile', both publicly and through less visible networks. The academic with the website, blog and string of adjunct positions and op-ed publishers is becoming more common. Policymakers, on the other hand, are first and foremost managers of teams, and typically large teams that build knowledge through highly collective processes, partly because of the speed in which they must produce. Social capital, collegiality and participatory leadership are essential components of well-oiled policy teams. These aspects are important in academia but not nearly as important. Rarely would academics be faced with briefing Cabinet on an important policy issue requiring input from multiple specialists with only a few hours' notice.

Whether it was by intellect or instinct (or both) we can reflect on the fact that Graeme Hugo was an exemplar of the highly intellectual scholar but he was also an exceptional team leader. Much in the same way as senior policymakers lead teams to manage complex issues, Graeme Hugo led a very large, perhaps global team of researchers committed to contributing to the ongoing pursuit of a better understanding of migration dynamics. At the heart of this global team was his faithful and talented University of Adelaide team, who worked tirelessly for Graeme by contributing to research and policy discourses, some of them over decades (most notably Janet Wall and Margaret Young). Graeme's style was participatory, he led by example and inspired respect and admiration all the while remaining accessible and open to new ideas. He worked very hard but many people also worked hard with him and for him, and his ability to lead teams and collaborate within a culture of extreme competition provides us some insights as to how and why he was so successful. But Graeme's passing has also left us with an important question to ponder: have we got the balance right between the individual pursuit of scholarship and the development of collective knowledge and wisdom?

There can be no doubt that Graeme Hugo's passing has left a substantial hole. Many of us continue to feel it. As Papademetriou has suggested, Graeme was a giant in the field of migration studies in a system that demanded very high standards of individuals. But perhaps there are ways that we can help soften such losses in the future by building on positive and participatory leadership styles in academia through instituting better systems and processes, such as team publication. The current system, for example, is not flexible enough to accommodate institutional team authorship in academic peer-reviewed journals but it shouldn't be discounted. For those who have grown up in the system, the mere suggestion would be so unorthodox as to be deemed ludicrous, and yet in non-academic publishing institutional team-based authorship is common (and in policy it is the norm). Reports and papers in the grey literature are often 'written' by institutional authors, such as UNHCR or IOM.[4] Encouraging and strengthening team-based approaches, including a place for academic institutional authorship, could assist in further developing knowledge hubs involving multiple stakeholders and scholars. It could be a useful complement to authorship by individuals (including co-authoring), which is undoubtedly a core value of academia, but only if such research outputs are able to be measured and counted in the academic reward system. There is no good reason, for example, why institutionally authored papers cannot be subjected to double-blind peer review.

It is but one small idea but I can't help reflect on the irony that it is likely that Graeme Hugo would have been a supporter of such a notion, seeing it positively as a way to build intellectual capital and knowledge transfer. He, like everyone else, worked within the system, and we should remain grateful for the ongoing benefits we can derive from his enormous contributions. I'm sure he would urge us all to explore ways to improve the system to enable more academics to contribute collectively to both research and policy discourses in the future. There are, after all, many talented minds and hearts willing to contribute.

Notes

1. University of Adelaide, 'Longer Term View Needed of Ageing Migrants', Media Release, August 9, 2013, https://www.adelaide.edu.au/news/news63721.html
2. The term 'policy scholars' draws on personal reflections of Dr Demetrios P. Papademetriou, Distinguished Senior Fellow, President Emeritus and Co-founder, Migration Policy Institute, Founder and President, Migration Policy Institute Europe, and is based on his 45 years' experience working with migration academics and policymakers globally.
3. Australian Research Council submission, 2014.
4. It is perhaps worth acknowledging that I have long argued that 'grey' literature could benefit from individual (named) authorship and more rigorous peer-review processes, specifically for research outputs (as opposed to policy outputs).

Acknowledgements

Thank you to Natascha Klocker and Olivia Dun for the invitation to contribute a Thinking Space piece to this Special Issue in Memory of Graeme Hugo. My appreciation also goes to Richard Bedford, Demetrios Papademetriou, Peter Hughes, James Raymer, Janet Wall and Jorgen Carling for commenting on an earlier draft of this manuscript. Special thanks to Janet Wall for the very useful unpublished background material she kindly provided.

Disclosure statement

No potential conflict of interest was reported by the author.

References

ABC (Australian Broadcasting Corporation). 2011. *Bowen Heckled over Malaysia Asylum Deal*, June 17. http://www.abc.net.au/pm/content/2011/s3246916.htm.
Bedford, R., and J. Nieuwenhuysen. 2015. "Graeme Hugo: Population and Migration Expert Spread his Message Far and Wide." *The Age*, March 1. http://www.theage.com.au/victoria/graeme-hugo-population-and-migration-expert-spread-his-message-far-and-wide-20150301.
Bodman, A. 1992. "Holes in the Fabric: More on the Master Weavers in Human Geography." *Transactions of the Institute of British Geographers* 17 (1): 108–109.
Carling, J. 2015. *Who is Who in Migration Studies: 107 Names Worth Knowing*, June 1. https://jorgencarling.wordpress.com/2015/06/01/who-is-who-in-migration-studies-108-names-worth-knowing/.

Cherney, A., B. Head, P. Boreham, J. Povey, and M. Ferguson. 2012. "Perspectives of Academic Social Scientists on Knowledge Transfer and Research Collaborations: A Cross-sectional Survey of Australian Academics." Evidence & Policy 8 (4): 433–453.

Colquhoun, D. 2011. "Publish-or-perish: Peer Review and the Corruption of Science." *The Guardian*, September 5.

Connell, J. 2015. "Graeme John Hugo, 1946–2015." *Australian Geographer* 46 (2): 271–279.

Foster, J., A. Rzhetsky, and J. Evans. 2015. "Tradition and Innovation in Scientists' Research Strategies." *American Sociological Review* 80 (5): 875–908.

Papademetriou, D. 2015. *The Field of Migration Studies Loses a Giant: Graeme Hugo.* Washington, DC: Migration Policy Institute. http://www.migrationpolicy.org/news/field-migration-studies-loses-giant-graeme-hugo.

Prosser Scully, R. 2015. "How the Pressure of Publish or Perish Affects us all." *The Medical Republic.* http://www.medicalrepublic.com.au/how-the-pressure-of-publish-or-perish-affects-us-all/.

Smith, R. 2006. "Peer Review: A Flawed Process at the Heart of Science and Journals." *Journal of the Royal Society of Medicine* 99 (4): 178–182.

Rethinking Australian migration

Stephen Castles

The Australian model of permanent immigration, based on easy access to citizenship plus multiculturalism, is under challenge. Changes in global and regional migration patterns question some of the key assumptions held by many Australians. Since the early 1990s, Australia's immigration policies have shifted from migration as population building to migration as an economic factor. The meaning of multiculturalism has changed, especially since the 1989 *National Agenda for a Multicultural Australia*, which defined multiculturalism as a system of citizenship rights and obligations. Moreover, the growth of temporary migration, which leads to new forms of personal identity and belonging (often referred to as 'transnational belonging'), in turn requires new solutions with regard to settlement policies, the meaning of citizenship and the ways in which social entitlements are recognised and delivered.

The major changes of recent years have received remarkably little public attention. Instead, the media and the politicians have focused on the relatively small number of 'irregular maritime arrivals' who come seeking protection from persecution, violence and discrimination. The definition of asylum seeking as a 'national crisis', along with the renaming of what was once the Department of Immigration and Multicultural Affairs (DIMA), and then the Department of Immigration and Citizenship (DIAC), as the Department of Immigration and Border Protection (DIBP) is emblematic of the securitisation—even the militarisation—of asylum. This essay seeks to go beyond the aggressive everyday political debates about 'migration as a threat' to explain some of the key underlying trends which will help shape the future of our multicultural society.

This Thinking Space piece was originally written in late 2014 for an international symposium that Graeme Hugo was organising in Canberra. He was unable to attend the meeting, as he was in hospital fighting against the illness that led to his untimely death. Graeme was deeply concerned about major shifts in official policies on immigration and multiculturalism that could undermine Australia's successful model for building an inclusive society that welcomed immigrants as people, not just as economic inputs. He was a fervent advocate of a socially just national approach to immigration, and worked tirelessly to achieve this (see, for example, Hugo 1994, 2006, 2011). His meticulous, evidence-based work has helped to inform both policy and public perceptions, and is sorely missed in the present situation in which immigration has become a key political theme, addressed all-too-often more at an emotional than a rational level.

Current trends in global and regional migration patterns

Figure 1 presents the most recent United Nations (UN) Population Division figures on global migration. The figures indicate that there were 244 million international migrants (defined as persons resident for at least 1 year in a country other than their country of birth) worldwide in 2015. The numbers of international migrants—especially of those going to developed countries—has increased sharply over the last two decades (UN Population Division 2015). However, the global population has increased at a similar rate, such that international migrants still make up only 3.3 per cent of the world's population—or to put it differently, nearly 97 per cent of people worldwide remain in their country of birth.

What makes international migration highly significant in economic and political terms is its global distribution and concentration: migrants form 11.2 per cent of the populations of 'developed regions' compared with only 1.7 per cent of the populations of developing regions (UN Population Division 2015). The Gulf oil-producing countries have massive shares of temporary migrant workers in their populations: ranging from 32.3 per cent of the population in Saudi Arabia through to 73.6 per cent in Kuwait and 88.3 per cent in the United Arab Emirates. Of the developed countries where permanent settlement of immigrants is common, Australia comes first: the UNDESA Population Division (2015) puts its migrant stock at 28.2 per cent of the nation's total population. The respective figures for Canada and the USA are 21.8 and 14.5 per cent. Western and northern European countries have migrant stocks of 13–14 per cent of their populations, while southern European countries have an average of 10.3 per cent. Eastern Europe is much lower at 6.7 per cent (UN Population Division 2015). Within these developed countries, the level of concentration of immigrants is highest in big cities such as Toronto and London—and, of course, Sydney and Melbourne.

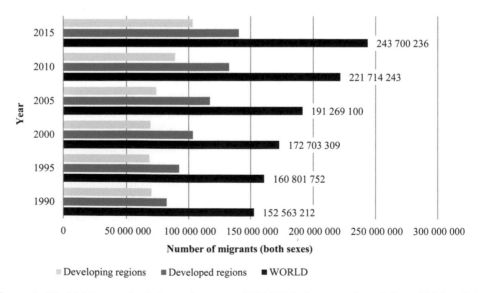

Figure 1. World Migrant Stock by major areas, 1990–2015. Source: adapted from UN Population Division (2015)

The Asia-Pacific region—home to over 60 per cent of the world's population—has always been a space of population movements (Castles, de Haas, and Miller 2014). Economic mobility is increasing today in response to better transport and communications, and patterns are becoming increasingly complex. Skilled migration to North America, Oceania and Europe (often seen as a 'brain drain') started in the 1950s. It frequently leads to family reunion and settlement, although some migrants return as new opportunities emerge. Most Asian migration is outwards to other regions. Asian countries attract few long-term settlers, with the exception of marriage migrants in East Asian countries (Lee 2008). The migrant stock is just 0.5 per cent of the total population in East Asia, 1.6 per cent in Southeast Asia and 0.8 per cent in South Asia (UN Population Division 2015).

Australia's new migration situation

Outward migration from Asia is especially important for Australia, since the region has become its main source of skilled migrants. According to data from the census round of 2000, only about 3 per cent of all Asian migrants moved to Oceania; while some 43 per cent went elsewhere in Asia; 24 per cent to the Americas; and 18 per cent to the Middle East (IOM 2010). As demographic surpluses decline and economic opportunities improve in Asia's emerging industrial centres, Australia may find it increasingly difficult to attract and retain migrants. This is where policymakers and analysts need to re-examine the assumptions upon which Australian immigration policies have been built. Table 1 summarises some of these traditional assumptions and the new realities.

Since the 1940s, Australia's model of immigration has been based on the principles of strict control of entries, predominance of permanent settlement migration and the expectation that immigrants would quickly become Australian citizens. This does not imply that there has been no change: migration and settlement patterns have been transformed by the ending of the White Australia policy, the shift away from assimilation and the emergence of multiculturalism, alongside the decline of British and European immigration and the rise of Asian arrivals. But the basic assumptions listed in Table 1 have remained. Now major adjustments to these assumptions are needed.

The opening of Australia's economy to global flows of trade and capital together with new means of transport and communication make border control much more complex,

Table 1. Challenges to Australian migration assumptions

	Long-standing assumptions about migration in Australia	Emerging situations that challenge these assumptions
1	Controllable borders	New modes of transport and increased economic ties make borders more porous
2	Australian governments decide who comes	Markets, families and individuals influence government decisions
3	Ready availability of skills and labour	Growing competition for skills and labour
4	Predominance of permanent settler migration	Growing complexity: permanent settlers, temporary migrants, students
5	One-way migration	Multi-directional and circular migration
6	Most migrants want to stay and become citizens	Diverse and changing motivations
7	Migrants adopt Australian identity (seen as multicultural identity)	Diverse forms of identity, including transnational identity

Source: author.

reducing the significance of geographical isolation. It has become easier for people to migrate on a temporary or circular basis. Migrants can keep in contact with families and communities in the homeland or other migration countries, facilitating not only further mobility but also transnationalism (i.e. regular cross-border relationships of an economic, political, social and cultural nature). Whether people move on temporary or permanent visas, the commitment to becoming permanent members of their destination community—and especially citizens—cannot be taken for granted.

This is shown most dramatically by the very rapid rise in temporary migration—particularly over the past decade (see also Breen, in this issue). Figure 2 shows that the 2014–15 target figures for permanent migration places under the Australian Migration Program were 190 000, of which 129 000 were expected to come through the Skilled Migration category and 61 000 through the Family Migration category (DIBP 2014, 2016). The Skilled Migration figure and the overall total were the highest ever, while permanent migration places under the Humanitarian category remained at 14 000, as they have for several years. But these entry figures are now dwarfed by the number of temporary entrants.

At the end of December 2015 there were 723 070 long-term temporary entrants (i.e. people granted leave to enter Australia for study or for limited terms of up to 4 years, rather than short-term visitors; see DIBP 2015). The main categories were:

- 159 910 holders of Temporary Work (Skilled) (subclass 457) visas: this is the main visa scheme for temporary employees, who can range from semi-qualified building workers to business executives;
- 328 130 international students: many of whom may stay on after completing their studies;
- 155 180 Working Holiday Makers: including students on gap years who provide a cheap source of labour (often 'off the books') for catering, retail and agriculture.

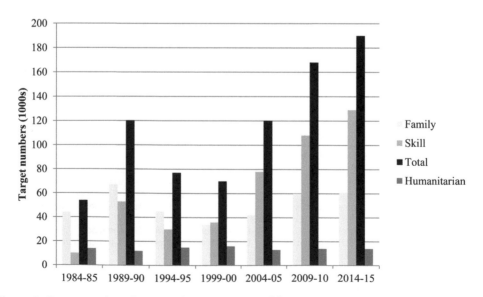

Figure 2. Target numbers for Australia's migration and humanitarian programs. Sources: based on Department of Immigration and Border Protection (DIBP 2014, 2016).

All of these categories have grown rapidly since about 2000. By 2013, there were 1.8 million people living in Australia with some sort of temporary permit. We should probably discount the 625 000 New Zealanders resident in Australia, since the majority would hold a Special Category Visa which, although a temporary visa, does permit New Zealand citizens to stay in Australia indefinitely under most circumstances (DIBP 2016). Even so, there were about 1.2 million temporary residents (equivalent to 5.6 per cent of Australia's population) in 2013, compared with about 4 million permanent immigrants. It is important to note that there is a new path to permanent residence that does not include selection prior to migration: many 457 visa holders and international students are able to apply for permanent residence while already in Australia. The very rapid growth of temporary migration, which fits well with the economic migration priority of both Coalition and ALP governments, represents an historic change away from permanent settler migration. Yet this change has gone almost unnoticed because the public debate has focused on the small number of asylum seekers coming to Australia (about 1.2 per cent of the world total; author's calculation based on UNHCR 2015).

Australia's comparative advantage

The days are over when governments could assume an almost endless supply of people who want to come to work and live permanently in Australia. In future, Australia may struggle in the global competition for skills if employers and policymakers emphasise high salaries only—highly-skilled migrants may prefer to go to countries such as South Korea, China and Singapore, which can offer higher salaries still. Knowledge-based industries in Australia may experience shortages of innovative personnel as a result. Australians need to think about other reasons why people might want to work and build their lives there. In fact, economic factors have never been the only—or even the main—motivation for migrants coming to Australia. Other factors have included:

1. Strong rights for permanent residents, including family reunion and quick and easy access to citizenship.
2. A good environment, facilitating a healthy and attractive lifestyle.
3. Democracy, security and political freedom.
4. Multiculturalism: especially recognition of cultural rights and government measures to combat racism and to ensure equal participation.

It is important to ask whether these draw-cards for immigrants are being maintained in the current context of securitisation and public scepticism about migration. In some ways, current attitudes and recent policy changes seem to be going in the wrong direction. For example, rights of family reunion have been restricted, so that it is now more difficult for migrants to bring in members of their extended families, such as parents. For many Asian migrants, it is very important to observe traditional obligations with regard to care of elderly parents, and also to look after the graves of parents who have passed away. Limiting family migration for parents might well put off potential migrants. Access to citizenship has also become more difficult since the lengthening of the waiting period from 2 to 4 years and the introduction of a 'citizenship test' through the *Australian Citizenship Act* of 2007. This legislation was introduced by the centre-right Coalition government led by Prime Minister John Howard in the wake of increased arrivals of asylum seekers by boat, as well as terrorist attacks in the USA, Europe and Bali. The more restrictive policy on citizenship was widely interpreted as implying that immigrants could not be trusted.

On the second point, Australia still offers a good environment and opportunities for a healthy lifestyle. However, the current denial of climate change, the rejection of renewable energy sources in favour of continued reliance on fossil fuels, and failure to provide adequate protection for forests, wildlife and the Great Barrier Reef may also make potential migrants question Australian governments' continuing commitment to these goals. Similarly, widespread public scepticism about political leadership and corruption in public life could make migrants question the strength of Australia's democracy and political freedom. Security discourses which appear to target specific origin groups (notably Muslims) can also have negative effects on the attractiveness of settlement in Australia.

Is multiculturalism still a draw-card for Australia? Since it was first introduced as a public policy in the early 1970s, multiculturalism has gone through a number of incarnations that cannot be discussed in detail here (see Jakubowicz and Ho 2013). Probably its high point as an attraction for potential migrants was reached in 1989, with the principles set out in the ALP government's *National Agenda for a Multicultural Australia* (OMA 1989):

- the right of all Australian citizens and residents to maintain their own culture, religion and language;
- the right to equal participation in all governmental services and benefits as laid down in the 'access and equity' strategy;
- the duty of government to combat racism and discrimination;
- the obligations of all Australian citizens and residents to accept democracy, the rule of law and gender equality.

These principles were never fully achieved, but they did set important benchmarks that have since been eroded. The idea of 'access and equity' to government services and benefits for all on the basis of need has been replaced by a critique of a so-called 'culture of entitlement'. Since the terrorist attacks of September 2001 security concerns have opened the door to Islamophobia, leading to the isolation and discrimination of Muslim and Middle Eastern migrants. The already-mentioned *Australian Citizenship Act* of 2007 has made citizenship less inclusive. Today the emphasis is no longer on multiculturalism as an aspect of citizenship, but rather on ideas of 'harmony' and 'social cohesion' as summed up in former Prime Minister Tony Abbott's image of a 'Team Australia', which implies that certain immigrants (again, those of Muslim and Middle Eastern background) should not be seen as loyal citizens.

In many ways, Australia seems to be moving away from the characteristics that made it so attractive to immigrants in the past. But we also need to ask whether new approaches are needed in view of the major shifts in migration patterns already outlined in this paper. As pointed out above, patterns of migration are getting much more complex. Some migrants do wish to settle permanently, others plan to stay just a few years and then return home, while yet others pursue international careers that may involve multiple movements of new migration, circular migration and return migration to various destinations. And in addition, some migrants do not have fixed plans with regard to migration;

they want to keep their options open, and to maintain cultural characteristics and citizenship rights that permit future flexibility. The emergent trend for Australia may be a shift from a multicultural notion of citizenship to a transnational model. What would that mean?

First, policymakers, and indeed the Australian public, need to ask whether temporary migrants actually want to become Australians. If people see their futures in flexible mobility, we may need to reshape our citizenship model. The answer cannot lie in denial of rights to people whose future plans are unclear, for the presence of a minority with inferior rights is always disadvantageous for all, as it can lead to a divided and conflictual society. Rather, we need to rethink the meaning of citizenship rights in an increasingly mobile world. Mobile people need to have rights where they live and work—even if they do not intend to stay there permanently.

Second, it is necessary to establish the legal and administrative preconditions for rights that transcend national borders. Clearly this requires international cooperation, since rights established unilaterally by a single nation could hardly gain regional or global recognition. Some such rights (such as the right to vote) could become dormant if people move on, and be reinstated upon return. Other rights need to be additive: accumulated pension and superannuation entitlements should be portable when a migrant moves to a new destination. This is already the case in the EU, which might serve as a model for regional portability of rights, as a precursor to global acceptance of the principle. Some rights should have global validity from the outset. This obviously applies to basic human rights such as protection from violence and to workers' rights such as equal pay for equal work. But it also applies to cultural rights, such as freedom of religion, language and cultural practices—provided these do not infringe the rights of others. Flexible and multi-layered identities are an intrinsic aspect of transnationalism.

Conclusion

The Australian model of immigration and multiculturalism has been highly successful. But major changes in the character of migration and in the aspirations and opportunities of migrants make it necessary to rethink the model and introduce new approaches. Just as Australia was a leader in the shift to multiculturalism from the 1970s to the 1990s, we now need to play a leading role by developing a new model of transnational citizenship. The challenge is to maintain the principles of equity and inclusion in new forms. The key issue we need to discuss, and on which Graeme Hugo was steadfastly fixed in his incomparable academic career, is what sort of society we want in the twenty-first century—and how immigration can contribute to achieving this.

Disclosure statement

No potential conflict of interest was reported by the author.

References

Castles, S., H. de Haas, and M. J. Miller. 2014. *The Age of Migration: International Population Movements in the Modern World*. 5th ed. Basingstoke: Palgrave Macmillan.

DIBP. 2014. *The People of Australia: Statistics from the 2011 Census*. Canberra: Department of Immigration and Border Protection, Australian Government.

DIBP. 2015. *Temporary Entrants and New Zealand citizens in Australia, as at 31 December 2015*. Canberra: Department of Immigration and Border Protection, Australian Government.

DIBP. 2016. *Fact Sheet Migration Programme Outcomes*. Canberra: Department of Immigration and Border Protection Accessed July 4, 2016. http://www.border.gov.au/about/corporate/information/fact-sheets/migrant

Hugo, G. 1994. Migration and the Family. Vienna: United Nations Occasional Papers Series for the International Year of the Family, no. 12.

Hugo, G. 2006. "An Australian diaspora?" *International Migration* 44 (1): 105–33. doi:10.1111/j.1468-2435.2006.00357.x

Hugo, G. 2011. *A Significant Contribution: The Economic, Social and Civic Contributions of First and Second Generation Humanitarian Entrants*. Canberra: Department of Immigration and Citizenship. http://www.immi.gov.au/media/publications/research/_pdf/economic-social-civic-contributions-booklet2011.pdf.

IOM. 2010. *World Migration Report 2010: The Future of Migration: Building Capacities for Change*. Geneva: International Organisation for Migration.

Jakubowicz, A. and C. Ho, eds. 2013. *For Those Who've Come Across the Sea: Australian multicultural theory, policy and practice*. Melbourne: Australian Scholarly.

Lee, H.-K. 2008. "International Marriage and the State in South Korea: Focusing on Governmental Policy." *Citizenship Studies* 12 (1): 107–123. doi:10.1080/13621020701794240

OMA. 1989. *National Agenda for a Multicultural Australia*. Canberra: AGPS.

UNHCR. 2015. *UNHCR Global Trends: Forced Displacement in 2014*. Geneva: United Nations High Commissioner for Refugees. http://www.unhcr.org/556725e69.html.

UN Population Division. 2015. *Trends in International Migrant Stock: The 2015 Revision*. New York: United Nations Department of Economic and Social Affairs.

Ageing research in Australia: reflecting on Graeme Hugo's four decades of contribution

Debbie Faulkner, Helen Barrie Feist and Judith Lewis

ABSTRACT

Globally, population ageing is one of the most pressing social and policy issues faced today. Over the next two decades, Australian society will face dramatic increases in the proportion of the population aged 65 years and over, as the baby boomers move into older age and fertility levels remain low. Yet population ageing is not a surprising or new trend—demographic changes in the age profile of a population tend to occur incrementally rather than suddenly. As a demographer and geographer, Graeme Hugo drew attention to this trend in Australia's population more than three decades ago. Throughout Graeme Hugo's vast breadth of work over the past 40 years, there has been a consistent thread of demographic analysis and academic thought associated with the ageing of Australia's population. This paper focuses on Hugo's contributions to academic thought and policy on Australia's ageing population and the challenges associated with this for both service delivery and health policy as Australian society moves into an unprecedented era of population ageing.

Introduction

Over the last few decades, the ageing of the world's population has been recognised as a process of major economic and social significance (Andrews 1999). The speed at which national populations are ageing is quite variable. In more developed countries including Australia, populations have been on an ageing trajectory for over a century and population ageing is well advanced. In developing regions of the world ageing is a more recent phenomenon but is occurring rapidly, requiring adaptation at a faster rate. In all major regions of the world, except Africa, the population aged 60 years and over is projected to reach 25 per cent by the middle of the century (United Nations 2015). In Australia, the numbers of older people will increase rapidly as the baby boomers[1] move past the age of 65 years, with the median Australian Bureau of Statistics (ABS 2013b) projections anticipating the numbers aged 65+ years in Australia to increase by 84.8 per cent, from 3.1 million in 2011 to 5.7 million in 2031. The proportion of the Australian population aged

65+ years will also increase. In 2011 people aged 65+ years made up 13.8 per cent of the total population; this will increase to 18.7 per cent by 2031. Graeme Hugo always emphasised the importance of anticipating population numbers as critical to planning ahead for aged care and health service needs, housing and infrastructure, and welfare costs. The proportion of the total population aged 65+ years allows us to consider the ratio of working to non-working people in the total population and the effects on intergenerational transfers of taxation, income and wealth.

Graeme Hugo's interests in population ageing, predominantly in Australia, began early in his career. Hugo was one of a limited group of researchers who first identified that structural ageing of populations could present challenges to society (e.g. Howe 1981; Rowland 1981, 1982a; Borrie and Mansfield 1982; Kendig et al. 1983; Howe and Preston 1984; Kendig and McCallum 1986). Over the decades, Hugo consistently reiterated the long-term, multifaceted complexity of the structural ageing of the population; a process shaped by multiple demographic phenomena including lower birth and death rates, increased longevity, and increased population mobility and migration. However, Hugo stressed that with acknowledgement, understanding and sufficient lead time to allow appropriate planning and policy development population ageing could be an opportunity for communities (Hugo 2013b, 2014a, 2014d). Through his research, Hugo was an advocate for promoting an understanding of the broader implications of population ageing and the need to set in place policies and practices that would create positive ageing outcomes—not only for the community but also for older individuals.

Throughout Hugo's vast breadth of work produced over the past 40 years, there has been a consistent thread of demographic and geographic analysis and academic thought associated with the ageing of populations. From his first publications in 1979 to his final contribution in 2014, he produced over 60 journal articles, reports, and book chapters and over 100 conference papers and presentations on population ageing in Australia and in developing countries. Additionally, many of Hugo's numerous reviews of overall population trends (see, for example, Hugo 1979, 1983a, 1986a, 1990, 2010a, 2010b, 2014a; Commonwealth of Australia 2010a) included sub-sections on ageing. Hugo's goals from early on were not simply to present detailed demographic and spatial information on ageing but also to elucidate the policy implications of these trends. Of central importance for Hugo was how the confluence of demographic change and policy would impact on communities and individuals. In recognition of the need for interdisciplinary collaboration to provide a holistic analysis of the impacts of demographic change he often worked collaboratively, not only with his demography and geography colleagues but also with colleagues across areas such as gerontology, social analysis, economics and health. Throughout his career, governments, academic institutes and national and international peak bodies and not-for-profit organisations sought Hugo's expertise and forethought in the field of human geography and geographical gerontology.

Human geographers' interest in, and contribution to, the field of ageing worldwide did not occur until the late 1970s and Hugo was at the forefront of this work. In a review of geography and gerontology by Warnes (1990, 26), Hugo was named as one of the 'very few geographers [who] have made gerontological questions their principal interest'. Similarly in Harper and Laws's (1995) review of the geography of ageing, the authors lamented the continuing low profile of geographical studies of ageing but, again, Hugo's enduring contributions were noted. Hugo's work is particularly noteworthy because of his attention to intra-urban analyses and his acknowledgement and study of the ethnicity of ageing populations in Australian cities (Hugo 1983b, 1998a, 2000a). This concentration on ethnically diverse populations has always been a focus of Hugo's work[2] yet this area remains, according to Skinner, Cloutier, and Andrews (2015), an area of population geography that has not attracted sufficient academic deliberation.

While ageing in developing countries and international migration of the elderly were areas of Hugo's interest and expertise (see Hugo 1986b, 1992a, 1993a, 1994, 2000b), the focus of this paper is on his work as it pertains to population ageing in Australia. In the 1980s and 1990s his focus was on the description of trends and characteristics, the recognition of ethnicity, spatial patterns of ageing and consideration of ageing in a national and world context. In the late 1990s, and into the first two decades of this century, Hugo concentrated more specifically on the pressing issues facing the new wave of older people (the baby boomers) in terms of health, migration and ethnicity, demand for aged care services and workforce issues. This paper begins with the genesis of his major work in these areas and highlights his perceptive grasp of various aspects of policy and population change that impact on the older population, as well as policy change driven by an ageing population. The paper argues that Graeme Hugo was an early thinker in the area of spatio-temporal demography in ageing and in elucidating the need for timely planning to accommodate the gradual changes in population, in particular the movement of the baby boom cohort through the age structure. It was finally at the turn of the century that ageing of the Australian population reached a prominent position on the government agenda and Graeme became highly sought and commissioned to provide his expertise in the ageing of the population and to be a member of various State and national committees.

Australia's ageing population: trends and characteristics

Demographically and from a policy perspective, the issue of ageing in Australia received little attention until the late 1970s (Howe 1981; Rowland 1997). The National Population Inquiry of the mid-1970s, known colloquially as the Borrie report (Borrie 1975), was the first national report of Australia's demographic position, and it did not regard ageing an issue of concern. Based on the assumptions with regard to population change in the Australian National Population Inquiry, Borrie predicted that the population aged 65 years and over would be between 9 and 15 per cent by the year 2030. However, the proportion of Australia's population aged 65 years and over reached 8.9 per cent by 1976, 14.6 per cent by 2011 and is projected to reach between 18.3 and 19.4 per cent by 2031 (ABS 2013a). As outlined by McDonald and Kippen (1999), the position taken by the report was justifiable based on the fact that it was only after its publication that birth and death rates began to fall significantly (and to lower levels than predicted in the report) and important and unforeseen social and economic changes began within society. Such changes included shifts in social attitudes, the changing role of women in society (for instance, to pursue education and employment opportunities), greater access to birth control and abortion, public health campaigns and significant advances in diagnostic, surgical and other medical advances. These advances in health were significant in reducing mortality and it was Hugo who appears to have first identified the significant and unprecedented improvements in life expectancy of older adults in Australia (Hugo 1986c). These changing conditions 'sparked', as reported by Howe (1981) and Rowland (1997), a renewed interest in ageing research that explored past and present patterns of population ageing, past and present experiences of older Australians and the implications of these patterns for the policies designed to assist older people (Howe 1981). This interest has continued unabated over the last 30 years, resulting in an ever-burgeoning interest in ageing research that today provides a multiplicity of perspectives.

One of the first areas attracting academic attention was the need for a comprehensive understanding of the growth and characteristics of the older population, generally defined as people aged 65 years or over. At the time, only limited use had been made of census data, with the main contributions being the Australian Bureau of Statistics (ABS) publication *Australia's Aged Population 1982* (ABS 1982), Pollard and Pollard's "The Demography of Ageing in Australia" (Pollard and Pollard 1981) and various works by Rowland (1981, 1982a, 1982b, 1983a, 1983b, 1984). For Hugo, this impetus to research, understand and identify the policy and planning implications of Australia's population was enhanced by funding available from the then Department of Immigration and Ethnic Affairs and the release of the 1981 census data. This enabled a more intensive examination of the growth and characteristics of Australia's population than had ever been available previously due to the more detailed age breakdown available in the published census data and, more importantly, because of the release for the first time of 1 per cent public use sample unit record tapes for both individuals and households (ADSRI 2009). As Hugo stated:

> This has opened up a wide range of research possibilities in allowing us to explore the characteristics of the aged population as well as their housing and living arrangements at a level of detail not previously possible. (Hugo and Wood 1984, 5)

Hugo in co-authorship with Wood (now Faulkner) (Hugo and Wood 1984), produced a 193 page Working Paper that extensively examined Australia's aged (65+) population at the time.[3] The conclusion of that Working Paper, and the chapter on ageing in the final report (Hugo 1986c),[4] identified a number of policy issues that continue to be relevant today: the need to re-examine health policy and financial policies; the need for innovative approaches in providing services in the context of an ageing population; the need for the planning process to be inclusive of older people themselves to ensure service quality and delivery, and the need for suitable housing which would encourage 'ageing in place' with the passage of time. The ability of the older population to lead a decent life was identified as a partnership between the individual, the government, family and the community.

With the availability of a census every 5 years and other sources of data (ABS, government departmental data and specific survey data, for example) Hugo frequently updated and refined his detailed analysis of the size, composition and spatial distribution of the population (e.g. Hugo and Rudd 1989; Hugo 1992b, 2010a, 2013a, 2014a; Commonwealth 2010a). For the older population he increasingly concentrated on the baby boom cohort as they moved towards, and into, the older age groups. Other authors also provided demographic insights into Australia's ageing population (see, in particular, Borowski and Hugo 1997; McDonald 2004; Borowski and McDonald 2007). There are a number of authored books and edited collections that reflect the breadth and depth of ageing research in the country and discuss in detail specific aspects of Australia's ageing population—in terms of employment, income, health, aged care, and housing, for example, as well as reflecting on major developments in policy at both national and State levels over time (Borowski, Encel, and Ozanne, 1997, 2007; Davies and James 2011; Butler 2015).

In a chapter on ageing in his book *Australia's Changing Population* (1986c), Hugo discussed that as the baby boomers were at this time moving into early and middle adulthood with less demand on public services, the difficulties society had endured in catering for this cohort (i.e. the heavy demand on service provision, in particular education) were significantly lessened. The baby boomer cohort was, according to Hugo (1986c, 158), 'entering a period where it will cease to be labelled as troublesome'. This optimism, he commented, had to be tempered by the understanding that this cohort would move into the older ages early in the next century and could potentially raise a number of challenges. However, as demographic change is often 'incremental and gradual', the movement of the baby boomers into older age in the first half of the twenty-first century could be anticipated well in advance and should not surprise policymakers and planners (Hugo 1991). Graeme continued to advocate the predictability of the changes in the aged population over time and the need to plan in advance. The ageing of Australia's population has been, and at times continues to be, seen as a crisis (Hunter 2014).

Over time, Hugo went on to examine, in greater detail, a number of aspects of Australia's ageing population. The following sections of this paper highlight some of the major issues foregrounded in his work: ethnicity and migration, the spatial dimensions of population ageing in Australia, the South Australian context, and the need for policymakers to respond to the movement of the baby boomer cohort into older age.

Ethnicity and migration: the diversity of Australia's 'older' population

The older population is not a homogeneous group but one of considerable diversity. In Australia, one of the defining features of this diversity is ethnicity. Substantial changes in immigration policy in the post-Second World War period meant that by the 1981 census, 42 per cent of Australia's total population had been born overseas or had a parent who was born overseas (Hugo 1984, 1986c). This post-war immigration would inevitably impact on the characteristics of the older population. By the 1970s, the overseas-born older population (aged 60+) was growing in size at a rapid rate, more than twice as fast as the Australian-born older population (Hugo and Wood 1984). The growth of the older population from non-English-speaking backgrounds (now referred to as Culturally and Linguistically Diverse—CALD—populations) was around four times faster than that of the Australian-born population (Hugo and Wood 1984). By the 2011 Australian census, one in five older Australians was born overseas and this will increase to one in four in the coming years (Hugo, Feist, and McDougall 2013). Further, while the Australian-born population aged 65 years and over increased by 68.1 per cent between 1981 and 2011, the overseas-born and CALD populations aged 65 years and over have increased by 191.5 and 339.5 per cent respectively over the same period (Hugo 2013b).

This rapid change in the characteristics of Australia's older population raised a number of issues in the 1980s that policymakers and service providers had not confronted previously, foremost of which were language and communication issues. Many among the older CALD population, and in particular women, had limited English-language skills and had little or no interaction outside their homes or outside their cultural communities (Hugo 1984).

Hugo's significant contribution to a geographic and demographic portrayal and understanding of the overseas born in Australian cities and regional areas (including the older cohort within this population) was through a series of analytical reports—the *Atlas of the Australian People* series—based on 1986 census data (Hugo 1989–92), and a second series of *Atlases* in collaboration with others, based on the 1991 census. These *Atlases*, commissioned by the Conference of Commonwealth, State and Northern Territory Ministers for Immigration and Ethnic Affairs, provided overwhelming detail on first- and second-generation immigrants to Australia, including details on the spatial distribution of major ethnic, birthplace, cultural and religious groups which comprise the population in each State and Territory. Hugo was of the strong opinion that to truly plan for the equitable and cost-efficient provision of appropriate services for these groups it was necessary to have a comprehensive understanding of their spatial distribution—not only at a single point in time but also of 'how and why the spatial pattern is evolving and changing' (Hugo 1989, 345). He argued that such understanding was essential to ensure that the provision of services could 'be planned in an "active" rather than "reactive", *post-hoc* way to ensure the well-being of the groups involved' (Hugo 1989, 345). Evidence of the acceptance of this shift can be seen in the steady engagement of policymakers and government departments with Hugo and his work, including commissioned reports and sitting on numerous advisory boards (Hugo 1989–92, 1990, 1998b, 2007; Hugo, Feist, and McDougall 2013).

In more recent years, Hugo's work on older CALD populations continued to highlight issues, especially access to information and aged care services, but also included more contemporary issues such as the CALD digital divide (Hugo, Feist, and McDougall 2013). Hugo's long history in researching this population resulted in Hugo being a keynote speaker at the inaugural 'Ageing in a Foreign Land' international conference in 2013. At this conference, Hugo highlighted issues such as language proficiency, home ownership, caring roles, access to services and pension sources for different birthplace groups. One of the final pieces of work Hugo undertook on CALD ageing was a comprehensive review of all Australian research on CALD ageing since the 1970s for the Federation of Ethnic Community Councils of Australia (Hugo, Feist, and McDougall 2013). This report provided a comprehensive overview of research to date and went on to identify key research gaps and define a research agenda for CALD ageing moving forward. This report highlighted the breadth of issues still not well understood, such as older people from new and emerging CALD communities, intergenerational care, the treatment and approaches to dementia and mental health for older CALD people and the complexities in providing an equitable aged care system that includes older people from CALD backgrounds. This report led to a parliamentary round table on CALD ageing issues in 2014.

Spatial distribution: Australia's uneven geographies of ageing

Being a geographer, Hugo had a strong passion for understanding the spatial distribution of populations. As time progressed, Australia's older population has exhibited a different and evolving geographic distribution when compared to the broader population, reflecting the settlement and development of regions and suburbs at particular points in time, as well as trends in residential choice. Nationally, there is considerable variation in the distribution and growth of the older population by State and even greater disparity at a regional and local government level (O'Brien and Phibbs 2011a). Understanding this spatial patterning and future changes is vitally important in understanding the contributions that older people bring to communities and the need for services and facilities for which much of the cost falls upon State and local governments (Hugo 1987, 2014b; O'Brien and Phibbs 2011b). As a result, government departments and numerous local councils have produced strategic plans and strategies focused specifically on the older population. While it can be difficult to directly link the work of academics to specific policy directions, Hugo was often commissioned to provide background papers and research focused on spatial issues for government. Such research, for example, included population projections for non-metropolitan areas in South Australia for the South Australian State Government Interdepartmental Forecasting Committee (Hugo, Rudd, and Cooper 1981), demographic profiles of local government areas in South Australia (Hugo et al. 1981), involvement in the NSW government's Ageing 2030 Working Groups to develop long-term priorities and strategies for an ageing population (Hugo 2007), and socio-demographic analysis of Australia's CALD older population to establish emerging aged care needs for the Department of Social Services (with Outcomes Plus) in 2014.

In Australia, most older people live in the capital cities of each State but the distribution of the older population within these cities has changed over time. Using Adelaide as an example, Hugo (2014a) described changes in the distribution of the older population over time, but the trends described reflect the patterns and dynamics of change in all major cities in Australia. For instance, in Adelaide in 1971, 44.5 per cent of people aged 65 years and over lived in the inner and coastal suburbs. By 2011, this ratio had declined to just 24.4 per cent as urban renewal, urban consolidation and gentrification changed these suburbs. Between 1971 and 1991 the proportion of older people living in the middle suburbs of Adelaide increased, but subsequently declined and the real growth in older populations is now in the outer suburbs, chiefly as a result of ageing in place. These areas 'with nucleated shopping centres, low density network[s] of public transport and low density of services for the elderly' pose important challenges for service provision (Hugo 2014a, 32).

Hugo's analyses not only reviewed metropolitan distributions of the older population but also examined regional and rural patterns, where the mismatch between the need for, availability of, and access to services is even starker than in urban areas (see, for example, Hugo, Rudd, and Downie 1984; Hugo 1987, 1990, 2003, 2014a; Feist et al. 2011). Regional areas of Australia in general have an older age structure than the capital cities; a result of ageing in place, the outmigration of young adults from non-metropolitan areas to the metropolitan areas for employment and educational opportunities, and the reverse movement of people aged 50 years and over from metropolitan areas to regional areas for lifestyle reasons (Hugo 2007, 2014a; Feist et al. 2011). This rural and regional structural ageing of populations has resulted in challenges for policymakers, planners and service providers to ensure that older people are given equitable access to housing, transport, aged care and health services and social opportunities, regardless of where they choose to live.

Hugo's recognition of the importance of spatial patterns in the distribution of Australia's older population, and his ability to justify and convey its importance to government, planners and policymakers from the late 1980s and early 1990s, was represented in much of the work undertaken by Hugo's National Centre for the Social Applications of GIS

(GISCA)[5] (see, for example, Hugo 1987, 1989, 2001, 2007; Hugo et al. 2009). Hugo recognised that understanding *where* older people lived was critically important to the provision of transport and other services because of the greater likelihood of limited mobility later in life creating a greater reliance on the local community and the provision of services to the home to enable ageing in place (Hugo et al. 2009).

South Australia's ageing population

Hugo was a champion of research into the population and social issues confronting his home State of South Australia. For over 40 years, ageing of the population in South Australia has been more pronounced and more rapid than for any other mainland State. This trend will continue over the coming decades because of the current older age structure of the population and the long-term pattern of a disproportionate loss of younger age groups through migration to elsewhere (O'Brien and Phibbs 2011a). Hugo's interest in ageing in South Australia was also sparked by the composition of the State's population. South Australia fared well in the 1950s and 1960s as a place of settlement for the post-Second World War migrants from the UK and Europe, with significant net migration gains (more than its proportionate share), consisting of young adults. These migrants now comprise a significant proportion of the older population of this State. In 2011, 38 per cent of South Australians aged 65 years and over were born overseas and just over half of this population had been born in countries where English is not the first language (Hugo 2014c). Hugo produced a number of papers and reports on South Australia's older population covering a range of topics including general demographic processes and characteristics, ethnic composition, rural and regional settlement, access to health and aged care services and the digital divide (Hugo 1983a, 1983b, 1984, 1989, 1998b; Hugo et al. 2009; Feist et al. 2011; Feist, Parker, and Hugo 2012).

One of the most comprehensive of these reports was the *State of Ageing in South Australia* report for the South Australian Office for the Ageing (Hugo et al. 2009). That report provided a comprehensive overview of demographic and social indicators for South Australia based on the State government's plan—*Improving with Age: Our Ageing Plan for South Australia*. At that time, Hugo described the ageing of South Australia's population as 'one of the most significant challenges facing the state during the next three decades' (Hugo et al. 2009, 8). As a State, he recognised the challenges an ageing population presented in terms of economic growth, service provision, and demands on the health and aged care systems. However, the report went on to emphasise that the ageing of South Australia's population also represented important 'opportunities which, if taken up, can contribute to the state's broader social and economic goals' (Hugo et al. 2009, 75). Hugo felt South Australia, and Australia more broadly, was in a position to lead the world in 'best practice' in terms of service provision, engagement of the older workforce, developing an 'ageing industry' and, most importantly, shifting the society 'mindset' and attitudes about older people and ageing. Hugo continued to emphasise this point in papers, reports and presentations throughout the remainder of his career; suggesting that such opportunities would be realised only if society undertook a significant conceptual change, especially among key stakeholders such as policymakers and employers. This challenge to see ageing as an opportunity for Australia, rather than a threat, continues to the present day (Commonwealth of Australia 2011; Kalache 2013; Butler 2015).

Baby boomers: Australia's future ageing demographic

The baby boomers, born between 1946 and 1965, are increasingly the focus of much attention by policymakers and researchers alike as they approach older age. Moving the impacts of the baby boomers growing older into a prominent position on the government agenda, Bishop (1999) argues persuasively that both the well-being of baby boomers and the well-being of the country as a whole would depend on innovative approaches taken by government to prevent significant reductions in national productivity, as baby boomers began to reach traditional retirement age, followed by an equally significant rise in demand for government income supports and services. The Department for Ageing's subsequent *National Strategy for an Ageing Australia* (Andrews 2002, v), which formalised government commitment to limiting the 'fallout' of population ageing, presented a strategic framework that it claimed would ' ... underpin the Government's leadership role in encouraging the development of appropriate economic and social policies'.

Amongst the slew of reports that were generated in the early 2000s that focused on the imminent ageing 'crisis' that would be initiated by the baby boomers were four *Intergenerational Reports* (IGRs) (Commonwealth of Australia 2002, 2007, 2010b, 2015[6]) that comprehensively investigated the potential outcomes of population ageing in Australia. The projections provided by the IGRs, adjusted over the first three editions, confirmed that, without policy interventions, the unfettered movement of baby boomers out of the workforce would result in dramatic reductions in the size of the labour force, reducing productivity and impacting economic growth. In addition, the significant increases in claims for government-funded income supports and in the demand for public health services would result in an unsustainable fiscal burden, posing a risk for living standards of all Australians.

Much of Hugo's work over the past three decades examining the demography and geography of ageing has involved an analysis of the changing nature of Australia's population as the baby boomers move through into older age. Hugo aptly pointed out that 'this growth is not crystal ball gazing but is able to be specified with a high degree of certainty' (Hugo 2014d, 1). While the ageing of Australia's population has been at the centre of much academic, social and political debate over the past decade or two, much of this work has focused on the numbers—the changing structure of the population and the impacts of this on Australian society. While Hugo himself contributed to this demographic analysis since the 1990s, in the last decade of his career he increasingly turned his attention to other dimensions of the baby boom generation, namely the diversity in social composition, health and well-being, the changing spatial distribution of this large population cohort, and the impacts these will have on society, planning, policy and service provision in the future.

Hugo argued that the changing composition of this population group, as it ages, will produce as many profound changes for society as the shift in the proportion and overall numbers of older Australians (Hugo 2013b). These changes in composition and social characteristics include greater cultural diversity; changing household composition (including increased divorce and separation leading to more people entering older age living alone, the rise of same-sex partnerships, blended families and increased rates of childlessness); greater use of new technologies, and different health issues (including increased rates of obesity and chronic disease than previous generations in later life) (Hugo, Taylor, and Dal Grande 2008; Feist et al. 2010, 2011; Buckley et al. 2013; Hugo 2013b; Taylor et al. 2014). Hugo (2013b, 2014d) also stressed that the geography of the baby boomers—in their older age—will be very different from previous generations. He argued that this needed to be explored further and taken into account when considering the location of future aged care-specific services as there would increasingly be a mismatch between current provision of aged care services, which tend to be heavily invested in location-specific capital such as nursing homes and retirement villages, and the location of the future older population.

Discussion

Graeme Hugo was an eminent scholar; by the 1990s he was considered Australia's leading population geographer (Stimson et al. 2015). While this paper can only briefly outline his academic achievements and influence, there are a number of themes that permeate his work in relation to ageing in Australia.

First, the numerical growth of the older population is highly predictable. There are only different scenarios with respect to the *proportion* of the total population aged 65 and over, due to the influence of fertility rates, migration intakes and so on. Thus, given this predictability, it is possible to plan ahead to make effective policy decisions. Second, structural ageing is complex, multifaceted and is a whole of population process influenced by social, medical and policy changes. Third, there is a need for policymakers and planners to be fully informed of the characteristics of the older population and how they change over time—population geographers and demographers have a great deal of insight to offer in this regard. The need to monitor and understand the spatial distribution of populations, and the likely spatial behaviour of the baby boomers is a fourth important theme. The opportunities and challenges that an ageing population presents to society were also key areas of discussion in Hugo's body of work, leading to his assertion that the ageing of the population is not a crisis but should be viewed as a positive achievement. However, Hugo was concerned that the potentially positive outcomes of population ageing would be compromised without adequate and timely policy development and preparation at all levels (family, community, all tiers of government) and from all angles (e.g. social, economic, health, infrastructure); but he believed that a window of opportunity to 'adapt' to an ageing population still exists.

Writing in the early 1980s about the older population in Australia, Hugo summarised the ability of older persons to have a fulfilling and successful older age:

> The ability of the elderly person to adapt and thrive is contingent upon his [sic] physical health, personality, earlier life experiences and on the societal support he [sic] receives in the way of adequate finances, shelter, medical care, social roles and vocation. These critical latter factors of societal support depend not only on older people themselves but on the rest of society committing resources to this end, not only public financial resources but family and private social investments of interest, time and caring. The demographic changes of the next two decades will place more pressure on society for this commitment than has

been the case at any time in our history, and it represents a most significant challenge. (Hugo 1986c, 187)

Whilst this statement was made over 30 years ago it is still relevant today. Income, housing, access to and cost of care and services, ethnicity, family support, workforce participation, housing, loneliness and social isolation remain significant issues of research and policy relevance.

Even though Graeme Hugo continually advocated for the need to plan well in advance for the ageing of the population, great strides have only been made since the turn of the century with the elevation of ageing onto the Australian government agenda. This has prompted a wider focus on the implications of ageing in Australia. As noted in the *Blueprint for an Ageing Australia* (Per Capita 2014, 8):

> Over the last decade the nation's top policy making bodies have started to focus our population's mind on the issue of ageing. Led by the Australian Treasury—most notably through its Intergenerational Reports and its report into the economic potential of senior Australians (*Turning Grey into Gold* Advisory Panel on the Economic Potential of Senior Australians 2011)—and by leading demographers like Professor Hugo, this process has increased our outstanding of the relevant issues substantially.

Since 2000 in particular, Hugo was highly sought for his expertise and insight into population ageing to move the policy agenda forward. He contributed to major national inquiries and debates and chaired a number of government bodies and committees focusing on the economic and social impacts of ageing. These included being a member of: the National Seniors Productive Ageing Centre Advisory Board, the Productive Ageing Centre Research and Education Advisory Committee, the National Housing Supply Council, the South Australian Aged Care Planning Advisory Committee and the Ministerial Advisory Board on the Ageing, South Australian Office for the Ageing. In addition, he was an advisor to the Council on the Ageing, Chair of the Demographic Change and Liveability Panel of the Ministry of Sustainability, Environment, Water, Population and Communities and most recently (from 2012 until his death) he was the Deputy Chair of the Aged Care Financing Authority of the Australian Government. Recently, the Productivity Commission (2013, 3) raised concerns that planning for an ageing population has lost its impetus:

> In 2005, the Commission reported that timely action to address the consequences of demographic change could avoid the future need for 'big bang' policy interventions later. Over eight years later, the discussion of the possible opportunities and policy challenges presented by an ageing population seems to have waned.

This lack of interest in ageing policy, as identified by the Commission, is an issue of concern as the movement of the baby boomer cohorts into the older age groups is no longer a future scenario. The need for 'big bang' or rushed policymaking was the precise situation that Graeme Hugo had been trying to avoid. It would be sad to think that after 40 years of investment by Hugo in raising awareness of the characteristics of the older population and the impetus that began at the turn of the century, in terms of government attention and policy development, that ageing would now slip from the policy agenda. Fortunately, there is ongoing recognition of the importance of research on Australia's ageing population amongst a sizeable group of scholars informed by Hugo's body of work (and who also influenced Hugo's work), who continue to push for policy development in this area (Davies and James 2011; Feist et al. 2011, 2012; Taylor et al. 2014; Butler 2015).

Concluding remarks

There are two reflections on Graeme Hugo's extensive research on ageing and his contribution to policy development that highlight not only his academic abilities but his inclusiveness and humanity. The first from Don Rowland (2015), a geographer who influenced Hugo's work, at the Graeme Hugo Colloquium in late 2015 who concluded: 'besides his academic achievements, he has probably done as much as anyone in taking demography to the people. This is especially evident in the study of population ageing.' And from a policy perspective, the last word belongs to the Premier of South Australia, the Honourable Jay Weatherill, who stated at Graeme Hugo's Memorial Lecture, 7 October 2015:

> I talked to him [Graeme], and read his research, on a regular basis, especially in regards to the social policy aspects of the ministerial portfolios I have held over the years. I invited him to join a number of advisory bodies, which he did. He supplied expert advice in relation to my responsibilities in local government, urban development, planning, housing, ageing, state development and treasury. My work in these portfolios, and indeed the work of others in all of these areas of public policy benefitted greatly from his findings and perspectives. And while Graeme had a remarkable command of the technical tools of his area of expertise, what impressed me most however was his ability and determination to decipher the larger story, the human story the data was telling us. It was this trait that I believe led ministers, parliamentarians and policy-makers right around the nation and right around the world to seek him out, to rate his analysis so highly …

Notes

1. The term 'baby boomers' is used to describe a generation of people born after the Second World War. The definition of the baby boomer period varies between countries, and over time the definition has been revised in Australia. In Hugo's early work (Hugo 1986c, 158) and the ABS (1986) it referred to people born between 1946–47 and 1960–61. The ABS now defines the baby boom generation as people born 1946–47 to 1965–66 (ABS 2003, 2015).
2. It continues as a major area of ongoing research in Hugo's research centre—the Australian Population and Migration Research Centre—under the direction of Helen Barrie Feist.
3. In this working paper the authors covered the changing age composition of the population, the spatial distribution of the aged (65+) population, migration, ageing in place, ageing and the family, housing and living arrangements, socio-economic differentiation (workforce participation, income, education), ethnicity, and health of the older population.
4. In a review of Australian demography at the millennium this book was seen as one of the major benchmarks in demographic analysis (Lucas 1994). This book and the working papers associated with it set the standard for all of Hugo's demographic analysis of processes, trends and characteristics of populations—highly in-depth but in a language that was easy to understand. This ability plus the applied nature or relevance of his work was why he was constantly sought as a conference speaker, by policy advisers and the media.
5. GISCA was established in July 1995, funded by the Australian Research Council as a joint Key Centre venture. When the ARC Key Centre funding ended in 2001, GISCA continued as an independent unit within the University of Adelaide. The work of GISCA continues under the Australian Population and Migration Research Centre, established by Graeme Hugo in 2012.

6. For reviews of the 2015 *Intergenerational Report*, see Duckett (2015); Kendig and Woods (2015).

Acknowledgements

The authors wish to thank Janet Wall for all her help in assisting with locating copies of Graeme's work and for her insight into his academic career. Thanks also to Dr Cecile Cutler for her comments on the paper.

Disclosure statement

No potential conflict of interest was reported by the authors.

References

Andrews, G. 1999. *Ageing Triumphantly, Dame Roma Mitchell Oration, Office of the Commissioner for Equal Opportunity and The Centre for Ageing Studies*. Adelaide, South Australia: Flinders University.

Andrews, K. 2002. *National Strategy for an Ageing Australia: An Older Australia, Challenges and Opportunities for all*. Canberra: (Department for Ageing), Commonwealth of Australia.

ABS (Australian Bureau of Statistics). 1982. *Australia's Aged Population 1982*. Catalogue no 4109.0. Canberra: Australian Bureau of Statistics.

ABS (Australian Bureau of Statistics). 1986. *Australian Demographic Trends*. Catalogue No 3102.0. Canberra: Australian Bureau of Statistics.

ABS (Australian Bureau of Statistics). 2003. "Baby Boomers and the 2001 Census." *Newsletter: Age Matters, Apr 2003* 4914.0.55.001 http://www.abs.gov.au/ausstats/abs@.nsf/7d12b0f6763c78caca257061001cc588/c9c319dd7e17542fca2572e300810a3f!OpenDocument.

ABS (Australian Bureau of Statistics). 2013a. *Reflecting a Nation: Stories from the, 2011 Census, 2012–2013*. Catalogue no. 2071.0. Canberra: Australian Bureau of Statistics.

ABS (Australian Bureau of Statistics). 2013b. *Population Projections Australia 201 to 2101*. Catalogue No. 3222.0. Canberra: Australian Bureau of Statistics.

ABS (Australian Bureau of Statistics). 2015. Talkin' 'Bout Our Generations: Where are Australia's Baby Boomers, Generation X & Y and I Generation? (Feature Article) *Population by Age and Sex, Regions of Australia, 2014*, Catalogue no 3235.0. Canberra.

Australian Demographic & Social Research Institute. 2009. *Integrated Microdata Access System International, General Notes on Australian Data*. Canberra: ADSRI, Australian National University. http://ipumsi.anu.edu.au/Documentation/AustralianData/gennotes.php.

Bishop, B. 1999. *A National Strategy for an Ageing Australia*. Canberra: Commonwealth of Australia.

Borowski, A., S. Encel, and E. Ozanne, eds. 1997. *Ageing and Social Policy in Australia*. Melbourne: Cambridge University Press.

Borowski, A., S. Encel, and E. Ozanne, eds. 2007. *Longevity and Social Change in Australia*. Sydney: UNSW Press.

Borowski, A., and G. Hugo. 1997. "Demographic Trends and Policy Implications." In *Ageing and Social Policy in Australia*, edited by A. Borowski, S. Encel, and E. Ozanne, 19–53. Melbourne: Cambridge University Press.

Borowski, A., and P. McDonald. 2007. "The Dimensions and Implications of Australian Population Ageing." In *Longevity and Social Change in Australia*, edited by A. Borowski, S. Encel, and E. Ozanne, 15–39. Sydney: UNSW Press.

Borrie, W. D. 1975. *Population and Australia: A Demographic Analysis and Projection*. 2 volumes, First Report of the National Population Inquiry, Parliamentary Paper No. 6. Canberra: Parliament of the Commonwealth of Australia.

Borrie, W. D., and M. Mansfield, eds. 1982. *Implications of Australian Population Trends, Academy of Social Sciences in Australia.* Canberra: ANU.

Buckley, J., G. Tucker, G. Hugo, G. Wittert, R. Adams, and D. H. Wilson. 2013. "The Australian Baby Boomer Population—Factors Influencing Changes to Health-related Quality of Life over Time." *Journal of Aging and Health* 25 (1): 29–55.

Butler, M. 2015. *Advanced Australia: The Politics of Ageing.* Victoria: Melbourne University Press.

Commonwealth of Australia. 2002. *The Intergenerational Report 2002.* Canberra: Commonwealth of Australia.

Commonwealth of Australia. 2007. *The Intergenerational Report 2007.* Canberra: Commonwealth of Australia.

Commonwealth of Australia. 2010a. *A Sustainable Population Strategy for Australia.* Canberra: Commonwealth of Australia.

Commonwealth of Australia. 2010b. *Australia to 2050: Future Challenges (The Intergenerational Report 2010).* Canberra: Commonwealth of Australia.

Commonwealth of Australia. 2011. *Realising the Economic Potential of Senior Australians: Turning Grey into Gold.* Advisory Panel on the Economic Potential of Senior Australians. Canberra: Commonwealth of Australia.

Commonwealth of Australia. 2015. *The Intergenerational Report 2015.* Canberra: Commonwealth of Australia.

Davies, A., and A. James. 2011. *Geographies of Ageing: Social Processes and the Spatial Unevenness of Population Ageing.* Surrey: Ashgate Publishing.

Duckett, S. 2015. "Intergenerational Report Misses the Point on Health Spending." *Australasian Journal on Ageing* 34 (4): 214–216.

Feist, H., K. Parker, N. Howard, and G. Hugo. 2010. "New Technologies: Their Potential Role in Linking Rural Older People to Community." *International Journal of Emerging Technologies and Society* 8 (2): 68–84.

Feist, H., K. Parker, N. Howard, and G. Hugo. 2011. "Rural Ageing-in-Place: Community Connectedness, Health and Wellbeing, an Opportunity for New Technologies?" *SA Public Health Bulletin* 8 (1): 24–29.

Feist, H., K. Parker, and G. Hugo. 2012. "Older and Online: Enhancing Social Connections in Australian Rural Places." *Journal of Community Informatics* 8 (1). http://ci-journal.net/index.php/ciej/article/view/818.

Harper, S., and G. Laws. 1995. "Rethinking the Geography of Ageing." *Progress in Human Geography* 19 (2): 199–22.

Howe, A. L., ed. 1981. *Towards an Older Australia.* Brisbane: University of Queensland Press.

Howe, A. L., and G. A. N. Preston. 1984. "Handicap in the Australian Aged Population: Part 1: Findings and Interpretations from the Handicapped Persons Survey." *Journal of the Australian Population Association* 1 (Spring): 66–81.

Hugo, G. 1979. "Some Demographic Factors Influencing Recent and Future Demand for Housing in Australia." *The Australian Quarterly* 51 (4): 4–25.

Hugo, G. 1983a. "South Australia's Changing Population, South Australian Geographical Society." *South Australian Geographical Papers* No. 1. Adelaide: South Australian Geographical Society.

Hugo, G. 1983b. *The Ageing of Ethnic Populations in South Australia.* Adelaide: National Institute of Labour Studies and South Australian Ethnic Affairs Commission.

Hugo, G. 1984. *The Ageing of Ethnic Populations in Australia with Special Reference to South Australia.* Occasional Paper in Gerontology No. 6. Melbourne: National Research Institute for Gerontology and Geriatric Medicine.

Hugo, G. 1986a. *Population Ageing in Australia: Implications for Social and Economic Policy.* Papers for the East-West Population Institute 98. Hawaii: East-West Centre.

Hugo, G. 1986b. "Demography of Ageing in Developed and Developing Countries." *Proceedings of the 21st Annual Conference of the Australian Association of Gerontology* 1: 52–55.

Hugo, G. 1986c. *Australia's Changing Population: Trends and Implications.* Melbourne: Oxford University Press.

Hugo, G. 1987. "Ageing in Australia: The Spatial Implications." *Urban Policy and Research* 5 (1): 24–26.

Hugo, G. 1989–92. *Atlas of the Australian People*. 8 volumes. Canberra: Australian Government Publishing Service.

Hugo, G. 1990. "Demographic Trends in Adelaide: Recent Trends and the Future Outlook." Paper prepared for the Metropolitan Adelaide Planning Review, Adelaide.

Hugo, G. 1991. "What Population Studies Can Do for Business." *Journal of the Australian Population Association* 8 (1): 1–22.

Hugo, G. 1992a. "Ageing in Indonesia: A Neglected Area of Policy Concern." In *Ageing in Newly Industrialising Countries of East and Southeast Asia*, edited by D. R. Philips, 52–55. London: Edward Arnold.

Hugo, G. 1992b. "Australia's Ageing Population: Patterns and Policy Implications." In *The Elderly Population in Developed and Developing World: Policies, Problems and Perspectives*, edited by P. Krishnan and K. Mahadevan, 151–185. Delhi: B.R. Publishing Corporation.

Hugo, G. 1993a. "Review of the Population Ageing Situation and Major Ageing Issues at Local Levels." In *Productive Ageing in Asia and the Pacific*, 23–45. New York: United Nations.

Hugo, G. 1994. "Ageing in Indonesia and Australia: Similarities and Differences." In *Future Directions in Aged Care in Indonesia*, edited by G. Hugo, Proceedings of the Joint Indonesia-Australia Seminar, 36–72. Adelaide: Department of Geography, The University of Adelaide.

Hugo, G. 1998a. "International Migration and the Spatial Distribution of the Elderly in Australia." In *Immigration, Internal Migration and the National Redistribution of the Elderly Population*, edited by A. Rogers. Boulder, Colorado: The Third Colorado Conference on Elderly Migration, University of Colorado at Boulder.

Hugo, G. 1998b. *Recent Trends in the Ageing of South Australia's Population*. Adelaide, SA: Department for Transport, Urban Planning and the Arts and the University of Adelaide.

Hugo, G. 2000a. "South Australia's Ageing Population and its Increasingly Multicultural Nature." *Australasian Journal on Ageing* 19 (1): 23–32.

Hugo, G. 2000b. "Lansia—Elderly People in Indonesia at the Turn of the Century." In *Ageing in the Asia-Pacific Region*, edited by D. R. Phillips, 299–321. London: Routledge.

Hugo, G. 2001. "Addressing Social and Community Planning Issues with Spatial Information." *Australian Geographer* 32 (3): 269–293.

Hugo, G. 2003. "Australia's Ageing Population: Some Challenges for Planners." *Australian Planner* 40 (2): 109–118.

Hugo, G. 2007. *Some Spatial Dimensions of Australia's Future Aged Population: A Demographic Perspective*, Background paper prepared for 'Ageing 2030—Creating the Future', New South Wales Government, October.

Hugo, G. 2010a. *Recent Trends in the Ageing of Australia's Population and Their Implications*, Paper prepared for the Inaugural National Council on the Ageing Congress, Older Australians: A Working Future?, The Changing Nature of Work and Retirement in the 21st Century, November.

Hugo, G. 2010b. *A Sustainable Population Strategy for Australia*, Issues paper and appendices, Sustainable Advisory Panel, Chair 2010.

Hugo, G. 2013a. "The Demography of Ageing in Australia." In *Positive Ageing: Think Volunteering*, edited by L. Rogers and J. Noble, 45–56. Adelaide: Volunteering SA Inc.

Hugo, G. 2013b. "The Changing Demographics of Australia over the last 30 years." *Australasian Journal on Ageing*, 32: 30 Year Anniversary Special Issue, 18–27.

Hugo, G. 2014a. The Demographic Facts of Ageing in Australia, Appendix Q for *Aged Care Financing Authority Second Annual Report*. Canberra: Department of Social Services, Australian Government.

Hugo, G. 2014b. The Changing Distribution of Population in New South Wales and its Drivers, Final Report to Department of Planning and Environment of the Government of New South Wales.

Hugo, G. 2014c. South Australia's Ageing Population and Some Implications for Eldercare, A report for Eldercare, The University of Adelaide, Adelaide.

Hugo, G. 2014d. "The Demographic Facts of Ageing in Australia: Patterns of Growth." *Policy Brief* 2 (2). Australian Population and Migration Research Centre. Accessed March 20 2016. www.adelaide.edu.au/aprmc.

Hugo, G., H. Feist, and K. McDougall. 2013. *What do we Know and What do we Need to Know? Compiling the Evidence, Identifying the Gaps and Making Research Accessible to Providers of Services for CALD Older People.* Canberra: Report for Federation of Ethnic Community Councils of Australia.

Hugo, G., M. Luszcz, E. Carson, J. Hinsliff, P. Edwards, C. Barton, and P. King. 2009. *The State of Ageing in South Australia.* Adelaide, South Australia: Summary Report for the Office for the Ageing.

Hugo, G., and D. Rudd. 1989. "Demography of Ageing in Australia." Paper for the 3rd Australian Rotary Health Fund International Conference on Alzheimer's Disease, Canberra: University House, Australian National University, October.

Hugo, G., D. Rudd, and J. Cooper. 1981. Population Projections for South Australian Non-Metropolitan Government Areas—A Methodology and Illustrative Projections, A Final Report to the South Australian State Government Interdepartmental Forecasting Committee, Flinders University of South Australia.

Hugo, G., D. Rudd, and M. C. Downie. 1984. "Adelaide's Aged Population: Changing Spatial Patterns and their Policy Implications." *Urban Policy and Research* 2 (2): 17–25.

Hugo, G., D. Rudd, M. Downie, A. Macharper, and A. Shillabeer. 1981. A Demographic Profile of the South Coast Region of South Australia with Particular Emphasis on the Aged, A Final Report to the South Australian Health Commission, School of Social Sciences, Flinders University of South Australia.

Hugo, G., A. Taylor, and E. Dal Grande. 2008. "Are Baby Boomers Booming Too Much? An Epidemiological Description of Overweight and Obese Baby Boomers." *Obesity Research and Clinical Practice* 2 (3): 203–214.

Hugo, G., and D. Wood. 1984. *Ageing of the Australian Population: Changing Distribution and Characteristics of the Aged Population*, 1981 Census Project—Paper 8. Bedford Park, S.A.: National Institute of Labour Studies, Flinders University.

Hunter, M. 2014. "Australia's Ageing Tsunami and the Coming Aged Care Catastrophe." *Independent Australia*, 16 January. https://independentaustralia.net/life/life-display/australias-ageing-tsunami-and-the-coming-aged-care-catastrophe,6070.

Kalache, A. 2013. *The Longevity Revolution-Creating a Society for all Ages*, Adelaide Thinker in Residence Final Report, South Australian State Government, Adelaide.

Kendig, H. L., D. M. Gibson, D. T. Rowland, and J. M. Himer. 1983. *Health, Welfare and Family in Later Life.* Sydney, NSW: New South Wales Council on the Ageing.

Kendig, H. L., and J. McCallum. 1986. *Greying Australia; Future Impacts of Population Ageing.* Canberra: National Population Council (Australia) Migration Committee.

Kendig, H. L., and M. Woods. 2015. "Intergenerational Report 2015: A Limited and Political view of our Future" *Australasian Journal on Ageing* 34 (4): 217–219.

Lucas, D. 1994. "Australian Demography at the Millennium." *Journal of the Australian Population Association* 11 (1): 33–54.

McDonald, P. 2004. "Getting a Little Older Each Year: The Demography of Ageing in Australia." In *Australia's Ageing Population: Fiscal, Labour Market and Social Implications*, edited by P. Smith and K. Henry, 31–37. Melbourne, Australia: Committee for Economic Development of Australia.

McDonald, P., and R. Kippen. 1999. "Ageing: The Social and Demographic Dimensions." In *Policy Implications of the Ageing of Australia's Population, Productivity Commission & Melbourne Institute of Applied Economic and Social Research*, 47–71. Melbourne: Productivity Commission & Melbourne Institute of Applied Economics and Social Research.

O'Brien, E., and P. Phibbs. 2011a. *Local Government and Ageing, Literature Review.* University of Western Sydney and Department of Families and Community services NSW. http://www.lgnsw.org.au/files/imce-uploads/35/local-government-ageing-2011.pdf.

O'Brien, E., and P. Phibbs. 2011b. *Local Government and Ageing, Final Report.* University of Western Sydney and Department of Families and Community services NSW. Accessed June

27 2016. https://www.adhc.nsw.gov.au/__data/assets/file/0004/250825/21_OBrien_and_Phibbs_ 2011_Local_Govt_Ageing_Literature_Review.pdf.

Per Capita. 2014. *Blueprint for an Ageing Australia*. Australia: Per Capita Australia Limited. Accessed June 24 2016. http://percapita.org.au/wp-content/uploads/2014/11/BlueprintForAnAgeingAustralia. pdf.

Pollard, A., and G. Pollard. 1981. "The Demography of Ageing in Australia." In *Towards an Older Australia: Readings in Social Gerontology*, edited by A. L. Howe, 13–34. St. Lucia, Queensland: University of Queensland Press.

Productivity Commission. 2013. *An Ageing Australia: Preparing for the Future*. Canberra: Commission Research Paper, Productivity Commission.

Rowland, D. T. 1981. *Sixty Five Not Out: Consequences of the Ageing of Australia's Population*. Sydney: Institute of Public Affairs.

Rowland, D. T. 1982a. "The Vulnerability of the Aged in Sydney." *Australian and New Zealand Journal of Sociology* 18: 229–247.

Rowland, D. T. 1982b. "Living Arrangements and the Later Family Life Cycle in Australia." *Australian Journal on Ageing* 1 (2): 3–6.

Rowland, D. T. 1983a. "Migration During the Later Life Cycles." Paper presented to Symposium on Migration in Australia, Royal Geographical Society of A/Asia (Qld. Branch) and Australian Population Association (Qld. Regional Group), Brisbane, 1–2 December.

Rowland, D. T. 1983b. "The Family Circumstances of the Ethnic Aged." In *Papers on the Ethnic Aged*, 171–214. Melbourne: Australian Institute of Multicultural Affairs.

Rowland, D. T. 1984. "Old Age and the Demographic Transition." *Journal of Population Studies* 38 (1): 73–87.

Rowland, D. T. 1997. "The Demography of Ageing and Families in Australia." *Australian Journal on Ageing* 16 (3): 99–104.

Rowland, D. T. 2015. *Graeme Hugo on Population Ageing*. Canberra: Australian Population Association, Graeme Hugo Colloquium, ANU. 2 Dec. https://www.apa.org.au/content/ seminar-highlights-graeme-hugo-colloquium.

Skinner, M. W., D. Cloutier, and G. J. Andrews. 2015. "Geographies of Ageing: Progress and Possibilities after Two Decades of Change." *Progress in Human Geography* 39 (6): 776–799.

Stimson, R., A. Maude, C. Forster, P. Smailes, D. Rudd, M. Bell, and D. Forbes. 2015. "Obituary, Professor Graeme John Hugo, BA, MA, PhD, FASSA, AO." *Geographical Research* 53 (2): 115–118.

Taylor, A., R. Pilkington, H. Feist, E. Dal Grande, and G. Hugo. 2014. "A Survey of Retirement Intentions of Baby Boomers: An Overview of Health, Social and Economic Determinants." *BMC Public Health* 14: 355. doi:10.1186/1471-2458-14-355.

United Nations. 2015. *World Population Ageing Report*. Accessed June 23 2016. http://www.un.org/ en/development/desa/population/publications/pdf/ageing/WPA2015_Report.pdf.

Warnes, A. M. 1990. "Geographical Questions in Gerontology: Needed Directions for Research." *Progress in Human Geography* 14 (1): 24–56.

Weatherill, J. 2015. *Professor Graeme Hugo AO Memorial Lecture*. The University of Adelaide, October 7, Accessed June 27 2016. https://www.youtube.com/watch?v=EGbLLf_H2yk.

Social Applications of Geographical Information Systems: technical tools for social innovation

Danielle Taylor and Jarrod Lange

ABSTRACT

The establishment of the National Key Centre for Social Applications of Geographical Information Systems (GIS) in 1995, under the directorship of Professor Graeme Hugo, was a turning point in the use of GIS in Australia. The field of GIS, previously dominated by environmental applications, now broadened its focus to include populations, services and the interactions between people and the environment. Social applications of GIS offered a unique opportunity to make service planning, reporting, funding allocations and research both smarter and fairer. Geography and geographic relationships as implemented in GIS became the integrating platform for social spatial information, invigorating social research, planning and policy. A key strength of this approach, recognised by Professor Hugo, was the ability to 'put people back into the planning process'. Further to being an integrating platform, GIS also offered the ability to generate new information and knowledge, which could facilitate evidence-based decision making. This paper focuses in particular on providing a written record of the development of the Accessibility/ Remoteness Index of Australia (ARIA) suite of spatial accessibility indices. The lasting legacy and continued relevance of this work in social applications of GIS is also reviewed in this paper, with reference to key examples of how social research and planning in Australia have been made both smarter and fairer through the contributions of Professor Hugo and his team.

Introduction

The geography of the Australian continent and its dispersed population poses many challenges for those seeking to provide services equitably to individuals and communities in need. Objective data about service need and service capacity at a local and regional level is critical to achieving resource allocations which are fair, equitable and efficient. With the capacity to integrate population, service, economic and environmental information within a spatial framework, Geographical Information Systems (GIS) have been recognised to offer great potential to contribute to social planning and research, although the uptake of this application of GIS in Australia was initially very slow. Garner (1990) observed that while GIS were well established in the areas of land information systems, natural resource and environmental management in Australia, applications of this technology to social, economic and planning activities were less well established, particularly when compared to other countries. With the information technology sector growing at a rapid rate, both within Australia and globally, the need for Australia to accelerate its use of spatially referenced social, demographic and economic data in the planning process was identified as essential by Professor Hugo and his collaborators when they applied to the Australian Research Council (ARC) for funding to establish a National Key Centre focused on social applications of GIS. Australia already had world-class social data collections; the challenge was to integrate them into a GIS and put them to better use (Hugo 1994). The overall aim of the Centre was to 'make public and private sector planning in Australia "smarter"' (Hugo 1994, 4). The Centre's research focus aimed to highlight the value of GIS within a social context, thereby facilitating 'a better understanding, description and explanation of demographic change at a range of geographic levels ... for achieving more efficient and equitable allocations of scarce resources' (Hugo 1994, 4).

A core component of the ARC funding proposal led by Hugo was the philosophical linkage between sound basic research and practical and innovative applied research. Amongst its other more detailed objectives, the Centre aimed to deliver training, education and new technologies in GIS, and to provide consultancy services and, importantly, opportunities for collaboration between planners and researchers (Hugo 1994). Potential economic benefits derived from making smarter planning decisions and a better use of scarce resources, together with the potential to generate export income from the delivery of regional expertise in the rapidly expanding Asia-Pacific region, formed a compelling argument for the establishment of the Centre.

The application for funding was successful, and in 1995 the National Key Centre for Social Applications of GIS was established. The Centre was led by Professor Hugo at the University of Adelaide, and involved close collaboration with both Flinders University and the University of South Australia. Key Centre government partners included the Australian Bureau of Statistics and the South Australian Departments of Housing and Urban Development and Environment and Natural Resources.[1] The Centre became known as GISCA and when the Key Centre concluded in 2001, GISCA became an independent entity within the University of Adelaide. In 2011, GISCA was incorporated into the Australian Population and Migration Research Centre (APMRC), also established by Professor Hugo at the University of Adelaide (APMRC 2015a).

While there have been numerous contributions to the field of social applications of GIS through the Centre's postgraduate education and its multitude of research projects, consulting reports and academic articles, a consistent conceptual theme of people and social justice has both driven and integrated the work of the Centre and is the key to many of its achievements and to the continued relevance of this research. The breadth of the research undertaken and its impacts have been substantial and include: world recognition received for building the first 3D city model, showcased at the United Nations Habitat Conference in Istanbul in 1996 (Hugo 1997); GIS-based influenza pandemic modelling conducted for the 2000 Sydney Olympics, ensuring a rapid and efficient mobilisation of resources in the event of an epidemic (Bryan and Bamford 2000); and early web mapping applications which were used to provide spatial information via a simple interface to a broad audience. Hugo (2001) identified these web mapping applications as having the potential to encourage individuals and communities to become actively engaged in planning and policy development, resulting in more considered and people-focused decisions. Research which directly aided policy and planning included: mapping to determine priority areas for fly-in fly-out female doctors; the production of the first 'Country Matters Social Atlas' with the Bureau of Rural Sciences (1999); a study of the provision of health services in non-metro-politan Australia (GISCA 2000); analysis of the accessibility of aged care services (Hugo and Aylward 1999); and the first GIS-based analysis of metropolitan walkability for Australia (Leslie et al. 2007). Arguably, one of the most significant and lasting contributions made by Hugo and his fellow researchers at GISCA and APMRC has been the development of the Accessibility and Remoteness Index of Australia (ARIA) and the use of ARIA+ (ARIA *plus*—an enhanced and refined version of ARIA) as the basis of the Australian Bureau of Statistics (ABS) Remoteness Structure. This has been incorporated into the ABS's Australian Standard Geographical Classification (ASGC) since 2001(ABS 2001b) and Australian Statistical Geography Standard (ASGS) since 2011 (ABS 2013).

Graeme Hugo's influence on critically appraising and shaping the Australian spatial data infrastructure has shaped Australian scholars' and planners' understanding of Australian geography, demography and service provision, and has delivered the means for improved social equity and better and smarter decision making. These achievements will be the focus of this paper. Following a brief description of GIS and social spatial data, we discuss the role of GIS in social spatial decision making and the capacity to leverage GIS for the allocation of resources. The paper will then review the origins, development and application of the ARIA suite of spatial accessibility indices and the ABS Remoteness Structure, drawing on published articles and Key Centre documents in addition to the personal experience of one of the authors (Taylor) as the spatial analyst responsible for the calculation and construction of the original ARIA and her subsequent involvement in the refinement and application of the ARIA methodology. Other changes to Australia's spatial data infrastructure advocated by Hugo are also discussed. The concluding sections highlight some of the many applications of ARIA+ and the ABS Remoteness Areas (RA) classification in policy and research, and discuss the expanding opportunities for future advances in the area of social spatial research.

Geographical Information Systems and social spatial data

GIS is the branch of information technology that supports the storage, management, retrieval, manipulation, display, modelling and analysis of geographic information (Delaney 1999). The important difference between a GIS and other computer-based information systems is the inclusion of a spatial reference (as simple as a postcode, or as precise as a geographic coordinate), which enables the consideration of various spatial contextual characteristics and the linking and integration of different datasets based on their location. Contextual factors (such as high or low socio-economic status neighbourhoods, or urban *vs* rural location) and measures of proximity to other features (such as the distance from someone's home to a hospital), can be included as variables for analysis within a GIS.

While mapping and visualisation are important functions of GIS which can greatly contribute to understanding problems and hypothesis generation, the analytic properties of a GIS—such as integrating different datasets, and quantifying multi-dimensional concepts such as socio-economic disadvantage, locational disadvantage and ecological vulnerability—offer great value to social and community planning (Hugo 2001). These analytical functions can be used to model and examine spatial distributions and outcomes that transform routinely collected administrative data into valuable information (Hugo 2001). The ARIA (Commonwealth Department of Health and Aged Care 1999), Metropolitan ARIA (Metro ARIA; see Taylor and Lange 2015), Pharmacy ARIA (PhARIA; see APMRC 2015b), Cardiac ARIA (Coffee et al. 2012) and walkability index (Leslie et al. 2007) are all examples of the analytical utility of GIS.

Equity in spatial decision making and the allocation of resources

The need for goods and services is not evenly distributed across all population groups and all geographic areas, and as populations and areas change, so does their level of need (Hugo 1997). Hugo argued (1997, 2) that 'people must be at the centre of all social and economic planning, whether this be in the private or public sector'. GIS were seen, by Hugo, as a vehicle to achieve more equitable resource allocation. Through their capacity to integrate, visualise and communicate information interactively, they could 'put people back into the planning process' (Hugo 1997, 18). This could be achieved through the consideration of people's needs, characteristics and the social conditions in which they live, in each stage of the planning process and through the involvement of people in the planning process itself (Hugo 1997). Hugo (1997, 2001) recognised the huge potential GIS offered to inform resource allocation, particularly by government agencies, noting that the provision of social and human services accounted for more than two-thirds of government spending and more than one-third of Australian Gross Domestic Product. Provided timely, accurate and relevant information was available, spatial information databases could be developed to enable a targeted needs-based approach to resource allocation which would underpin the more equitable, effective and efficient use of scarce resources (Hugo 1997). Hugo (1997) advocated using GIS to inform needs-based resource allocations, rather than relying on other methods of resource allocation such as: fixed allocation (per area or per capita); response-based allocation (in reaction to lobbying or agitation); demand-based allocation (where funding is allocated based on current demand and usage rates); or historically-based allocation (whereby funds are allocated based on how they have been allocated in the past). GIS could also be used to evaluate the effectiveness of resource allocations (Hugo 1997). This philosophy underpinned the development of customised accessibility indices such as PhARIA (APMRC 2015b), and General Practice ARIA (GPARIA) (APMRC 2014), which together with additional service information facilitated the distribution of Commonwealth government service retention funding to ensure remote populations had access to nearby pharmacy and medical services.

The early origins of ARIA—Australian Standards Geographical Classification review

One of the seminal pieces of work produced by Hugo et al. (1997) was the review of the Australian Standards Geographical Classification (ASGC), commissioned by the ABS. The ASGC is the framework used by the ABS between 1984 and 2006 for the collection and dissemination of geographically classified population statistics (ABS 2011a). This review critically appraised the ASGC, examining both conceptual and practical issues relating to the criteria used to describe remoteness and to delimit urban areas, and identified ways in which the current classification could be improved. The report (Hugo et al. 1997) paved the way for the creation of the Accessibility and Remoteness Index of Australia (ARIA) (Commonwealth Department of Health and Aged Care 1999) and the subsequent adoption of a classified version of ARIA+ to form the ABS Remoteness Structure (ABS 2001b). Hugo et al. (1997) also argued that Census Collectors Districts (CDs), at the time the smallest available statistical unit for the release of census data, were unsatisfactory census output units. Their report advocated for more flexibility to be built into the ASGC and suggested that, where possible, data collections should be geocoded (Hugo et al. 1997). The report supported the adoption of the New Zealand 'mesh block units' and suggested the ABS establish its own GIS to facilitate the management and dissemination of census data (Hugo et al. 1997). Looking back, we see many of these suggestions have been implemented. Since 2011, mesh blocks have been the base spatial unit underpinning the ASGS (ABS 2010). The ABS continues to implement more flexibility into its spatial data products, such as the release of the national population grid, a 1 km grid of the national population (ABS 2011b), and the ability to generate custom area statistics through their TableBuilder census product (ABS 2015). The 2016 census will be the first that is geocoded, although the data will be aggregated to preserve confidentiality for dissemination purposes (Brady and Van Halderen 2013).

Defining accessibility and remoteness in Australia

'Accessibility to services is a crucial element of the economic and social well-being of Australians' (Hugo 2001, 276). Accessibility is a complex and multi-dimensional concept which has both spatial and aspatial dimensions. The spatial component, often termed 'geographic accessibility', refers to the physical separation between people and the services they use, and may also include consideration of mode of transport. Aspatial measures of accessibility refer to the affordability, acceptability, availability and adequacy of services (Ngamini Ngui and Vanasse 2012). Accessibility indices can be used for a range of applications including academic research, planning and policy development. Measures of accessibility are useful because they offer a means by which access inequalities can be measured. In a planning and policy context, accessibility measures can be a valuable source of information to identify areas of need where new services, improved transport infrastructure, or additional funding could be beneficial. They also provide a useful way to report data and examine relationships between accessibility and other factors, and can be used to monitor changes in accessibility brought about by the introduction or removal of services.

The ASGC review appraised a range of methods that had been used to produce remoteness classifications, including those by Holmes (1977), Faulkner and French (1983), Griffith (1996) and the Rural, Remote and Metropolitan Areas (known as RRMA) by the Department of Primary Industry and Energy and the Department of Human Services and Health (cited by Hugo et al. 1997). However, Hugo et al. (1997) recommended that for the purposes of establishing a general-purpose remoteness index for incorporation into the ASGC, a geographic approach to defining remoteness would be the most appropriate. This recommendation did not deny the importance of aspatial factors in contributing to remoteness and isolation, but rather recognised that for a national general-purpose index it was better to keep these components separate. Within a single location, it was considered likely that some individuals would face the 'double jeopardy' of both geographic remoteness and socio-economic disadvantage (Hugo et al. 1997). To be able to develop policy or program responses to overcome disadvantage, it was important that the sources of disadvantage could be distinguished. While the ASGC review (Hugo et al. 1997) recommended the development of a geographic index of remoteness using GIS methods, no single approach to developing the index was recommended; rather, three alternative methods were proposed as the starting point for further analysis and investigation.

Following the ASGC review, the Commonwealth Department of Health commissioned GISCA (under Hugo's leadership) to develop a geographic remoteness index for Australia, using GIS technologies, that would supersede the dated RRMA index. The ASGC review had identified a number of limitations of the RRMA approach, including the use of Statistical Local Areas (SLAs) as the basic building block of the index, which were criticised for being too large and heterogeneous. The use of straight-line distance measurements also came under critique (Hugo et al. 1997). The result of this work was the development of the Accessibility/Remoteness Index of Australia (ARIA) (Commonwealth Department of Health and Aged Care 1999), which became widely accepted as a standard national measure of geographic remoteness in Australia and was recognised internationally for its contribution to health geomatics (Kamel Boulos, Roudsari, and Carson 2001).

A detailed account of the development of the ARIA methodology and its calculation, in addition to a summary of the previous approaches to measuring remoteness, is given in the Commonwealth Department of Health and Aged Care (2001) occasional paper *Measuring Remoteness: Accessibility/Remoteness Index of Australia (ARIA)*. ARIA conceptualised geographic remoteness as a continuum which describes the ease or difficulty with which individuals can access a range of services, some of which are available in smaller centres and others in larger centres (Figure 1). ARIA quantified accessibility by measuring the road distance people travel to access service centres. Distance measurements from 11 340 populated towns to 201 service centres (defined as larger populated towns and centres with more than 5000 people) classified into four service levels (A to D; see Table 1) were used as the basis of the calculation of the index. The resulting ARIA values for each of the 11 340 populated towns were interpolated to a 1 km grid, producing a continuous measure of remoteness covering Australia with values ranging from 0 to 12 (with 0 being highly accessible and 12 being very remote). The ARIA grid could be used to determine the relative remoteness of any location in Australia or, alternatively, grid values could be aggregated and averaged to produce minimum, maximum and average ARIA values for any spatial unit.

Figure 1. Accessibility/Remoteness continuum. *Source*: Taylor, Bamford, and Dunne (1999).

Table 1. Aria and ARIA+ service centre classes

ARIA	ARIA+	
Level A	Level A	More than 250 000 persons
Level B	Level B	48 000–249 000 persons
Level C	Level C	18 000–47 999 persons
Level D	Level D	5000–17 999 persons
	Level E	1000–4999 persons

Source: Commonwealth Department of Health and Aged Care (1999, 13) and APMRC (2015c).

ARIA+ and ABS remoteness structure

Prior to the 2001 Australian census, an update and refinement of the original index was developed, and named ARIA+. Rather than four levels of service centres, ARIA+ included five (as shown in Table 1), making the index more sensitive to smaller service centres with populations between 1000 and 4999 people. As a result of the inclusion of the additional service class, the range of values for ARIA+ increased by three to a range of 0–15 (with 0 being highly accessible and 15 being very remote). In addition to this change, some minor refinements to the methodology were incorporated into the calculation of the ARIA+ including the calculation of road distances from the edge of a service centre as opposed to its centroid. The index continued to be calculated for a 1 km grid across Australia (Figure 2).

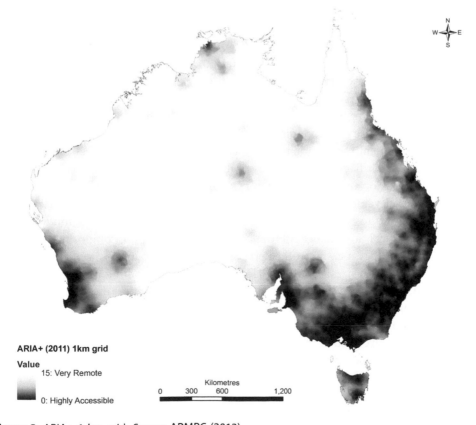

Figure 2. ARIA+ 1 km grid. *Source*: APMRC (2013).

ARIA+ was evaluated by the ABS and later included into the ASGC in 2001, as a national measure of geographic remoteness. It underpinned the development of the ABS RA classification, and continues to be used to this day. The purpose of the RA classification was to define 'broad geographical areas which share common characteristics in terms of physical distance from services and opportunities for social interactions' (ABS 2001a, 5). The RA classification comprises five remoteness classes,[2] which were based on a classification of ARIA+ values aggregated to census CDs. With each Australian census since 2001, ARIA+ has been updated and recalculated using current census population data to define the service centre classes. These updated versions of ARIA+ have been used to determine the RA classification for each census year. The ABS continues to produce the RA classification from ARIA+, which is now included in the updated ASGS and is based on average ARIA+ values aggregated to Statistical Area 1 (SA1) spatial units (ABS 2013). The procedure for classifying ARIA+ to produce the ASGS RA is further described in ABS (2013), although it should be noted that there is a degree of subjectivity involved. For example, the ABS uses a minimum population percentage to draw the boundary between remote and very remote categories. Thus, for some applications, the broad RA classes may not provide sufficient discrimination between areas of differing accessibility. In such cases the ARIA+ values (0–15), or a customised finer classification of the ARIA+ values, may provide better differentiation between areas.

Derivatives of the ARIA methodology

The development and refinement of the ARIA methodology was the beginning of a large body of work by Hugo and his team at GISCA, which is being continued by the APMRC. While both ARIA and ARIA+ are useful general-purpose indices of geographic accessibility and remoteness, for some applications more precise information regarding the distance to particular service types is desirable. Customised accessibility indices were developed to provide information regarding the remoteness and isolation of particular services and service providers. These custom indices have been used to inform funding allocations targeted to assist and retain services in remote locations. PhARIA (APMRC 2015b), GPARIA (APMRC 2014), Cardiac ARIA (Coffee et al. 2012) and more recently Metro ARIA (AURIN 2015; Taylor and Lange 2015) are examples of customised ARIA-type indices.

PhARIA was developed by GISCA for the Commonwealth Department of Health in 2000 and continues to be used to assist in determining the allocation of funding under the Rural Pharmacy Maintenance Allowance scheme. PhARIA is a composite accessibility index which combines measures of general accessibility with more specific measures of access to pharmacy services (APMRC 2015b). By combining both a general accessibility measurement, represented by ARIA+, and a professional isolation measure (the average road distance to five pharmacies), the index provides a sensitive tool to assess not only the general remoteness of a pharmacy from populated services centres but also how isolated a pharmacy is from other pharmacies. This provides a basis for ensuring that funding can be directed to those services that are both remote and isolated from other pharmacy services, having the effect of supporting the equitable distribution of pharmacy services across Australia.

A further customisation of the ARIA methodology is Metro ARIA, which quantifies accessibility within urban areas, recognising that even within cities, areas of locational disadvantage exist. The methodology for Metro ARIA was developed in 2001, although it was not until 2015 that a national Metro ARIA for all capital cites was completed and released via the Australian Urban Research Infrastructure Network (AURIN; see AURIN 2016). The index reflects the ease or difficulty people face accessing five basic service types: educational, financial and postal, health, public transport, and shopping. Similar to ARIA, road distance measures form the basis of the index calculation, although the resolution for the analysis is much finer, being calculated for each land parcel within the metropolitan area before being aggregated to SA1 ABS ASGS spatial units and classified into five access classes. In this way, Metro ARIA complements both ARIA+ and the ASGS RA classification, which quantify accessibility for non-metropolitan areas. Also available on the AURIN portal are each of the five service sub-indices used to create the composite Metro ARIA, offering a further five measures of service-specific accessibility.

Advantages and limitations of ARIA

The ARIA methodology offers a mechanism for: (1) measuring geographic access inequality and identifying areas of poor accessibility; (2) informing service planning through the integration of census data and other spatial datasets, such as socio-economic variables, to understand those factors contributing to areas of low accessibility (both geographic and socio-economic) and the best approaches to ameliorate these; and (3) reporting data and identifying whether there are relationships between accessibility/remoteness and other factors, such as health outcomes.

The key advantages of the ARIA methodology are: the use of road distance measures rather than straight-line distance; the pure focus on measuring geographic remoteness; the conceptual simplicity of the methodology; the relative stability of the index over time; the national coverage of the index (it provides uninterrupted coverage across the entire country, with a weighting factor for islands); it quantifies remoteness as a continuous variable from 0 to 15; the continued support and update of the index every 5 years in line with the census; and its flexibility (ARIA values can be assigned to particular locations or can be aggregated to any spatial unit). The ARIA methodology can also be used to quantify service access to particular services.

Despite being broadly supported as a national general measure of remoteness, several criticisms of the ARIA methodology and the RA classification have been made. As a general measure of remoteness, ARIA and ARIA+ can lack specificity with regard to access to particular services, since the indices use population size to represent different levels of service centres (ABS 2001a). To address this issue, particularly with regard to service funding models, it was always envisaged that ARIA and ARIA+, as measures of general remoteness, would be combined with more specific measures of access to particular services, to make the index sensitive to both general remoteness and accessibility factors specific to the service type of interest. It was this approach that resulted in the development of the PhARIA and GPARIA indices. The wider availability of geocoded service information in recent times has also paved the way for accessibility measures to be built based on particular service locations rather than having to rely on using population centre sizes as a surrogate measure of service availability. This approach underpins the Metro ARIA.

The use of road distance rather than travel time has been raised as a further concern with regard to the ARIA methodology (ABS 2001a). The decision to use road distance was, at the time of the development of the original ARIA, the only viable method to calculate accessibility. The available Australian Surveying Land and Information Group (AUSLIG) 1:250 000 road network did not contain information regarding travel speeds, and, while there was a distinction between sealed and unsealed roads, road type would (on occasion) change across map sheets, leading to questions regarding the accuracy and currency of this component of the dataset. Also, there was no indication of road condition, which was considered to be an important determinant of travel time. It was for these reasons that the decision to use road distance was made. Recent improvements to commercial road networks may make the incorporation of travel time possible, although the costs of such datasets are extremely high and generally prohibitive for research purposes. In addition, any change of this type would require careful evaluation to ensure the network was accurate enough not to introduce bias into the index.

Changes to the accessibility of areas due to road access being cut by flooding at certain times of the year is not captured by the ARIA+, and forms another area of criticism (ABS 2001a). However, analysis conducted by the ABS found that communities in the Northern Territory that had no road access for a large part of the year were already in the 'very remote' remoteness class, so additional consideration of the access constraints in these areas would have no influence on their final remoteness class (ABS 2001a).

The cost of purchasing ARIA values has been perceived as a barrier to a greater level of use of these data. However, the APMRC relies on the sale of ARIA products to cover the costs associated with the development of the ARIA products every 5 years to correspond to ABS census releases. The APMRC provides the ARIA+ 1 km grid to the ABS for the creation of the RA classification, which can be downloaded and combined with census data at no cost. The sale of the detailed ARIA+ scores (ranging from 0.00 to 15.00) for individual towns, different ABS geographies, and for custom areas (on request) is viewed as necessary so that the APMRC can ensure the maintenance and continuation of this valuable piece of Australia's spatial data infrastructure. Markham (2016) has produced an alternative version of ARIA termed the Open Accessibility and Remoteness Index of Australia which is freely available. However, the name is somewhat misleading since the index departs from the ARIA methodology and, among its variations, uses straight-line (Euclidian) distances rather than road network distances as the basis of the calculation. This notable difference gives rise to numerous variations in measures of accessibility between the two indices. For example, the road distance between Adelaide and Port Lincoln is 650 km, but 250 km by Euclidian distance.

Other areas of concern have arisen from the use of the RA classification as a basis for calculating retention payments for general practitioners by the Commonwealth Department of Health (Commonwealth Department of Health 2013). The use of the RA classification was introduced following a decision by the Department of Health to move to a more general measure of remoteness that could be used across the Department rather than using more area-specific indices such as GPARIA, which were created specifically to support general practice retention payments. The application of the RA classification for this purpose was found to be too broad within the RA classes Inner Regional and Outer Regional Australia, resulting in towns of different population sizes falling within the same remoteness class and thus receiving the same payment incentives. This resulted in some small towns within these two classes being disadvantaged in their ability to recruit and retain doctors. To address this problem the Modified Monash Model (MMM) (Commonwealth Department of Health 2016b), based on work by Humphreys et al. (2012), has been developed. For a complete description of the MMM the reader is referred to the Department of Health's website (Commonwealth Department of Health 2016). The MMM uses the RA classification of ARIA+, but combines the Inner Regional and Outer Regional classes, and within this broad band applies population size categorisations to determine payment categories. While this approach has resolved some of those previously identified anomalies in the index, it has created others. By ignoring the influence of remoteness within this broad band, towns close to large cities, and in some cases within commuting distance of a large city, are allocated the same payment category as towns hundreds of kilometres from a large city. For example, Tanunda, which is located within the Barossa Valley region of South Australia, has a population of 4672 and a road distance of 72 km from the South Australian capital city of Adelaide. In comparison, the town of Naracoorte (also within South Australia), which has a similar sized population of 4888, is located 336 km by road from Adelaide, but shares the same payment category of MMM5 with Tanunda.

Coombes and Raybould (2001) identify population size, remoteness and population concentration as key dimensions of modern human settlement patterns, and argue that these dimensions should be measured individually and that one dimension cannot replace the others. Thus, an improvement to the use of the RA classification for retention payment purposes may have been to overlay population size onto the classification, rather than replacing remoteness measures with population size, as has been done with the MMM.

Policy, planning and research applications of ARIA

The ARIA approach to quantifying geographic accessibility and all of the developed derivative ARIA indices (ARIA+, PhARIA, GPARIA, and Metro ARIA) have provided a valuable means to identify and measure spatial inequities between and within the city and the bush. ARIA+ and the other custom ARIA indices have served to raise our national consciousness of the need to be mindful of spatial inequalities, and the need to act to ameliorate them. Many reports of national data collections now routinely use the ASGS RA classification (Major Cities, Inner Regional, Outer Regional, Remote and Very Remote) to highlight spatial inequalities, drawing attention to areas where outcomes need to be improved. The *Social Health Atlas of Australia* (Glover and Tennant 1999; PHIDU 2016) and the various reports compiled by the Australian Institute of Health and Welfare are examples of standard national reporting processes that routinely use the RA classification to report health outcomes.

The Australian *Education Act 2013* (Department of Education and Training 2015) incorporates ARIA+ as a component of the Australian government's funding of schools (Australian Education Act 2013). The National Climate Change Adaptation Research Facility has used ARIA+ in policy guidelines at the Commonwealth and State level to emphasise the challenges of adapting to climate change for Indigenous communities and associated agencies (NCCARF 2013). The Australian Institute of Family Studies has used ARIA+ to highlight 'how the characteristics of families differ between the "city", and the "country" or "bush'" (Baxter, Gray, and Hayes 2013, 2). ARIA+ has also featured in the reporting of police statistics (OESR 2012) and other regional profile analyses which seek to differentiate varying levels of accessibility/remoteness (OESR 2013). To this end, it has been noted that ARIA+ has become and continues to be a de facto Australian government agency standard for defining remoteness across Australia (Alizadeh 2013).

From a research perspective, the Australian Longitudinal Study on Women's Health, a survey of over 58 000 Australian women, currently in its 20th year (ALSWH 2016), has utilised ARIA scores to characterise the geographic access context of survey participants. The use of ARIA in this context offers researchers the capacity to focus analysis on regional and remote participants if desired (Byles et al. 2010), in addition to providing an avenue to compare health outcomes with respect to remoteness, in conjunction with other Australian census data. ARIA+ has also provided a visual representation of geographic accessibility/remoteness that has been used to illustrate the geographic and demographic landscape of Australia. Maps of ARIA+ feature in Australian geography and humanities and social sciences textbooks (see, for example, Butler et al. 2013; Bliss, Reid, and Chaffer 2014; McInerney et al. 2014).

GIS and social research—advances, limitations and opportunities

The last 20 years have seen numerous advances in the spatial industry in Australia. Not only has the usability and functionality of GIS improved, including huge advancements in mobile mapping technology, but access to small-area spatial and demographic datasets though the ABS and some other government departments has been greatly facilitated by reduced data costs, including many free datasets and online data portals such as those provided by the ABS, Location SA (Government of South Australia 2016), AURIN (AURIN 2016) and National Map (Australian Government 2016a). The ABS is supporting the extraction of statistics for custom defined aggregates of ASGS units in TableBuilder (ABS 2015), allowing more flexibility for community-focused spatial analysis. The Australian Commonwealth government's national innovation and science agenda aims to promote innovation through publishing and sharing non-sensitive public data (Australian Government 2016b). This national initiative has led to the release in February 2016 of the Australian Geocoded National Address File (GNAF) by the Public Sector Mapping Agency (PSMA) (PSMA 2016), which will undoubtedly generate much interest in the geospatial community as it further enhances its ability to conduct analysis and plan at the small-area level.

While some of the impediments to the application of GIS to social research documented by Hugo in 1997 and 2001 are not such a concern today, data confidentiality and the limited number of trained personnel are still factors that limit the development of sophisticated social spatial applications in Australia. Methods to enable researchers to access unit record data, which is often the most detailed for research, planning and policy development, are still required. The currently lengthy ethics processes often result in researchers choosing to use less detailed aggregate data to avoid lengthy delays in data access. A method of researcher accreditation which would allow endorsed researchers access to confidential data under strict ethical guidelines could offer one solution. This type of approach would facilitate the best use of routinely collected administrative data and accordingly may produce higher quality research outcomes.

Despite the proliferation of data and websites, there is still a need for considered and careful analysis of the available spatial data which is underpinned by a good geographic understanding of spatial processes, analysis methods, scale and the limitations of the spatial datasets. The availability of appropriate spatial units for community analysis (Hugo 1997) remains an area that is often overlooked, but crucial to achieving reliable analysis results. Hugo (2002) argued for the need to debate and re-examine structures used to represent settlement systems, particularly in light of major global changes to population distributions and advances in GIS which can enable more sophisticated conceptualisation and measurement of these systems. This ongoing critical review process and debate needs to be maintained and progressed since advancements in this area can have important implications for our understanding of the influence of geography and our ability to communicate this. Certainly ARIA+ and the RA classification are examples of the benefits of this type of critical spatial thinking. The modifiable areal unit problem and issues around the interpretation of ecological data (taking care to avoid the ecological fallacy) are additional factors that continue to warrant consideration when interpreting results of spatial analyses which use area-level data.

The growing availability of web-based GIS and spatial data offers many of the opportunities that Hugo (1997) urged for, with these tools providing a means to integrate GIS into the community and offer pathways for people to be more involved in the planning process. The utilisation of these systems to empower communities and provide evidence on which to base decision making to enhance the planning process should continue to be encouraged. While GIS is not a silver bullet for all planning and policy development, given the identified advances in both data availability and technology, GIS can make an increasingly important contribution to an equitable evidenced-based planning process, to ensure that the most effective allocation of scarce resources across all areas and communities is realised.

Summary and conclusion

A strong focus on people and social justice was the driving force behind much of Professor Graeme Hugo's work and is a key characteristic of the projects and research of the Key Centre, GISCA and APMRC. The Key Centre's mission, to demonstrate the use of GIS as an analytic tool, integral to strategic decision making and policy development, led to research that was both innovative and applied across a range of application areas. The operationalisation of the concept of geographic accessibility and remoteness is arguably one of the most influential and lasting contributions made by Professor Hugo and his team. ARIA, ARIA+ and the customised accessibility indices that have been developed continue to assist the equitable delivery of health care, contribute to the identification of spatial inequalities, and underpin the geography used for the dissemination of national statistics.

The changing nature of populations, information and technology requires ongoing review and innovation to ensure methods of resource allocation and the conceptualisation of social and spatial processes remain current and relevant, and make the best use of available data and technology. To this end, Professor Hugo's vision remains. It is our job as geographers to continue this work and to communicate to planners, policy makers and the public what is possible.

Notes

1. Private-sector partners included ESRI Australia, Silicon Graphics Corporation, Communica Systems, Space Time Research Pty Ltd, Cole Associates, MAPTEK Pty Ltd, Formida Computer Associates, Kinhill Engineers, Daedalus SA Pty Ltd, Spatial Concepts Pty Ltd and Coded Australia.
2. ABS Remoteness Area classification

RA category	RA name	Average ARIA+ value ranges for CDs and SA1s
0	Major Cities of Australia	0 to 0.2
1	Inner Regional Australia	Greater than 0.2 and less than or equal to 2.4
2	Outer Regional Australia	Greater than 2.4 and less than or equal to 5.92
3	Remote Australia	Greater than 5.92 and less than or equal to 10.53
4	Very Remote Australia	Greater than 10.53

Source: (ABS 2001b, 2013).

Acknowledgements

We are grateful to the two anonymous referees and Natascha Klocker for their helpful comments. We would also like to thank and acknowledge the past and current employees of the National Key Centre for Social Applications of GIS, GISCA and APMRC who under the direction of Professor Hugo have contributed to this vast body of work and to its legacy.

Disclosure statement

No potential conflict of interest was reported by the authors.

References

Alizadeh, T. 2013. *Towards the Socioeconomic Patterns of the National Broadband Network Rollout in Australia*. Proceedings of State of Australian Cities (SOAC) conference, Sydney. Australian Policy Online. Accessed June 10, 2016. http://apo.org.au/.

ALSWH. 2016. *Australian Longitudinal Study on Women's Health*. Accessed February 16, 2016. http://www.alswh.org.au/.

APMRC. 2013. *Accessibility/Remoteness Index of Australia Plus (ARIA+) 2011*. Adelaide, SA: Australian Population and Migration Research Centre (APMRC), The University of Adelaide.

APMRC. 2014. *General Practice ARIA—GPARIA*. Accessed December 12, 2015. https://www.adelaide.edu.au/apmrc/research/projects/gparia.html.

APMRC. 2015a. *Australia Population and Migration Research Centre*. Accessed January 20, 2016. https://www.adelaide.edu.au/apmrc/.

APMRC. 2015b. *Pharmacy ARIA (PhARIA)*. Accessed February 8, 2016. https://www.adelaide.edu.au/apmrc/research/projects/pharia/.

APMRC. 2015c. *ARIA (Accessibility/Remoteness Index of Australia)*. Accessed July 14, 2016. https://www.adelaide.edu.au/apmrc/research/projects/category/about_aria.html.

AURIN. 2015. *Metropolitan Accessibility / Remoteness Index of Australia (Metro ARIA)*. Accessed December 10, 2015. http://aurin.org.au/projects/data-hubs/metro-aria/.

AURIN. 2016. *Australian Urban Research Infrastucture Network*. Accessed March 1, 2016. http://aurin.org.au/.

Australian Bureau of Statistics (ABS). 2001a. *Information Paper—Outcomes of the ABS Views on Remoteness Consultation, Australia.* Catalogue No. 1244.0.00.001. Canberra: Australian Capital Territory.

Australian Bureau of Statistics (ABS). 2001b. *Australian Standard Geographical Classification (ASGC).* Catalogue No. 1216.0. Canberra: Australian Capital Territory.

Australian Bureau of Statistics (ABS). 2010. *Australian Statistical Geography Standard (ASGS): Volume 1—Main Structure and Greater Capital City Statistical Areas.* Catalogue No.1270.0.55.001. Canberra: Australian Capital Territory.

Australian Bureau of Statistics (ABS). 2011a. *Australian Standard Geographical Classification (ASGC) July 2011.* Catalogue No. 1216.0. Canberra: Australian Capital Territory.

Australian Bureau of Statistics (ABS). 2011b. *Australian Population Grid, 2011.* Accessed February 10, 2016. http://www.abs.gov.au/ausstats/abs@.nsf/Lookup/1270.0.55.007main+features12011.

Australian Bureau of Statistics (ABS). 2013. *Australian Statistical Geography Standard (ASGS): Volume 5—Remoteness Structure.* Catalogue No. 1270.0.55.005. Canberra: Australian Capital Territory.

Australian Bureau of Statistics (ABS). 2015. *TableBuilder.* Accessed January 25, 2016. http://www.abs.gov.au/websitedbs/censushome.nsf/home/tablebuilder?opendocument&navpos=240.

Australian Education Act. 2013. *Commonwealth Government of Australia. Act 67 of 2013.* Accessed March 18, 2016. http://www.legislation.gov.au/.

Australian Government. 2016a. *National Map.* Accessed March 1, 2016. https://nationalmap.gov.au/.

Australian Government. 2016b. *National Innovation and Science Agenda—Data Sharing for Innovation.* Accessed March 1, 2016. http://www.innovation.gov.au/page/data-sharing-innovation.

Baxter, J., M. Gray, and A. Hayes. 2013. *Families in Regional, Rural and Remote Australia: Facts Sheet 2011.* Melbourne: Australian Institute of Family Studies (AIFS). Accessed June 10, 2016. http://aifs.gov.au/.

Bliss, S., G. Reid, and L. Chaffer. 2014. *Geo World 7: For the Australian Curriculum.* South Yarra: Macmillan Education Australia.

Bureau of Rural Sciences. 1999. *Country Matters—Social Atlas of Rural and Regional Australia.* Canberra: Australian Capital Territory.

Brady, M., and G. Van Halderen. 2013. *A Sstatistical Spatial Framework to Inform Regional Statistics. World Statistics Congress.* Canberra: Australian Capital Territory.

Bryan, B. A., and E. Bamford. 2000. *The Spatial Distribution of Health Outcomes of an Influenza Pandemic in Australia.* Brisbane: Proceedings of the International Society for Ecosystem Health.

Butler, D., R. Cooke, D. Lergessner, S. Miller, and M. Robertson. 2013. *Geography for the Australian Curriculum Year 9.* Port Melbourne: Cambridge University Press.

Byles, J., A. Dobson, N. Pachana, L. Tooth, D. Loxton, J. Berecki, R. Hockey, D. McLaughlin, and J. Powers. 2010. *Women, Health and Ageing: Findings from the Australian Longitudinal Study on Women's Health: Major Report E.* Accessed June 10, 2016. http://www.alswh.org.au/.

Coffee, N., D. Turner, et al. 2012. "Measuring National Accessibility to Cardiac Services Using Geographic Information Systems." *Applied Geography* 34: 445–455.

Commonwealth Department of Health. 2013. *Review of Australian Government Health Workforce Programs—Mason Review.* Accessed July 14, 2016. http://www.health.gov.au/internet/main/publishing.nsf/Content/review-australian-government-health-workforce-programs.

Commonwealth Department of Health. 2016. *General Practice Rural Incentives Programme.* Accessed June 10, 2016. http://www.ruralhealthaustralia.gov.au/internet/rha/publishing.nsf/Content/General_Practice_Rural_Incentives_Program.

Commonwealth Department of Health and Aged Care. 1999. *Measuring Remoteness: Accessibility/Remoteness Index of Australia (ARIA) Occasional Papers.* Canberra: Australian Capital Territory.

Commonwealth Department of Health and Aged Care. 2001. *Measuring Remoteness: Accessibility/Remoteness Index of Australia (ARIA).* Rev. ed. Occasional Papers New Series Number 14. Accessed May 12, 2016. http://www.health.gov.au/internet/main/publishing.nsf/Content/health-historicpubs-hfsocc-ocpanew14a.htm.

Coombes, M., and S. Raybould. 2001. "Public Policy and Population Distribution: Developing Appropriate Indicators of Settlement Patterns." *Environment and Planning C: Government and Policy* 19 (2): 223–248.

Delaney, J. 1999. *Geographical Information Systems: An Introduction*. South Melbourne: Oxford University Press.

Department of Education and Training. 2015. *Australian Education Act 2013*. Accessed February 28, 2016. https://www.education.gov.au/australian-education-act-2013.

Garner, B. J. 1990. "GIS for Urban and Regional Planning and Analysis in Australia." In *Geographic Information Systems: Development and Applications*, edited by L. Worrall, 41–64. London: Belhaven Press.

GISCA. 2000. "A Study of Provision of Health Services in Non-Metropolitan Australia." Report prepared for The Department of Health and Aged Care. Adelaide, The University of Adelaide. Adelaide, South Australia.

Glover, J., and S. Tennant. 1999. *A Social Health Atlas of Australia*. Adelaide: Openbook Publishers.

Government of South Australia. 2016. Location SA, Adelaide. Accessed March 1, 2016. http://www.location.sa.gov.au/.

Hugo, G. 1994. *Australian Research Council, Grant Submission for: Key Centre for Teaching and Research in Social Applications of Geographical Information Systems*. Adelaide: The University of Adelaide.

Hugo, G. 1997. "Putting People Back into the Planning Process: The Changing Role of GIS." *Monograph Series 4*. National Key Centre for Social Applications of Geographical Information Systems. The University of Adelaide. Adelaide, South Australia.

Hugo, G. 2001. "Addressing Social and Community Planning Issues with Spatial Information." *Australian Geographer* 32 (3): 269–293.

Hugo, G. 2002. "Changing Patterns of Population Distribution in Australia." *Journal of Population Research. Issue Special ed.* 2002: 1–21.

Hugo, G., and R. Aylward. 1999. "Using Geographical Information Systems (GIS) To Establish Access to Aged Care Residential Services in Non-Metropolitan Australia." *5th National Rural Health Conference*. Adelaide, South Australia.

Hugo, G., D. Griffith, P. Rees, P. Smailes, B. Badcock, and R. Stimson. 1997. "Rethinking the ASGC: Some Conceptual and Practical Issues." *Monograph Series 3*. Adelaide, South Australia: National Key Centre for Social Applications of Geographical Information Systems, The University of Adelaide.

Humphreys, J., M. McGrail, C. M. Joyce, A. Scott, and G. Kalb. 2012. "Who Should Receive Recruitment and Retention Incentives? Improved Targeting of Rural Doctors Using Medical Workforce Data." *Australian Journal of Rural Health* 20: 3–10.

Kamel Boulos, M. N., A. V. Roudsari, and E. R. Carson. 2001. "Health Geomatics: An Enabling Suite of Technologies in Health and Healthcare." *Journal of Biomedical Informatics* 34 (3): 195–219.

Leslie, E., N. Coffee, L. Frank, L. N. Owen, A. Bauman, and G. Hugo. 2007. "Walkability of Local Communities: Using Geographic Information Systems to Objectively Assess Relevant Environmental Attributes." *Health & Place* 13 (1): 111–122.

Markham, F. 2016. *The Open Accessibility and Remoteness Index for Australia*. Accessed June 7, 2016. https://figshare.com/articles/2001_2006_and_2011_raster_data_The_Open_Accessibility_and_Remoteness_Index_for_Australia/1574190.

McInerney, M., A. Woollacott, M. Jeffery, G. Somers, and J. Cain. 2014. *Humanities & Social Sciences for the Australian Curriculum Year 9*. Port Melbourne: Cambridge University Press.

NCCARF. 2013. *NCCARF Policy Guidelines Brief 6: Adaptation and First Australians: Lessons and Challenges*. Alice Springs: National Climate Change Adaptation Research Facility (NCCARF). Accessed June 10, 2016. http://www.nccarf.edu.au/.

Ngamini Ngui, A., and Vanasse, A. 2012. "Assessing Spatial Accessibility to Mental Health Facilities in an Urban Area." *Spatial and Spatio-Temporal Epidemiology* 3: 195–203.

OESR. 2012. *Queensland Police Service Profile*. Brisbane: Office of Economic and Statistical Research (OESR), Queensland Treasury and Trade. Accessed June 10, 2016. http://www.police.qld.gov.au/.

OESR. 2013. *Department of Communities, Child Safety and Disability Services (DCCSDS) Profile: Far North Queensland DCCSDS Region—Selected Key Indicators*. Brisbane: Office of Economic and Statistical Research (OESR), Queensland Treasury and Trade. Accessed June 10, 2016. http://statistics.qgso.qld.gov.au/.

PHIDU. 2016. *Social Health Atlases of Australia: Remoteness Areas*. Accessed March 1, 2016. http://www.publichealth.gov.au/phidu/maps-data/data/#remoteness.

PSMA. 2016. *G-NAF & Administrative Boundaries Now Available from data.gov.au*. Accessed March 1, 2016. https://www.psma.com.au/.

Taylor, D. S., E. Bamford, and L. Dunne. 1999. "Quantifying Access to Services in Australia using Network Analysis. OZRI 99." The 13th Annual Australian ESRI and ERDAS Users Conference. 31 August to 1 September. Glenelg, South Australia.

Taylor, D. S., and J. Lange. 2015. *Metro ARIA and Metro ARIA Health: New Accessibility Indices for Australian Capital Cities*. Institute of Australian Geographers Conference 2015. Canberra, Australian Capital Territory.

Responding to Negative Public Attitudes towards Immigration through Analysis and Policy: regional and unemployment dimensions

Kate Golebiowska, Amani Elnasri and Glenn Withers

ABSTRACT

This paper examines two key dimensions of the impact of immigration for Australia and related policy aspects. One is sub-national and the other is national. They are, first, the regional location aspects of immigration and, second, the aggregate unemployment implications of immigration. These are chosen so as to focus on two important issues that condition public attitudes towards immigration. In relation to the first, there is a common positive view that channelling migration towards regional areas assists regional development and reduces pressure on metropolitan areas. The paper reviews regional concepts embodied in Australian immigration policy and the ways in which visa arrangements have implemented policies geared towards the regional dispersal of immigrants. Using official data, it discusses the demographic impacts of these policies and, in particular, considers the extent to which immigrants to regional Australia remain there over the longer term. In relation to unemployment, a common concern is that immigrants take jobs from local workers. The paper examines —using statistical regression methodology—the relationship between immigration and national aggregate unemployment in Australia. It evaluates the net consequences of immigration for both existing residents and new arrivals together. The paper concludes that, with good policy design in each case, regional location encouragement can be effective for immigrants and that immigrants need not take more jobs than they create. The analysis demonstrates that mixed-methods approaches to important social science issues can be productive, and helpful also for policy. Evidence, such as that presented in this paper, offers a powerful basis from which to counter negative public and political discourses surrounding immigration in contemporary Australia.

Introduction

Within the wide ambit of Graeme Hugo's work, immigration had a prominent place. This paper examines two key dimensions of the impact of immigration for Australia. One is sub-national and the other is national. They are, first, the regional location aspects of immigration and, second, the aggregate unemployment implications of immigration. These are chosen so as to focus on two important issues that help condition public attitudes to the phenomenon of immigration.

Markus (2011) concluded that majority opinion—within the Australian population—typically supports the view that immigration unduly pressures provision of city infrastructure. There is thus a common positive view that if more immigration towards regional areas of Australia can be enabled, this will assist regional development and reduce pressure on metropolitan areas. Markus (2011) also found that attitudes to immigration are closely correlated with the unemployment rate. A common concern relating to Australia's immigration intake is that immigrants take jobs from local workers (see also Davis and Deole 2015; Goot and Watson 2011; Markus 2016). This paper addresses both of these issues in two main parts.

The first half of the paper reviews regional concepts embodied in Australian immigration policy and the ways in which visa arrangements have influenced the dispersal of immigrants to regional locations. Using official data, we examine evidence of the impacts of these dispersal efforts and find that such policy can have some significant redistribution effects. In the second part we analyse the relationship between immigration and national aggregate unemployment in Australia, using statistical regression methodology based on causality and co-integration. This provides insights into the net consequences of immigration both for existing residents and for new arrivals. We show that Australia's large-scale immigration program has *not* been significantly associated with any overall increase in unemployment rates.

Immigration to regional Australia

Regional immigration policy: background and evolution

In Australia, immigration matters are the constitutional responsibility of the Commonwealth government. However, the Commonwealth can be conscious of regional matters in its policy formulation and work with States and Territories at its discretion. In recent decades, formal regional immigration policy for Australia has centred on the State-Specific and Regional Migration (SSRM) Scheme, which was instituted by the Commonwealth in 1996–97. This scheme includes a suite of skilled and business visas for individuals interested in settling and working outside Australia's major cities.

The SSRM Scheme is intended as one mechanism to support population growth in slower growing and stagnant regions, alleviate environmental pressures resulting from sustained immigration flows to major cities, and respond to skills shortages outside these cities (Hugo 2008a, 2008b; Withers and Powall 2003). Former New South Wales (NSW) Premier Bob Carr put the issue of population pressure around Australia's major cities starkly:

> Right down the east coast of Australia, you'd see the end between the coast and the mountain range, you'd see the end of any farming. You'd see the end of any conservation, open space. You'd have cities … a totally urbanised east coast. (Carr 2000)

Specific regional visas have been the vehicle chosen to address some of these concerns, short of reductions in total immigration levels. In the case of the SSRM Scheme, in order for these visas to be granted, prospective migrants must be explicitly supported by a State/Territory government, an employer, or a family member living in a regional area.

The State/Territory role operates under ideas of 'co-operative federalism'. This consultative process is ongoing. State and Territory governments are interested in increasing skilled and business immigration to their respective jurisdictions and distributing these migrants where they can best contribute to the labour market. Each government has a skilled and business migration unit that promotes and facilitates such migration. Some local councils also have operational roles: they are gazetted as 'regional certifying bodies' to assist with the administration of the Regional Sponsored Migration Scheme (RSMS) visas. RSMS visas are one of a number of visas that fall under the broader SSRM Scheme—as shown in Table 1.

Since the inception of the SSRM Scheme 20 years ago, the regional immigration visa structure has been modified and the visa criteria revised. The main visa categories and characteristics are outlined in Table 1. Regardless of these modifications, as federal immigration visas they do have universal criteria relating to age, English-language proficiency, skill levels and relevant work (or business/investment) experience. Differences in criteria then relate to the skilled occupations in demand, which may vary between State and Territory labour markets, and between large and smaller area labour markets within States and Territories. The skilled occupations in demand are determined and periodically reviewed by the State and Territory governments. While the universal criteria are generally identical to those adhered to in the independent skilled immigration program, regional visa applicants can avail themselves of bonus points[1] for a family or State/Territory government nomination.

Table 1 summarises the characteristics of the core SSRM skilled visas, including those that were available until the first major immigration policy reform in 2007. The reason for including the now ceased visas is that some of them were in operation for a decade and contributed to the regional immigration policy outcomes that are considered in this paper. The holders of those visas were surveyed (by the Department of Immigration and by State and Territory governments), and the results of some of these surveys led to policy modifications discussed later in this paper.

There are also regional dimensions to temporary visas such as the Working Holiday Maker visa, including as part of the recent Northern Australia Development Agenda (DIBP 2015; Hugo 2008b). Further, since 2009, Australia has been operating the Pacific Seasonal Workers Scheme (PSWS). This Scheme has permitted employers in horticulture—and since 2016 agriculture more broadly—and the tourism sector in northern Australia[2] to recruit temporary, low-skilled and unskilled labour from Pacific Island countries and Timor Leste. The PSWS and the Working Holiday Maker visa are separate from the SSRM Scheme.

The visas offered under the SSRM Scheme started off from a small base. They represented just 4 per cent of Australia's annual skilled migration stream in 1996–97 (Golebiowska 2007), but grew to account for 38.8 per cent of this stream by 2013–14 (DIBP (Department of Immigration and Border Protection) 2014). Region-linked international immigration has therefore become a major element of Australian immigration.

Regional immigration policy: operation, definitions and transition

In the development of regional policy for Australian immigration under the SSRM Scheme, two key issues emerge as crucial to the operation of the visa programs: how regions are defined for regional visa purposes and access to permanent residency.

Table 1. Key characteristics of the SSRM Scheme skilled visas

Visa name	Stay	Points test	Nomination	Job offer	Concessional criteria	Areas eligible
Regional Sponsored Migration Scheme (RSMS)[a]	Permanent, minimum 2-year stay with the nominating employer	No	Employer	No	Concessions are available for age, skills and English-language ability (also for the non-regional version of this visa, the Employer-Nomination Scheme (ENS) visa)[b]	Regional or low population growth areas excluding: Sydney, Wollongong, Newcastle, Melbourne, Brisbane and Gold Coast[c]
Skilled-Designated Area Sponsored (SDAS). Ceased in 2007	Permanent until 2006 then a two-step visa (temporary to permanent)	No	Eligible family member residing in a Designated Area who provided an assurance of support	No, but occupation from the Skilled Occupation List (SOL)	Concessional minimum period of work experience and lower English-language standards than under the non-regional family-nominated visa	All Australia was Designated except: Sydney, Newcastle, Wollongong, Brisbane, Sunshine Coast, Gold Coast and Perth
State/Territory-Nominated Independent (STNI). Ceased in 2007	Permanent, minimum 2-year stay in the nominating State/Territory	Yes	State/Territory government	No, but occupation from a State/Territory List of Occupations in Demand (some occupations may be in demand only in some regions of a State/Territory)	Concessional points to qualify (pool mark not pass mark)	Jurisdictions were joining STNI progressively. Initially, this visa was offered in Tasmania, Victoria, South Australia, from 2005 Western Australia, then followed by other jurisdictions
Skilled-Independent Regional (SIR) introduced in 2004 and ceased in 2007	Temporary leading to permanent after meeting minimum residency and work conditions in the jurisdiction/Designated Area for which the nomination was made. E.g. for SIR 2 years of residence and 1 year of employment before applying for permanent residence	Yes	State/Territory government which attracted bonus points		Concessional points to qualify (SIR pass mark)	Regional or low population growth areas excluding: Sydney, Newcastle, Wollongong, NSW Central Coast, Brisbane, Gold Coast, Perth, Melbourne and ACT
Skilled-Regional, prior to 2012–13 known as Skilled-Regional Sponsored (SRS). The SRS was an amalgamation of SDAS and SIR visas		Yes	State/Territory government or eligible family member—both nominations attract bonus points	States/Territories determine if job offer required. In any case, occupation from a State/Territory List of Occupations in Demand (some occupations may be in demand only in some regions of a State/Territory)	Competent English (i.e. score of 6 in each of the four components of International English Language Testing System) acceptable but attracts no points	For State/Territory government nomination regional or low population growth areas excluding: Sydney, Newcastle, Wollongong, NSW Central Coast, Brisbane, Gold Coast, Melbourne and ACT. For nomination by a family member, Designated Areas which cover all Australia except: Sydney, Newcastle, Wollongong and Brisbane
Skilled-Nominated, prior to 2012–13 known as Skilled-Sponsored, which was an amalgamation of STNI and Skilled-Australian Sponsored visas. Both were ceased in 2007	Permanent, minimum 2-year residency and work in the jurisdiction for which the nomination was made	Yes	State/Territory government which attracts bonus points			All States and Territories

Notes: IELTS, International English Language Testing System.

[a]The RSMS was the first explicit regional visa piloted in 1995 and expanded in 1996 (Parliament of Australia 2001).

[b]Prior to 1 July 2012 the RSMS visa required lower English-level ability than the Employer-Nomination Scheme (ENS) visa and had more generous concessions for skill levels than currently available.

[c]In September 2011 Perth became an eligible location for the following visas: RSMS, Skilled-Regional Sponsored (SRS) temporary and Skilled-Regional permanent visas (DIBP n.d.b).

Sources: DIBP (n.d.a, n.d.b; various websites); Golebiowska (2007); Parliament of Australia (2001).

With regards to the former, Australia has adopted a flexible and substantially delegated approach. Areas eligible for regional settlement vary between visas and not all areas eligible would be intuitively considered 'regions' by many. This is partly because State/Territory governments were given authority, by the Commonwealth, to determine where within their jurisdictions regional immigrants could settle (Parliament of Australia 2001). Initially under the SSRM Scheme, 'regional' Australia covered (a) areas with fewer than 200 000 residents and (b) low population growth metropolitan areas. The latter were those that, in the last intercensal period (preceding the launch of the policy), had posted an average population growth rate below 50 per cent of the national average population growth rate (DIMIA (Department of Immigration and Multicultural and Indigenous Affairs) 2005c). Under these criteria all non-metropolitan and some metropolitan areas, including certain capital cities, were included in the definition of 'regions'. Sydney, Brisbane and Perth were excluded, but Adelaide, Hobart and Darwin were included. Indeed, applying the above criteria meant that the entire States of South Australia and Tasmania, and the Northern Territory, became eligible locations. Adelaide and Hobart qualified because of their trends of net out-migration and low population growth rates. Darwin qualified mainly due to its population being below 200 000 and due to its geographic isolation from the rest of Australia. Even Melbourne has been eligible for some regional visas under the SSRM Scheme. This situation arose from its low average population growth in the first part of the 1990s, preceding the launch of the policy. Its eligibility has been contested (Parliament of Australia 2001) but as Hugo (2008b, 555) observed, the strong pro-immigration stance of the Victorian government has been critical in retaining eligibility.

Switching to the present, a combination of demographic, economic and political factors can explain why some areas remain eligible for regional immigrant settlement and others do not. Specifically, the average intercensal population growth rates (2001–06, 2006–11) have exceeded the 50 per cent benchmark in Adelaide, Hobart and Melbourne (ABS (Australian Bureau of Statistics) 2012; Golebiowska 2012) and, technically speaking, these cities have ceased to meet the low growth metropolitan area criterion of the SSRM Scheme. However, if Adelaide and Hobart were excluded accordingly, South Australia and Tasmania would lose one key mechanism by which they can support their small and stagnant populations. In the south-west, lobbying by the Western Australian government and industry has resulted in re-classifying Perth as an eligible city for some regional skilled visas, despite its not meeting the aforementioned criteria. The primary motivation was acute skills shortages in Perth (DIBP n.d.b; Trenwith 2011).

The second key issue that is crucial to the operation of the SSRM Scheme is permanency. In Australia, most regional skilled and business visas are now two-step visas whereby meeting the temporary (usually 2 years) residency and employment (or business in the case of business immigrants) requirements in the area for which an immigrant is nominated permits a subsequent application for a permanent residence visa. The two-step process is intended as a retention measure, aimed at supporting population and economic growth in areas of initial settlement. It is anticipated that after a period of working and living in a regional area, immigrants may be less prone to relocate to a major city upon attaining permanent residence. Families, especially, acquire accommodation, have schooling arrangements, build social networks, acquire employment and so on (DIMIA 2005a, 2005b; Wulff and Dharmalingham 2008).

For the regional temporary visa holders, common routes to permanent residency include applying for other visas under the SSRM Scheme. For example, holders of the temporary Skilled Regional visa can apply for a permanent Skilled Regional visa. This visa does not require a State/Territory government nomination but operates on the premises that immigrants may be interested in staying in the original area of settlement after they have lived there for the minimum 2 years required by their temporary Skilled Regional visa. Holders of the temporary 457 visas who are working in regional areas can apply for the RSMS visa through their employer. The RSMS visa requires a minimum 2-year stay with the nominating employer, meaning effectively at least 2 additional years' stay in a regional area. RSMS visas may be cancelled if migrants do not see through the 2-year employment period with their employers (DIBP 2016). A further example of a permanent visa which also carries a minimum 2-year stay condition in the nominating State/Territory is the Skilled-Nominated visa (see Table 1). Holders of this visa may relocate to another jurisdiction if they advise the original nominating State/Territory government (Migration Western Australia n.d.).

In relation to the question of permanency, the history of the Skilled-Designated Area Sponsored (SDAS) visa provides an excellent example of how documentation and analysis of immigrant settlement behaviours can lead to constructive policy change. SDAS was initially a permanent visa (see Table 1). However, in the early 2000s a survey conducted by DIMIA (2005b) revealed that SDAS migrants were not settling in the Designated Areas where their nominators lived and were choosing to live in major cities instead. This was contrary to the objectives of this visa (Parliament of Australia 2001) and, with a view to assist retention, in 2006 SDAS became a two-step visa (Phillips and Spinks 2012). This case illustrates how demographic and geographical research can provide insights into immigrant behaviours and outcomes and be an important tool for informing policy development. Likewise, the Skilled-Independent Regional (SIR) transitional visa (Birrell, Hawthorne, and Richardson 2006) was introduced following a new research analysis of regional migration and associated visa policy reform suggestions (Withers and Powall 2003). Policy measures that could ensure more sustained regional residence by immigrants, such as transitional visas, were therefore a partial antidote for then NSW Premier Carr's criticisms of high Commonwealth immigration intake levels.

Given the broad objectives of the SSRM Scheme—to support population growth in regional areas and to alleviate population pressures in major cities—it seems clear that transitional conditional visas are crucial to ensuring that internal mobility does not undermine, from the beginning, any strong regional settlement experience and outcomes.

Assessing the benefits of regional immigration: a review of the literature

Given this history and the importance of regional immigration growth, it is appropriate that significant research has been conducted on regional immigration matters. In the initial years of the SSRM Scheme, the federal and State/Territory governments routinely commissioned or conducted surveys to understand how regional visa holders were settling and performing economically and if there were grounds therein for policy adjustments (Cully and Goodes 2000; DIMIA 2005a, 2005b). There has also been complementary literature such as parliamentary reports and reviews of regional immigration (Parliament of Australia 2001, 2015) and commissioned reports and research written by academics

(Hugo 2008a, 2008b; Institute for Social Science Research 2010; Khoo, McDonald, and Hugo 2005; Withers and Powall 2003; Hugo 1999) or consultants (Piper and Associates 2009).

As a phenomenon, international migration to regions also happens in other countries. Hugo and Morén-Alegret (2008) have argued that international migration to regional areas of high-income countries has recently become an integral element of the economic, demographic and social change in these areas and has been an outcome of longer-term trends affecting them (e.g. out-migration of youth and labour shortages). Argent and Tonts (2015) similarly adopted an international perspective and placed their considerations of regional immigration in Australia in the context of 'the global countryside', a concept developed by Woods (2007), which refers to rural spaces engaging with globalisation in multiple ways and undergoing a transformation as a result.

To date, scholarly analyses of regional immigration policy in Australia have looked at the governance of the policy, and its economic and demographic impacts, in particular retention rates and labour market participation (Cameron et al. 2012; Golebiowska 2012, 2015; Hugo 2008a, 2008b; Massey and Parr 2012). They have generally documented good participation in the labour market and noted that retention of immigrants in regional areas depends on a combination of factors, including job satisfaction and career prospects, quality of the local services and infrastructure (e.g. schooling, health, transport, and recreation), and attachment to the local community. Social adaptation of immigrants in regional Australia has been more specifically studied, for example, by Wulff and Dharmalingham (2008) and Krivokapic-Skoko and Collins (2016). These studies have found that social connectedness in regional centres is strong for those immigrants who have lived in Australia for longer periods, for families with children and for immigrants from certain countries. They found that South African, Zimbabwean and the Filipino-born develop particularly strong local connections. Krivokapic-Skoko and Collins (2016) have also observed that the existence of 'meeting places' that cultural groups can use plays a role in developing a sense of belonging locally and attracts immigrants to specific regional centres.

Understanding, and in turn influencing, immigrants' mobility motivations is clearly important for the success of regional immigration policies. In addition to some of the directly related works mentioned above, these motivations have been considered, for example, by Hugo, Khoo, and McDonald (2006), Goel and Goel (2009) and Taylor, Bell, and Gerritsen (2014). Broadly speaking, these studies have found that economic, lifestyle and social factors are reasons both for moving into and out of a regional area. An emerging stream of research has looked specifically at immigrant settlement in remote and peripheral regions of Australia (Golebiowska et al. forthcoming; Institute for Social Science Research 2010; Taylor, Bell, and Gerritsen 2014). It has found that immigrant mobility to and away from these regions is motivated by the same set of factors as above, and that sufficient stock of quality, accessible and affordable housing is one of the critical facilitators of longer-term settlement in remote and peripheral regions (perhaps even more so than in larger regional urban centres).

The studies reviewed above have analysed the demographic, economic and social contributions that regional immigrants make, identified the conditions that should be in place to support retention and what factors contribute to mobility. The next section enhances this knowledge base by analysing, in chronological order, the results of selected surveys of regional immigrants. This makes it possible to explore the demographic outcomes of regional immigration policy at different 'touchpoints'. It reveals variations in the rates of actual and intended continued residence (retention) in the areas of original settlement, depending on the visa type and visa conditions at the time of the survey.

Appraisal of Australia's regional immigration policy: immigrant retention in regional areas

Full formal evaluation of the demographic and economic or other impacts of Australia's regional migration policy is not straightforward. This is due especially to modifications of the visa criteria over the years, visa amalgamations in recent years, imperfect comparability of statistical data across years and some large capital cities such as Melbourne or Adelaide being eligible locations for settlement. Also, with the recently 'refreshed' DIBP website, not all surveys of regional visa holders previously available on the website are accessible now. With these limitations in mind, in terms of the demographic impacts of regional immigration, there is greater knowledge of what happens when immigrants arrive, where they intend to settle and do settle, than of what happens *after* they have fulfilled their minimum residency and work visa obligations.

However, useful findings are available from the results of surveys that the Department of Immigration commissioned, conducted or otherwise supported and also, as a case study, from the results of a survey commissioned by the Northern Territory government (Taylor, Bell, and Gerritsen 2014). The specific surveys discussed in the remainder of this paper include: the Cully and Goodes survey of the RSMS migrants (2000) commissioned by the (then) Department of Immigration and Multicultural Affairs (DIMA) and held in 2000; the DIMIA surveys of RSMS (2005a) and SDAS (2005b) migrants held in 2004 and involving migrants residing in all States and Territories; and Taylor, Bell, and Gerritsen's (2014) survey of RSMS and State and Territory Nominated migrants in the Northern Territory in 2012.

Starting with the issue of retention, the DIMIA (2005b) survey of SDAS migrants was conducted prior to SDAS becoming a two-step visa. It was revealed then that retention outside metropolitan areas ranged between less than 50 per cent for NSW and Queensland, and 36 per cent for Victoria. Melbourne was a strong magnet: 9 per cent of all SDAS migrants with a sponsor from outside Melbourne lived there in addition to 58 per cent of SDAS migrants who had a sponsor from Melbourne and lived in the city. On arrival in Australia, 10 per cent of all SDAS migrants by-passed Designated Areas and settled directly in non-Designated Areas. Furthermore, 16 per cent of SDAS migrants who had resided in Australia for more than 3 years (at the time of the 2004 survey), lived in non-Designated Areas such as Sydney or Brisbane. These 16 per cent included most of the 10 per cent of all SDAS migrants, who had never resided in a Designated Area. These retention and dispersal outcomes were poor and the introduction of a two-step visa process was thus intended to assist with reversal of such findings. However, no later surveys of SDAS or SRS visa (which replaced SDAS and SIR) holders are available to ascertain the exact extent of improvements.[3]

Conversely, assuming little change of intentions, the RSMS visa (which now requires a minimum 2-year stay in a regional area) may have delivered good retention rates. The overall retention rates of the RSMS migrants (that is, for those settled and working in an eligible metropolitan area such as Adelaide, as well as in non-metropolitan areas) have fluctuated over the years. In 2000, Cully and Goodes reported a 70 per cent retention rate. Meanwhile, in 2005, DIMIA reported an 85 per cent retention rate for those who had spent the minimum 2 years in their original location, and 91 per cent for those still on the original 2-year contracts (DIMIA 2005a). In the Northern Territory, Taylor, Bell, and Gerritsen (2014) reported an 84 per cent retention rate for those past their original 2-year contracts. These fluctuations are partially affected by the fact that the minimum 2-year stay with the original employer (or else a visa cancellation) was introduced after Cully and Goodes's (2000) survey, which found a 30 per cent separation rate from the original employer, before the conclusion of the initial contract.

In Taylor et al.'s (2014) survey conducted in 2012, 93 per cent of the RSMS respondents (all of whom arrived in 2008–11) were still in the Northern Territory, and 78 per cent intended to continue living there because of employment opportunities, a liking for the lifestyle and the climate. The dominant region of origin was Southeast Asia and it is likely that familiarity of these Southeast Asians with the tropical lifestyle and climate has contributed to this outcome. The 7 per cent of RSMS visa holders who left the Northern Territory had nevertheless stayed for a median period of 38 months. Another 22 per cent intended leaving, thus resulting in a likely overall leakage of around 30 per cent of RSMS migrants from the Northern Territory.

Looking briefly to the economic contributions of regional migrants, the surveys above have reported high rates of employment of the principal visa holders (DIMIA 2005a, 2005b; Taylor, Bell, and Gerritsen 2014). The effects of regional migrants' employment will naturally vary between States and Territories. In the early 2000s, employment of regional visa holders either supported additional annual State/Territory labour force growth or was helping to offset a more general pattern of labour force decline (Golebiowska 2007). The contributions were between 0.23 per cent and 10 per cent of annual labour force growth in large States, but larger still in the less populous jurisdictions such as South Australia, the ACT, Tasmania and the NT with smaller labour forces (Golebiowska 2007).

Overall, between 1996–97 and 2013–14, the regional skilled and business immigrant intake has grown nearly 10-fold as a share of Australia's annual skilled migration program (from 4 per cent to 38.8 per cent). Given this expanded share, alongside evidence of improved immigrant retention and strong economic contributions in regional areas found in the reviewed surveys, the policy of encouraging regional settlement of immigrants can be considered effective. These surveys suggest that even with the unavoidable secondary mobility of some regional immigrants, the overall positive population effects from their settlement in Australia (on non-metropolitan areas, and on smaller and/or slower-growth capital cities like Darwin, Hobart or Adelaide) are now strong, certainly for the short to medium term. This applies because of policy design, in particular the role of the 2-year visa condition. Such evidence, particularly of improved retention rates, can serve to counter negative public attitudes and discourses about immigration to Australia. The evidence presented here shows that, with well-designed policy mechanisms, immigrants can be channelled into, and then retained in, regional areas—rather than adding to the populations of Australia's largest metropolitan areas. In the remainder of this paper, we use statistical evidence to counter another prominent argument that is regularly used to foster anti-immigration sentiment: the idea that immigration creates unemployment for the Australian-born.

Immigration and unemployment in Australia

As discussed above, the capacity of Australian (local, State, Territory, and federal) governments to ensure that immigration flows can help meet regional development objectives can be said to enhance public support for immigration policy (Markus and Arnup 2010). Likewise, an ability of government to convincingly reassure electors that immigration will not be taking jobs from 'locals' will also likely improve immigration support (Markus and Arnup 2010). At the very least, if it can be maintained that immigrants create as many jobs (or more jobs) than they fill, that will be an important political and economic proposition.

To an average citizen it may seem obvious that immigrants add to labour supply and hence take jobs. And in a direct sense this is true. What is also true is that immigrants can also cause the demand for labour to increase—both through their own spending (many bring financial assets with them from sale of businesses or property and investment funds and savings) and through others' spending in response to their arrival. The latter includes businesses and governments, community organisations and local communities and family networks, with all of them increasing outlays in response to the settlement and living needs of new arrivals. What matters is the balance of these contending supply and demand influences. Determining their precise balance is an empirical issue. Conveying the findings and facts from the empirical evidence—on the balance between jobs taken and created by immigrants—is a political leadership and communication issue. The remainder of this paper focuses on the empirical economic analysis of how immigration impacts employment. Such analysis can underpin the capacity for governments to communicate effectively to the public regarding the employment effects of immigration. The empirical evidence presented in this paper provides a solid basis from which to allay public concerns, since it affirms that overall the modern Australian immigration experience is such that new arrivals create at least as many jobs as they take.

For Australia, among a large number of studies which have examined the impact of immigration on Australian unemployment, the earliest was Withers and Pope (1985), later extended to cover a much longer time period in Pope and Withers (1993). More recent studies using different definitions, data periods and statistical techniques have found similar results, as, for example, with Shan, Morris, and Sun (1999), Kónya (2000) and Boubtane, Coulibaly, and Christophe Rault (2013). There is, in fact, a clear consensus across these studies, and from Australian immigration research more broadly, that increases and decreases in immigration have *not* been associated with net increases or decreases in the aggregate unemployment rate. Here, this consensus is tested further so as to include the latest immigration experience and to test the earlier findings using more advanced statistical methodology. The finding is to affirm the previous conclusions. An accumulation of such findings, tested rigorously, provides a strong basis for the aspiration to evidence-based policy in this publicly contentious field.

The force of this conclusion can be seen descriptively by a simple graphical investigation in which a measure of the unemployment rate is plotted against the immigration rate, using data for 1960 to 2013 (Figure 1).

Figure 1 shows that the Australian unemployment rate was relatively high during the mid-1980s and early 1990s. However, it recorded a consistent decline through the middle and towards the end of the 1990s, and throughout the early 2000s. Starting

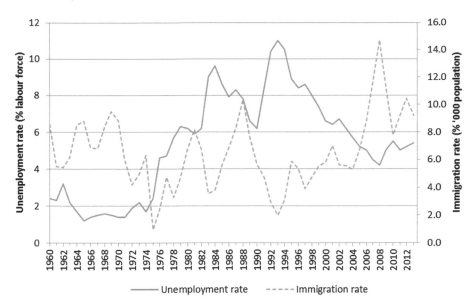

Figure 1. Immigration and unemployment rates in Australia, 1960–2013.
Source: Elnasri (2015).

from 2008, the unemployment rate began to increase again until the end of the period (2013). On the other hand, the immigration rate shows considerable variation during this period. As shown in Figure 1, it was relatively high during the mid-1960s, end of the 1970s and end of the 1980s. However, during the 1990s and the early 2000s it recorded lower levels, but then increased substantially in the middle of the first decade of the twenty-first century, reaching the highest level of all over the entire sample period in 2008. Afterwards, there was a sharp fall over 2009 and 2010. A slight increase is observed towards the end of the period.

By comparing the trend of the two series together, Figure 1 indicates that there is no co-movement between immigration and unemployment. In particular, movement in the Australian immigration rate appears to be inconsistent with observed movement in the unemployment rate during most of the last five decades. That said, this straightforward graphical analysis is not sufficient to draw conclusive evidence about the nature of the relationship. In the following sections we report on new formal investigations conducted using the so-called 'Granger causality' test designed for examining such relationships. These tests are named after Clive Granger, who received the Nobel Prize for this work in economics. Full technical details of data and procedures are available in Elnasri (2015).

Co-integration analysis: methods

The relevant co-integration analysis is presented here through two stages. Such analysis seeks to ensure that any relationship found is not coincidental or 'spurious', i.e. the variables of interest are 'causally' related and hence truly closely linked or 'co-integrated'. This is accomplished here, first, by implementing a simple bivariate framework and, second, by then applying a multivariate framework. In the two frameworks, we test the hypothesis as to whether changes in the Australia immigration rate, M_t, cause changes in Australia's

Table 2. Granger causality test results: bivariate model

Dependent variable	Causal variable	Causal lag	Chi^2	P-value
U_t	M_t	1	1.125	0.289
U_t	M_t	2	0.001	0.974
U_t	M_t	3	0.062	0.804
U_t	M_t	4	0.083	0.773
U_t	M_t	5	0.055	0.814
M_t	U_t	1	0.197	0.657
M_t	U_t	2	0.033	0.856
M_t	U_t	3	0.065	0.799
M_t	U_t	4	0.069	0.794
M_t	U_t	5	0.019	0.888

unemployment rate, U_t, or vice versa. Table 2 presents the results of chi-squared statistics and the corresponding P-values of the test.

It is well recognised that the results from such causality testing may be sensitive to the lag structure, especially the length of time allowed for the effects to flow through. Accordingly, results are presented for several lag lengths (i.e. 1–5 lags). In the upper half of Table 2, the null hypothesis tested is whether M_t (changes in Australia's immigration rate) does *not* Granger-cause U_t (changes in Australia's unemployment rate). In the lower half of the table, the null hypothesis tests whether U_t does not Granger-cause M_t.

As can be seen from Table 2, the above-stated null hypotheses are not rejected, and accordingly we can conclude that, within such a bivariate framework, there is no causality running from immigration to unemployment, or vice versa. However, there is argument in the previous literature, commencing with Pope and Withers (1993), that the simple causality method undertaken above can suffer from an omitted variable problem. Thus, to check the robustness of the results from the bivariate model, a more general model is specified to represent the relationship between immigration and unemployment, by including further explanatory variables. This is stage two of the analysis.

In particular, we have adopted the model of Pope and Withers (1993), which is based on a general disequilibrium framework of unemployment. More specifically, a four-dimensional vector autoregressive model is represented by the following equation:

$$z_t = \alpha_\circ + \sum_{t=1}^{k} \beta_t\, z_{t-i} + \epsilon_t,$$

where z_t is a vector consisting of four non-stationary variables beyond the variables looked at in the bivariate analysis: per capita real wages (W_t), real per capita GDP (Y), change in industrial structure of employment measured by the Stoikov index, (STO_t), and unemployment benefits proxied by the number of persons receiving unemployment benefits (BR_t). These join the unemployment rate (U_t) and immigration rate (M_t), for the wider stage two analysis.

More discussion on this disequilibrium model is provided in Pope and Withers (1993) and Shan, Morris, and Sun (1999). But in line with more recent literature (e.g. Islam 2007), the present study can further analyse the Pope–Withers model itself within the more advanced co-integration framework proposed by Johansen and Juselius (1990, 1994). This is therefore the full wider test sought to re-examine the immigration–unemployment relationship even more authoritatively and with more up-to-date statistics.

Table 3. Unit root test

Variable	ADF test statistic	p-value
U_t	−1.367	0.5978
ΔU_t	−5.073	0.0000
M_t	−2.493	0.3314
ΔM_t	−4.785	0.0005
W_t	−1.052	0.7338
ΔW_t	−5.747	0.0000
Y_t	1.245	0.9963
ΔY_t	−5.501	0.0000
STO_t	−2.930	0.1528
ΔSTO_t	−4.153	0.0053
BR_t	−2.002	0.2856
ΔBR_t	−4.400	0.0003

Note: The Augmented Dickey-Fuller (ADF) tests whether a unit root is present in a time series. The null hypothesis is that the series contains a unit root, and the alternative hypothesis is that the series was generated by a stationary process.

Before testing for the co-integration relationship between immigration and unemployment in this further way, it is important to determine whether all variables of interest are integrated of order one, I(1), so that it can be affirmed that the results are not biased. Thus, Augmented Dickey–Fuller (ADF) tests were carried out on the time series in levels and differences. Three lags were chosen to determine the stationarity of the variables. As shown in Table 3, the tests suggest that the series are indeed integrated of order one.

Co-integration analysis: findings

First, the Johansen tests for co-integration were applied. The trace statistic at rank = 0 is found to be 207.74, which exceeds the critical value 94.15. Thus, the null hypothesis of no co-integrating equations is rejected. The evidence of co-integration between variables in the VAR model tests the possibility of Granger non-causality. However, this does not provide information on what the co-integration equation or the direction of the causal relationship could be. To examine this issue, the vector error correction model (VECM) was applied. VEC has two advantages: it reveals the direction of causality, and it distinguishes between the short-run and long-run Granger causality. A VECM was estimated for time series covering the period 1985–2013. Following the estimation, short-run and long-run Granger causality tests were performed and their results are reported in Table 4. As seen in the table, the results indicate that the null hypotheses of Granger non-causality from immigration to unemployment, and Granger non-causality from unemployment to immigration, cannot be rejected at conventional significance levels.

Because the time series available for BR_t (the number of persons receiving unemployment benefits) and STO_t (change in industrial structure of employment measured by the Stoikov index) start only from 1982 and 1985 respectively, while the series of other variables start from 1960, it is of interest to explore the information available in the longer time series. Thus another specification of VECM, which excludes STO_t and BR_t, is estimated to cover the period 1960–2013. Results of the subsequent short-run and long-run Granger causality tests are reported in Table 5. Similar to the previous models, there is no evidence here either that immigration causes Australia's unemployment. There is no co-integration. The results are sustained even with alternative additional variables included and different time periods of analysis.

Table 4. Granger causality test results from VECM short-run causation test, 1985–2013

Equation	EC	U_t Wald F-statistics ΔU_t lags	M_t Wald F-statistics ΔM_t lags	W_t Wald F-statistics ΔW_t lags	Y_t Wald F-statistics ΔY_t lags	STO_t Wald F-statistics ΔSTO_t lags	BR_t Wald F-statistics ΔBR_t lags
U_t	0.01 (0.9410)	1.33 (0.7227)	5.54 (0.1361)	0.38 (0.9443)	1.31 (0.7262)	4.34 (0.2270)	1.77 (0.6225)
M_t	10.91 (0.0010) ***	6.13 (0.1056)	10.69 (0.0135)**	11.60 (0.0089)*	8.87 (0.0310)**	14.35 (0.0025)***	7.95 (0.0471)**
W_t	0.26 (0.6090)	0.13 (0.9883)	2.24 (0.5235)	4.19 (0.2416)	2.75 (0.4316)	2.23 (0.5264)	5.35 (0.1477)
Y_t	0.00 (0.9600)	0.97 (0.8089)	0.85 (0.8364)	0.42 (0.9358)	0.29 (0.9618)	0.49 (0.9203)	1.26 (0.7376)
STO_t	0.08 (0.7840)	6.40 (0.0937)	4.82 (0.1856)	2.70 (0.4398)	7.90 (0.0481)**	8.47 (0.0372)**	1.90 (0.5941)
BR_t	0.18 (0.6674)	0.81 (0.8467)	2.57 (0.4634)	0.09 (0.9925)	1.91 (0.5913)	2.96 (0.3982)	0.77 (0.8567)

Notes: the short-run causality tests are conducted by testing whether all the coefficients of the first difference of each variable are statistically different from zero as a group. The long-run causality is tested by the significance of the error term EC.
Terms *, **, and *** denote significance at the 10 per cent, 5 per cent and 1 per cent levels respectively.

Table 5. Granger causality test results from VECM 1960–2013 short-run causation test

Equation	EC	U_t Wald F-statistics ΔU_t lags	M_t Wald F-statistics ΔM_t lags	W_t Wald F-statistics ΔW_t lags	Y_t Wald F-statistics ΔY_t lags
U_t	1.25 (0.2638)	2.48 (0.4782)	2.21 (0.5301)	1.13 (0.7697)	2.50 (0.4745)
M_t	0.12 (0.7300)	1.63 (0.6518)	6.75 (0.0802)	0.20 (0.9774)	0.34 (0.9522)
W_t	3.14 (0.0764)	1.73 (0.6304)	0.56 0.9046	1.65 (0.6482)	2.61 (0.4552)
Y_t	3.37 (0.0663)	0.51 (0.9161)	1.12 (0.7722)	0.38 (0.9444)	0.33 (0.9541)

Notes: the short-run causality tests are conducted by testing whether all the coefficients of the first difference of each variable are statistically different from zero as a group. The long-run causality is tested by the significance of the error term EC.
*, **, *** denote significance at the 10 per cent, 5 per cent and 1 per cent levels respectively.

These results overall confirm that there is no causal relationship, in either direction, between immigration rates and unemployment rates in Australia. Therefore, this empirical evidence does not support any belief that, in aggregate, immigrants rob jobs—at least across the last three decades in Australia. The migration policy settings in place in Australia have therefore allowed significant immigration intakes to be received, without substantial aggregate adverse impacts for Australian unemployment rates resulting. This is in spite of Australia having one of the higher shares of overseas-born in its population across the OECD (Organisation for Economic Co-operation and Development) countries. Similar findings exist in relation to female workforce participation. Australia's labour market has expanded enormously through increased female workforce participation over recent decades, without related changes in aggregate unemployment.

Factors in Australia's migration policy settings that may have assisted with these positive employment outcomes are not directly examined in the statistical analysis here. But they could include such factors as the high skill share that typifies Australian immigration program management, relative to other countries. The predominant roles accorded to the points-selection process for independent migrants, plus the employer nomination and State/Territory nomination schemes for permanent visa entry, both provide selection mechanisms that favour entry of immigrants with skills. There are also strong skill, or human capital, elements embedded explicitly or implicitly for temporary entry through the 457 and the Working Holiday Maker visa schemes, as well as for student visa entry with associated (capped) work rights. These entry administration arrangements are especially possible to enforce for an island continent such as Australia. Together they ensure that regulated visa entry (except for visa-free entry from New Zealand, which is itself a high-wage country) can reassure the Australian public that working opportunities for the least skilled are *not* unduly disadvantaged by the immigration numbers experienced.

Conclusions

This paper has examined two key dimensions of the impact of immigration for Australia. One was sub-national and the other was national. We have argued that policy design— particularly the two-step visa process for regional migrants—has allowed substantial and effective regional location encouragement for immigrants over the past decade. Equally, the Australian immigration program's emphasis on skilled migration has helped to ensure that there have been no net job losses for the Australian economy as a consequence of the overall immigration program. Putting the two together, the job-creation dimensions of immigration may mean that regional policies seeking greater population growth away from the metropolitan locus of much Australian demography can be benefited by the use of targeted migration visa entry conditions. If there are economies of scale and scope in such regional areas, as will often be the case, then, in economic terms, this may be a net advantage economically compared to metropolitan settlement. Skill requirements for entry can also ensure that immigrants do not disadvantage less-skilled resident workers and, indeed, combined with the regional encouragement element, can productively up-skill regional workforces.

However, whether this potential has been fully realised to its optimum under the Australian immigration model, with the growth over time of both a 'skilled worker' and 'regional location' emphasis, remains as a future research project that looks for an analyst with the multi-disciplinary capabilities and ceaseless intellectual curiosity of a Graeme Hugo. Further, more detailed research is needed, as ever. Key research questions remain, including whether the skills that are prioritised are the right ones for Australian labour market needs, whether regional areas are selected well in allowing for critical minimum mass in retaining and benefiting from immigrant skills, whether wage and income effects diverge from employment impacts, and more. But the potential seems clear from the cases examined here for carefully focused social science research to inform policy advance for the national benefit, even in somewhat contested areas where seemingly self-evident propositions can be shown to require more nuanced understanding. Widely understood benefits can be enhanced and seeming negatives can be shown to be otherwise, or mitigated with well-designed policy. The public discourse around immigration can become better informed accordingly.

Notes

1. Permanent skilled and business migration to Australia operates on a points-based system. Prospective migrants must reach a minimum number of points to become eligible to apply for a visa.
2. Including the Northern Territory (NT), in its entirety, as well as Western Australia (WA) and Queensland (QLD) above the Tropic of the Capricorn.
3. There is one later survey of regional immigrants (Institute for Social Science Research 2010) but it contains only a minuscule number of the SRS immigrants, which does not permit evaluating how much retention has improved.

Acknowledgements

The authors are grateful to the editors and referees for very constructive comments. Responsibility for the contents remains that of the authors.

Disclosure statement

No potential conflict of interest was reported by the authors.

Funding

This work was partly supported by the Australian Council of Learned Academies from the Australian Research Council Grant [LS120100001 "Securing Australia's Future: Project 1 – Australia's Comparative Advantage"].

References

ABS (Australian Bureau of Statistics). 2012. "Australian Demographic Statistics December Quarter 2011." cat.no. 3101.0. Dated June 20. Accessed February 11, 2016. www.ausstats.abs.gov.au/ausstats/subscriber.nsf/0/66175C17C773120DCA257A2200120F63/$File/31010_Dec%202011.pdf.

Argent, N., and M. Tonts. 2015. "A Multicultural and Multifunctional Countryside? International Labour Migration and Australia's Productivist Heartlands." *Population, Space and Place* 21 (2): 140–156. doi:10.1002/psp.1812.

Birrell, B., L. Hawthorne, and S. Richardson. 2006. "Evaluation of the General Skilled Migration Categories." *Report*. Accessed February 3, 2016. www.flinders.edu.au/sabs/nils-files/reports/GSM_2006_Full_report.pdf.

Boubtane, E., D. Coulibaly, and Ch. Christophe Rault. 2013. "Immigration, Growth, and Unemployment: Panel VAR Evidence from OECD Countries." *Labour* 27 (4): 399–420. doi:10.1111/labr.12017.

Cameron, R., T. Dwyer, S. Richardson, E. Ahmed, and A. Sukumaran. 2012. "Skilled Migrants and their Families in Regional Australia: A Gladstone Case Study." e-book, CQUniversity, Gladstone, Qld. http://hdl.cqu.edu.au/10018/923857.

Carr, R. 2000. "Bob Carr on Immigration." *The World Today Archive*, March 6, Sydney ABC. Accessed July 15, 2016. http://www.abc.net.au/worldtoday/stories/s107556.htm.

Cully, M., and R. Goodes. 2000. "Evaluation of the Regional Sponsored Migration Scheme: Final Report Prepared for the Department of Immigration and Multicultural Affairs and the Department of Employment, Workplace Relations and Small Business."

Davis, L., and S. Deole. 2015. "Immigration Attitudes and the Rise of the Political Right: The Role of Cultural and Economic Concerns Over Immigration." CESIFO Working Paper. No. 5680, December.

DIBP (Department of Immigration and Border Protection). 2014. "2013–14 Migration Programme Report. Programme Year to 30 June 2014." Accessed November 4, 2015. www.border.gov.au/ReportsandPublications/Documents/statistics/report-migration-programme-2013-14.pdf.

DIBP (Department of Immigration and Border Protection). 2015. "Working Holiday Maker Visa Programme Report." Accessed February 23, 2016. http://www.border.gov.au/ReportsandPublications/Documents/statistics/working-holiday-report-june15.pdf.

DIBP (Department of Immigration and Border Protection). 2016. "Regional Sponsored Migration Scheme Visa." Accessed July 1. 2016. www.border.gov.au/Trav/Visa-1/187-.

DIBP (Department of Immigration and Border Protection). not dated a. "Fact Sheet 26. State-specific Regional Migration." Accessed January 27, 2016. https://www.border.gov.au/about/corporate/information/fact-sheets/26state.

DIBP (Department of Immigration and Border Protection). not dated b. "Migration Blog." Accessed February 10, 2016. http://migrationblog.border.gov.au/category/regional-skilled-migration-scheme-rsms/.

DIMIA (Department of Immigration and Multicultural and Indigenous Affairs). 2005a. *Analysis of the Regional Sponsored Migration Scheme Subclass.* Canberra: Commonwealth of Australia.

DIMIA (Department of Immigration and Multicultural and Indigenous Affairs). 2005b. *Analysis of the Skilled Designated Area Sponsored Subclass.* Canberra: Commonwealth of Australia.

DIMIA (Department of Immigration and Multicultural and Indigenous Affairs). 2005c. *Population flows: Immigration aspects, 2003–04 edition.* Canberra: Commonwealth of Australia.

Elnasri, A. 2015. "Immigration and Unemployment in Australia." In *Commissioned Statistical Studies for the Australia's Comparative Advantage Project*, 60–64. Melbourne: ACOLA, July. Accessed June 5, 2016. http://www.acola.org.au/PDF/SAF01/13.%20Commissioned%20Statistical%20Studies%20-%206%20July.pdf.

Goel, K., and R. Goel. 2009. "Settlement of Immigrants in Regional South Australia—Role of Socio-economic Determinants." Conference proceedings of 10th National Rural Health Conference, May 17–20. Accessed March 7, 2016. http://ruralhealth.org.au/10thNRHC/10thnrhc.ruralhealth.org.au/papers/docs/Goel_Kalpana_D1.pdf.

Golebiowska, K. 2007. "Regional Policy for Skilled Migration in Australia and Canada." PhD thesis passed at the Australian National University, http://hd.l.handle.net/1885/15782.

Golebiowska, K. 2012. "Intergovernmental Collaboration in Immigration, Settlement and Integration Policies for Immigrants in Regional Areas of Australia." *e-Politikon* 1: 121–152.

Golebiowska, K. 2015. "Are Peripheral Regions Benefiting from National Policies Aimed at Attracting Skilled Migrants? Case Study of the Northern Territory of Australia." *Journal of International Migration and Integration*, online June 5. doi:10.1007/s12134-015-0431-3.

Golebiowska, K., A. Taylor, T. Carter, and A. Boyle. Forthcoming 2016. The International Migration and the Changing Nature of Settlements at the Edge. In *Settlements at the Edge: Remote Human Settlements in Developed Nations*, edited by A. Taylor, D. Carson, P. Ensign, L. Huskey, G. Eilmsteiner-Saxinger, and R. O. Rasmussen, 75–97. Cheltenham, UK: Edward Elgar.

Goot, M., and I. Watson. 2011. "Population, Immigration and Asylum-seekers: Patterns in Australian Public Opinion." Parliament of Australia, Parliamentary Library, May.

Hugo, G. 1999. "Regional Development through Immigration? The Reality Behind the Rhetoric." Research Paper 9 (1999–2000), Parliament of Australia, www.aph.gov.au/library/pubs/rp/1999-2000/2000rp09.htm, accessed 5 July 2016.

Hugo, G. 2008a. "Australia's State-specific and Regional Migration Scheme: An assessment of its Impacts in South Australia." *Journal of International Migration and Integration* 9 (2): 125–145. doi:10.1007/s12134-008-0055-y.

Hugo, G. 2008b. "Immigrant Settlement Outside of Australia's Capital Cities." *Population, Space and Place* 14 (6): 553–571. doi:10.1002/psp.539.

Hugo, G., S-E. Khoo, and P. McDonald. 2006. "Attracting Skilled Migrants to Regional Areas: What does it Take?" *People and Place* 14 (3): 26–36.

Hugo, G., and R. Morén-Alegret. 2008. "International Migration to Non-metropolitan Areas of High-income Countries: Editorial Introduction." *Population, Space and Place* 14 (6): 473–477. doi:10.1002/psp.515.

Institute for Social Science Research. 2010. "Factors that Influence Skilled Migrants Locating in Regional Areas. Final report." Prepared for DIAC. Accessed May 21, 2015. www.dss.gov.au/sites/default/files/documents/01_2014/factors-influence-skilled-migrants-locating-regional-areas.pdf.

Islam, A. 2007. "Immigration Unemployment Relationship: The Evidence from Canada." *Australian Economic Papers* 46 (1): 52–66.

Johansen, S., and K. Juselius. 1990. "Maximum Likelihood Estimation and Inference on Cointegration: With Applications to the Demand for Money." *Oxford Bulletin of Economics and Statistics*, 52(2), 169–210. doi:10.1111/j.1468-0084.1990.mp52002003.x.

Johansen, S., and K. Juselius. 1994. "Identification of the Long-run and the Short-run Structure: An Application to the ISLM Model." *Journal of Econometrics* 63 (1): 7–36. doi:10.1016/0304-4076 (93)01559-5.

Khoo, S-E., P. McDonald, and G. Hugo. 2005. "Temporary Skilled Migrants in Australia: Employment Circumstances and Migration Outcomes." Report to the Department of Immigration, Multicultural and Indigenous Affairs. Accessed February 29, 2016. www.sapo.org.au/binary/binary3662/457s.pdf.

Kónya, L. 2000. "Bivariate Causality between Immigration and Long-term Unemployment in Australia, 1981–1998." Victoria University Applied Economics Working Paper 18/00.

Krivokapic-Skoko, B., and J. Collins. 2016. "Looking for Rural Idyll 'Down Under': International Immigrants in Rural Australia." *International Migration* 54: 167–179. doi:10.1111/imig.12174.

Markus, A. 2011. "Immigration and Public Opinion." In Productivity Commission, A Sustainable Population: Key Policy Issues, Proceedings, Canberra: Productivity Commission, Chapter 10, pp. 93–110.

Markus, A. 2016. *An inventory of Australian public opinion surveys. Mapping Australia's population*. Melbourne: Monash University. accessed June 5, 2016. https://www.monash.edu/mapping-population/public-opinion/surveys/inventory-of-surveys.

Markus, A. and J. Arnup. 2010. *Mapping Social Cohesion: The 2009 Scanlon Foundation surveys*. Melbourne: Monash University Institute for the Study of Global Movements.

Massey, S.L.J., and N. Parr. 2012. "The Socio-economic Status of Migrant Populations in Regional and Rural Australia and its Implications for Future Population Policy." *Journal of Population Research* 29: 1–21. doi:10.1007/s12546-011-9079-9.

Migration Western Australia. not dated. "Frequently Asked Questions about State Nomination." Accessed July 1, 2016. www.migration.wa.gov.au/services/skilled-migration-wa/frequent-questions-about-state-nomination.

Parliament of Australia. 2001. "New Faces, New Places: Review of State-specific Migration Mechanisms (September 2001)." Joint Standing Committee on Migration. Accessed February 3, 2016. www.aph.gov.au/Parliamentary_Business/Committees/Joint/Completed_Inquiries/mig/report/ssmm/index.

Parliament of Australia. 2015. "Report of the Inquiry into the Business Innovation and Investment Programme." Joint Standing Committee on Migration, March 2015. Accessed February 7, 2016. www.aph.gov.au/Parliamentary_Business/Committees/Joint/Migration/BIIP/Report.

Phillips, J., and H. Spinks. 2012. "Skilled Migration: Temporary and Permanent Flows to Australia." Background note. Accessed February 3, 2016. www.aph.gov.au/About_Parliament/Parliamentary_Departments/Parliamentary_Library/pubs/BN/2012-2013/SkilledMigration#_Toc342559467.

Piper, M., and Associates. 2009. "Regional Humanitarian Settlement Pilot Ballarat." Report of an evaluation undertaken by Margaret Piper and Associates for the Department of Immigration and Citizenship. Accessed July 29, 2016. www.dss.gov.au/sites/default/files/documents/12_2013/settlement-pilot-ballarat_access.pdf.

Pope, D., and G. Withers. 1993. "Do Migrants Rob Jobs? Lessons of Australian history, 1861–1991." *The Journal of Economic History* 53 (4): 719–742.

Shan, J., A. Morris, and F. Sun. 1999. "Immigration and Unemployment: New Evidence from Australia and New Zealand." *International Review of Applied Economics* 13: 253–260.

Taylor, A., L. Bell, and R. Gerritsen. 2014. "Benefits of Skilled Migration Programs for Regional Australia: Perspectives from the Northern Territory." *Journal of Economic and Social Policy* 16 (1): 35–69. Accessed September 3, 2014. http://epubs.scu.edu.au/jesp/vol16/iss1/3.

Trenwith, C. 2011. "Perth Branded Regional City of Hardship in Bid to Attract Foreign Workers." Accessed February 11, 2016. www.watoday.com.au/wa-news/perth-branded-regional-city-of-hardship-in-bid-to-attract-foreign-workers-20110719-1hmad.html, dated 11 July.

Withers, G., and D. Pope. 1985. "Immigration and Unemployment." *Economic Record* 61 (2): 554–564. doi:10.1111/j.1475-4932.1985.tb02010.x.

Withers, G., and M. Powall. 2003. *Immigration and the Regions: Taking Regional Australia Seriously.* Sydney: Chifley Research Centre.

Woods, M. (2007). Engaging the Global Countryside: Globalization, Hybridity and the Reconstitution of Rural Place. *Progress in Human Geography*, 31(4), 485–507. doi:10.1177/0309132507079503.

Wulff, M., and A. Dharmalingham. 2008. "Retaining Skilled Migrants in Regional Australia: The Role of Social Connectedness." *Journal of International Migration and Integration* 9 (2): 147–160. doi:10.1007/s12134-008-0049-9.

Understanding Ethnic Residential Cluster Formation: new perspectives from South Australia's migrant hostels

Karen Agutter and Rachel A. Ankeny

ABSTRACT

Throughout Australia's history, successive governments have lamented the clustering of non-English-speaking migrants in 'ethnic enclaves' or 'ghettos'. From the early Chinatowns of the 1800s till today, urban concentrations of ethnic groups have raised concerns and fears in local populations and authorities alike, despite decades of international research which suggests that ethnic residential clusters actually aid long-term assimilation and adjustment. Many of the ethnic residential clusters in contemporary Australia have been claimed to be a direct consequence of the migrant hostels and reception centres which operated between 1948 and the 1990s. This paper traces migrant settlement patterns in South Australia in rich detail, revealing the complexities of lived experiences that shape migrant settlement decisions. Against the background of public and scholarly debates over 'ethnic enclaves', and drawing on quantitative and qualitative historical research on the lived experiences of former hostel migrants, it analyses how migrant hostels and reception centres contributed to the settlement experiences of diverse migrants. We conclude that migrant hostels were just one among various factors that led to the growth and maintenance of ethnic residential clusters.

Introduction

Hostels were an essential component of Australia's migrant intake for over 40 years, from 1948 to the mid-1990s. We use the term 'hostels' generically to refer to all forms of communal State and Commonwealth-funded accommodation for migrants, covering reception and training centres, workers' hostels, and holding centres as well as institutions formally called hostels. Hostels provided not only relatively low- or no-cost temporary accommodation for refugees and assisted migrants but also a variety of services that were essential to the host nation's attempts to assimilate, integrate, and acclimatise new arrivals. Past studies have claimed that the hostels of the 1970s and 1980s, and particularly those used to house Vietnamese refugees, directly contributed to the formation of ethnic residential clusters, or even enclaves and ghettos (e.g. Jupp, McRobbie, and York 1990; Birrell 1993).

In this paper, we first briefly consider the concept of ghettos and ethnic enclaves as applied, in other immigrant societies, to large and segregated ethnic residential concentrations. We discuss evidence that such concentrations did not really occur in Australia in the same way despite frequent popular use of this terminology. Nevertheless, there have been and continue to be residential clusters of ethnic populations in Australia that are worthy of scholarly exploration.[1] We contend that it is critical to continue to research segregation and clustering because of the effects such patterns may have on migrants. Also, it is critical to investigate these types of concentrations because they continue to be seen as problematic in the wider population and popular media, and these perceptions often can become 'self-fulfilling prophesies'. As noted by scholars of migration, 'concentration neighbourhoods [i.e. neighbourhoods with large populations from the same ethnic group] can turn into breeding grounds for misery because they are so perceived' (Kempen and Özüekren 1998, 1634). In other words, communities can develop in problematic ways if those around them have low and negative expectations and fail to provide support for their positive growth.

We do not seek to cover old ground by undertaking a demographic analysis of ethnic residential clusters. Rather, we seek to understand the factors surrounding immigration policy and practices that may have contributed to (or worked against) the formation of such clusters, drawing on our findings about the operation of the hostel system in South Australia based on an Australian Research Council funded Linkage Project 'Hostel Stories: Toward a Richer Narrative of the Lived Experiences of Migrants'. We consider a range of factors including the location of migrant hostels, government policies around assimilation and placement of migrants into holding centres and work camps, and the impacts of the external community and the internal migrant community. We show that the hostel system alone did not create ethnic residential clusters: in fact no single factor alone can explain the establishment or growth of these clusters. Rather, a combination of factors must be considered when seeking to understand migrant settlement patterns.

Background: ghettos and enclaves

Across the world, particularly in nations of immigration, researchers from various disciplines have long sought to understand the formation, maintenance and significance of ethnic enclaves and ghettos. They have tended to explore these issues in the context of specific host nations and individual cities[2] and with reference to particular ethnic origins.[3] There has always been strong political, public, and media interest in ethnic settlements and concentrations, more often than not arising from the fear and apprehension of migrants thought to be notably different from the majority population of the respective host nation. So what is a ghetto, and how is a ghetto different from an ethnic enclave? Does either exist in any recognisable form in Australia?

Historically the word 'ghetto' has been associated primarily with European Jewry,[4] though today we tend to equate the term with a slum area occupied primarily by a minority group; it is the pejorative associations with the word that tend to arise in popular discourse. By contrast, the term 'ethnic enclave' is used to describe a geographic area that has a high ethnic concentration both living and conducting cultural and economic activity within its boundary (Abrahamson 1996) . Unlike a ghetto, an enclave is generally not considered to be a slum, although enclaves continue to have negative associations in Australia.

Social commentator Bernard Salt recently asked: 'At what point in a suburb's development does the presence of a demographic group stop being cultural diversity and start being an enclave?' (Salt 2010, n.p.). Scholars have attempted to place a numerical figure on the proportion of a particular ethnic population needed in a given area for it to create segregation (in terms of the representation of non-Australian birthplace groups), and thus to be considered a ghetto or an enclave. Generally, the figure is considered to be around 90 per cent to be a ghetto (Peach 1996) and 30 per cent to be an enclave (Dunn 1998). However, such simplistic definitions are problematic where there are multiple ethnicities living within an area, where the dominant ethnicity changes over time (see, for example, McKenzie 1999 on the Melbourne suburb of Footscray), or where the particular ethnic concentration is considered to be acceptable to the host society as a whole, as with heavy concentrations of British migrants (Ang et al. 2006; Salt 2010, n.p.) in Elizabeth to the north of Adelaide, or Burns Beach north of Perth (one of the largest ethnic residential clusters in Australia).[5] As Jupp, McRobbie, and York (1990, 1) explain, 'white English-speakers have always been excluded from any discussion of ghettos or enclaves even when they have numerically dominated large suburban areas', as policy and public opinion have been concerned only with those from non-English-speaking backgrounds.

In Australia, there is evidence of ethnic residential clustering, most notably, for example, in the so-called 'little Italies', 'Chinatowns', and so on. However, the general consensus is that Australia has not experienced the formation of ghettos and major ethnic enclaves (e.g. Burnley 1999; Ang et al. 2002; cf. Poulsen, Johnston, and Forrest 2002, 2004) as commonly found in the USA, Britain and Canada. The lack of ghetto formation in Australia has been attributed largely to a series of specific policies that controlled and severely limited non-Anglo-Saxon arrivals, including pre-Federation anti-Chinese legislation on poll and landing taxes, and entry controls implemented as a result of the *1901 Immigration Restriction Act* (the so-called 'White Australia' policy). By contrast, when mass migration occurred in the post-Second World War period as Australia sought to increase its population and grow its economic base, European refugees and assisted-migrant arrivals were discouraged from forming ethnic clusters through a strict policy of assimilation and mandatory 2-year work contracts. The latter generally allowed authorities to disperse these 'New Australians' across the country.[6] However, the dismantling of the White Australia policy and the subsequent adoption of an official policy of multiculturalism in 1972 allowed for a much wider range of non-European-born arrivals to enter Australia. As refugees from Southeast Asia were joined by refugees and migrants from other parts of Asia, the Middle East and Africa, many Australians were, and continue to be, outspoken about their fears of, and opposition towards, ethnic enclave and ghetto formation. The threat of ethnic concentrations has been used frequently in wider debates on Australian multicultural policy, including in historian Blainey's (1988) statement that Australia must break down the walls of the ghettos, and the infamous claims of the right-wing politician Pauline Hanson[7] (1996, n.p.) that the nation was being swamped by Asians who 'have their own culture and religion, form ghettos and do not assimilate'. Although history reveals that ethnic residential clusters are temporary phenomena, that in fact play important roles in the process of assimilation and integration of migrants into the wider community, isolated events such as the Cronulla riots[8] serve to stimulate ongoing fears in the wider community about concentrations of migrants and to prompt scholarly research.

Australian studies of ethnic residential segregation have relied heavily on models formulated overseas. However, our immigration patterns and experiences, particularly post-Second World War, have been unique for a number of reasons. Important differences include control of the movement of migrants through policies such as work contracts, careful overseas selection of future citizens, recent use of detention centres and offshore processing, and, as highlighted in this paper, the use of hostels as initial accommodation for new arrivals. If we cannot simply use the models and analysis conducted in other nations, then what approach should we take?

Australian studies of ethnic residential concentration and community diversity have relied primarily upon census data and resultant mapping of the residential distribution of immigrant groups to indicate their separation and segregation from the host society (e.g. Poulsen, Johnston, and Forrest 2002). As Grimes (1993, 104–105) argues, these 'mapping exercises are a useful first step' but they also have their limitations as they may 'hide significant heterogeneity in immigrant populations'. There are dangers in making assumptions about ethnic groups based on birthplace statistics as provided in census tables. For example, many post-War displaced children who came to Australia were born in Germany and yet their familial ethnicities were not German. Frequent movement prior to immigration is common among the displaced and refugee arrivals who are prominent in Australia's immigration history, and similar patterns continue today particularly among refugees from the Middle East and parts of Asia. As Burnley (1994) has argued, there is a need for a more extensive analysis of immigrant groups. Furthermore, questions regarding ethnicity/ancestry and birthplace within the Australian census have changed over time, making longitudinal and comparative study problematic.[9] Hence, given these Australian census data limitations and the aforementioned unique immigration patterns and policies associated with the movement of migrants within Australia, we contend that a new approach to research on the formation of ethnic residential clusters is required.

Methods

The Hostel Stories project traces the lived experiences of thousands of migrants who passed through South Australia's 14 government-operated migrant hostels and reception centres and various work camps from the late 1940s to the 1990s. Ethics clearance was obtained from the University of Adelaide's Human Research Ethics Committee (approval H-2012–120). Participant recruitment for the project typically begins with a registration of interest (available via the project in several languages, and paper and electronic forms) asking for basic information on migrants' backgrounds and hostel experiences. As of July 2016, over 600 registrations of interest had been received from former hostel residents; these allow us to gather qualitative data including migrants' initial places of residence and type of accommodation upon departure from the hostel, and any free-hand comments on these issues.

Based on sampling from the registrations of interest to represent diverse countries of origin, time and age of migration, reasons for migration, and hostels of residence, approximately 95 oral history interviews have been performed. Interviews are conducted using a set script with open-ended questions, and including questions about how migrants decided where to live upon leaving the hostel.

Among the archival materials sourced (which cover all of the government-operated hostels in South Australia throughout the time period in which they operated and numerous work camps), the hostel arrival and departure registers and accommodation records[10] have proven to be extremely useful sources of information. They provide insights into the initial movement of migrants upon leaving the hostels, which is a critical stage in migrants' lives and settlement in Australia, as they note the first destination address. Data from these records, in combination with other archival sources including alien registration documents[11] and applications for naturalisation and the qualitative data described above, have allowed us to establish settlement patterns over time from first departure from the hostel, thus providing more nuanced insights into larger-scale migrant settlement patterns and factors impacting choice of locale and type of housing than can be generated through use of census data.

Factors related to ethnic residential cluster formation

Studies of Vietnamese settlement frequently claim that the hostels which housed these refugees and migrants in the 1970s and 1980s were directly connected with the development of Vietnamese residential concentrations in Australian cities. Jupp, McRobbie, and York (1990, 79) contend that 'the use of hostels was the most important single factor in creating such concentrations'. However, our evidence from the Hostel Stories project is that there were numerous factors that contributed to settlement patterns following hostel residence. In the remainder of this paper, we analyse a number of these, considering the extent to which each may have contributed to the formation of ethnic residential concentrations in South Australia.

Hostel location and length of stay

Salt, writing in *The Australian* in 2010, observed that:

> The largest group of any non-Australian-born population in Australia is to be found in Sydney's southwestern suburb of Cabramatta. Here, Vietnamese comprise 27 per cent of the population in postcode 2166. That's because, when the Vietnamese arrived en masse just 25 years ago they were channelled through migrant hostels at Springvale in Melbourne and Villawood in Sydney. On moving into the general community they gravitated to nearby Cabramatta and to Springvale, where they are 21 per cent of the population in the local postcode district. (Salt 2010, n.p.)

Implicit in this statement is an argument that hostels can act as a 'seed' for the growth of ethnic residential clustering in the surrounding community. However, studies of earlier migrant arrivals and settlement do not support strong links between the hostels and migrant clustering upon settlement. Jupp, McRobbie, and York (1990, 30) claim that clustering did not occur in the earlier periods (late 1940s and 1950s) because migrants did not stay in hostels long enough to establish local links; furthermore, hostels 'had been located in remote rural areas'. By the 1970s and 1980s (the period when Vietnamese refugees arrived in Australia), hostels were 'located in the suburbs of metropolitan centres', contributing to the formation of residential concentrations. Testing this hypothesis against the

South Australian hostel system over many decades, we can find examples that both support and contradict this argument.

While it is certainly true that displaced persons (DPs) and early assisted migrants (under the requirements of the mandatory 2-year work contracts) were often moved out of the hostel system quickly to be sent to work in remote, rural, and sometimes urban locales while their dependent wives and children were housed in hostels around the country, other assisted migrants and family groups in particular often stayed in hostels for long periods of time. Those who had long stays created considerable anxiety for the Commonwealth Government, leading to rule changes in 1958 and evictions of migrants who had been in hostels for over 5 years.[12] Based on our research, while many earlier arrivals (those who arrived in the late 1940s and 1950s) were dispersed across the country, many others stayed in hostels, some by circumstance and some by choice, for many years and actually far longer than those of Vietnamese origin who migrated in the later period (arriving in Australia in the 1970s and 1980s). Thus length of stay was on average much longer in the earlier periods during which clustering did not typically occur, as compared to later periods which are often associated with clustering.

With respect to Jupp, McRobbie, and York's (1990) claim about the remoteness of early hostels in locales that were less attractive to migrants, we note that in South Australia there was a mixture of urban and suburban locations. Many hostels were purposely located near transport,[13] so that establishments such as that at Smithfield (30 km from the Adelaide Central Business District (CBD)) were not so distant as to prevent men from travelling to work in Adelaide suburbs while living at the hostel. There were also a number of hostels located in outer suburbs and in newly developing areas on the fringes of metropolitan Adelaide, including Rosewater (14 km from the CBD) and Semaphore (22 km from the CBD) which were not subsequently correlated with significant ethnic residential concentrations. One reason for this, at least in the South Australian case, is that there was significant movement of people between hostels, and therefore migrants did not necessarily put down roots in the surrounding communities. Analysis of hostel arrival/departure registers confirms this ongoing movement between hostels. For example, when the Gawler Hostel closed after only 3 years of operation, residents were sent to Finsbury (later renamed Pennington) in the first instance and then, with the arrival of significant numbers of British migrants to that hostel, on to Glenelg Hostel.

Considering the applicability of the 'seed' effect as described above, although the Elder Park Hostel in the Adelaide CBD catered primarily for English migrants, the concentration of this migrant group occurred in the satellite suburb of Elizabeth some 30 km to the north of the city (due primarily to State and private housing schemes targeted specifically at migrants). In contrast, Finsbury (located 11 km from the Adelaide CBD), which was the largest and longest operating hostel in South Australia (1949–90s), housed numerous waves of migrants including DPs, assisted British and European migrants, and other refugees from Europe, Asia, and South America. Today, the community profile of the Pennington district indicates some notable clusters of ethnicities, including 16 per cent who claim Vietnamese ancestry, 13 per cent Italian, 6 per cent Greek and 6 per cent German, as well as 26 per cent Australian and 28 per cent English (City of Charles Sturt 2011, n.p.). This distribution could be considered to reflect the ethnic background of the migrants who passed through the hostel. However, it must be noted that the concentrations of each ethnicity are still far below the numbers required to constitute an enclave using the international standards discussed previously, even for the Vietnamese, the ethnic group most closely associated with Pennington in the public imagination. Settlement patterns thus are much more complex than a simple theory of hostel proximity generating ethnic enclaves or residential clusters might suggest.

Availability of work and housing

The Finsbury/Pennington case may indicate that some migrants preferentially settled in the vicinity of the hostel; however, there were other factors at play in the residential decision-making process, most notably the availability of affordable housing and work, particularly for migrants who were largely working class (see also Zang and Hassan 1996). In the early years, the area around this hostel was rich in job prospects, providing easy access to large manufacturing companies such as General Motors Holden, Phillips, and Kelvinator. According to historian Marsden (1977, 228), by the early 1950s Finsbury housed 'almost all the heavy industry (mostly engineering, auto-motive and "whitegoods") the government had hoped to see established in South Aus-tralia' and so the availability of jobs may well have been a significant contributor to patterns of migrant settlement. In contrast, several of our participants noted that in the 1970s and 1980s, when work was less readily available in the area surrounding Fins-bury/Pennington, many Vietnamese and other refugees moved elsewhere in Adelaide or even interstate in search of better job opportunities. Thus job prospects often critically contributed to settlement decisions, which often were not well-correlated with hostel location.

Additional evidence of the association between housing and work can be found by tracking hostel residents through alien registration documents, where we found many instances of change of abode and of employment. For example, a Polish-born migrant and his wife arrived in Melbourne on 28 March 1950 and were sent initially to Finsbury Hostel. Between their arrival in 1950 and his naturalisation in Enfield (8 km from the hostel) in 1956 (National Archives Australia [NAA], D4878), the couple moved, appar-ently to be closer to his work. Under the 2-year work contract, he was employed at Fauld-ing and Co. in Southwark (current-day Thebarton) and on leaving the hostel in September 1950 he and his wife moved to Allenby Gardens, just 3 km from this employer. On release from his 2-year contract, the couple moved between several homes as he changed jobs, having four addresses in as many years, including a few months in Lobethal (in the Ade-laide Hills) while he was employed at the Onkaparinga Woollen Mills. While all of their moves, with the exception of Lobethal, were within a 10 km radius of the original hostel, they were also within easy commuting distance to his respective places of employment (NAA, D4881).

The influence of work availability on choice of residential location was evident in many of the oral histories conducted. A Lithuanian DP (JD, 26/2/15),[14] having worked out his 2-year contract in railway camps across South Australia, was prompted to request a transfer to Adelaide because of the prospect of being part of the growing Lithuanian community. However, when seeking accommodation, proximity to his job at Mile End influenced his choice to lodge in an Australian household. A second Lithuanian-born DP (JV, 26/2/15) found rooms with an Australian man in Enfield because that location was near his work in the cold storage facilities at Mile End. Although assisted by a fellow Lithuanian DP to find accommodation, an Estonian couple (LG, 26/2/15) ultimately settled in a room in a large house near Wakefield Street in Adelaide as it was near to the wife's work in a belt factory and the husband's as a mechanic. The hostel registers are also revealing on this point: at Gawler Hostel, out of 1231 departures, only 203 migrants/refugees settled in Gawler and surrounding towns. Despite there generally being enormous local support for them, most moved elsewhere probably as a consequence of the limited work opportunities. The large numbers of migrants who went to Woomera, Whyalla, and Renown Park (all of which had a large number of jobs) reveal availability of work as an obvious causal factor for choice of abode upon leaving a migrant hostel.

The importance of affordable housing for migrants was another key factor, and is particularly evident in the cases of those who lived at Glenelg Hostel. Examination of the arrival and departure registers for this centre reveals that very few families moved from the hostel to the surrounding area despite most of our participants describing its attractiveness given its proximity to the beach, easy transport to the city, and schools accustomed to educating migrant children. While their destinations were diverse, many moved north towards Finsbury. For example, a sampling taken from Glenelg Hostel registers during the month of July 1959[15] shows that, aside from one family who returned to Austria, the remaining families moved to houses in Beverley, Croydon, and Prospect, all areas which offered lower housing costs and better work opportunities at that time. Evidence from our oral histories of these patterns includes the statement by a Polish-born DP (IGM, 7/5/15) who noted that the Ukrainians settled largely around Seaton and Seaton Park (in the western suburbs) 'because the land was cheap, they could buy the land, live on it in car crates until they actually built their own homes'. Similarly, an English family (ILD, 28/10/14) noted that ideally they would have stayed in Glenelg after their time there in the hostel but 'the cost of buying a house there was half as much again as the one that we bought in Fairview Park [a suburb to the northeast] so it was a matter of … we just couldn't afford to go where we really wanted to'. Similarly the Gawler Hostel registers reveal large numbers of departures to live in new and affordable areas such as Enfield.

Assimilation to multiculturalism: the impact on hostels and communities

From 1947 through to the 1960s, the official Commonwealth Government policy was one of assimilation, whereby new arrivals were actively encouraged to adopt the 'Australian way of life'[16] and to disperse into the community. Newspaper reports from that time highlight the policy aims of assimilation, and the education and involvement of the wider host society:

> Many of the new Australians are from non-English speaking lands. They need in the first place to learn something of the language, and it is pleasing to know that provision is being made for elementary instruction, and that some 20,000 are being taught. This is one of the aids which no one thought about in other decades, with the result that foreign immigrants were greatly frustrated and this led to their forming national groups apart from the general community. This is something far from desirable, especially as the numbers of foreign migrants are far greater than ever before. We must take action which will prevent the formation of national enclaves. (*Advocate* 1950, 4)

Members of the broader Australian community were actively encouraged to visit the hostels and under the umbrella of the Commonwealth Government funded and initiated Good Neighbour Movement, innumerable Australian organisations and individuals (including religious denominations and volunteer groups such as the YWCA, the CWA, the Girl Guides, and Boy Scouts, and service organisations including Rotary and Apex), entered the hostels in order to introduce new arrivals into the Australian way of life.

This emphasis on assimilation, and the fact that there were a large number of hostels geographically dispersed around the Adelaide metropolitan area, meant that ethnic residential concentrations were minimal during this period. Even over time, as ongoing arrangements with the Intergovernmental Committee for European Migration (ICEM), evolving assisted-passage agreements between Australia and various other countries, and the realities of shipping meant that hostels contained increasingly large numbers of migrants from the same ethnic groups, the focus on assimilation still restricted the formation of ethnic enclaves in the community.

By the 1970s, when significant numbers of Vietnamese migrants arrived, the official government policy had changed to multiculturalism, where retention of ethnic identity was not only tolerated but encouraged. At the same time, the numbers of migrant hostels had dramatically reduced (for example, in South Australia the number of hostels declined from 13 in the 1950s and 1960s to a single hostel, Pennington, by the mid-1970s) so that there was naturally a higher concentration of particular ethnic populations in those remaining hostels. The method of service provision for new arrivals also changed: groups of new arrivals now learned English and received social assistance through more specifically trained individuals often linked to, or situated within, the hostel itself, rather than 'well-intentioned' Anglo-Australian volunteers.[17] Furthermore, new arrivals were now joining an Australian society composed of significant numbers of settled migrants. By 1981, 21 per cent of Australia's population was overseas-born, and 20 per cent of Australian-born children had a least one migrant parent, and these numbers would continue to increase (Collins 1992, 103–104). Many of these established migrants provided links for new arrivals into the wider community, acting in a similar way to the chain migration that had been occurring concurrently outside of the hostel system and stimulating the movement of new arrivals into existing and evolving ethnic residential clusters.

For those ethnic groups for whom there was little existing community in South Australia (such as those from South America), accommodation in large hostels like Pennington along with hundreds of other migrants from diverse ethnicities was particularly isolating, as noted by many of our interviewees. Although many of these migrants arrived in family groups, they often sought solace and support from each other in the hostel, which in this case provided a place for nascent ethnic group formation that would later flourish in the wider community and draw individuals of the same ethnic origins together.[18] Thus changing policies helped to shape settlement patterns across the long period in which hostels existed in South Australia.

Other influencing factors

Oral histories gathered during the Hostel Stories project also provide evidence of other factors that influenced migrants' residential decision-making processes. As an example, consider a Czechoslovakian family who arrived in South Australia in 1972 and left

Pennington Hostel after approximately 3 months in residence to move to a rented flat in Brompton. While its proximity to the hostel was noted, the interviewee stated that the location of the school that the children had been attending was a key factor in the decision. She explained:

> we got a newspaper and went around and had a look for a flat ... close to the city, not far from the hostel ... and close to Brompton School ... we could walk there. That's what we were looking for, especially the school, shops weren't too close, had to take bus to go shopping but school was walking distance which was good. (LC, 3/7/13)

Of course, on many occasions there was no single reason for migrants' place of settlement after leaving the hostel system, but rather a combination of influencing factors that then evolved with changing circumstances. This process is well illustrated by the testimony of an Italian assisted migrant who (along with her family) was housed at Pennington Hostel. Upon leaving the hostel, they moved to nearby Woodville to stay with an Italian family. A sampling of Italians from Pennington in 1973–74 supports this oral testimony about small-scale clustering around Woodville and Ottoway. When asked whether they chose this location because of the family being Italian or near to the hostel, our interviewee (BR, 1/5/13) replied it was because 'of my husband's work, he was working at Woodville ... it was closer'. However, she also explained that the area was affordable and housing was cheap. As the family became more established, the area gained additional appeal because of the growing Italian community: 'our church was there and so that was the first area we get familiar with'. The family eventually chose to settle at Aberfoyle Park 'close to the sea'.

Although many British migrants clustered in particular communities due to the availability of housing, notably in Elizabeth to the north of the Adelaide metropolitan area, others actively sought to resist joining these communities. For instance, following a stay at the Glenelg Hostel, a British interviewee recounted the thought processes that caused her to avoid these communities:

> A number of years later I'd read about Elizabeth. One or two books, and I was dead determined I wasn't going to live there ... Elizabeth didn't appeal to me at all. And I'd read about long waiting lists for sporting clubs and that kind of thing and I was given some idea already about the distances between things. And I hadn't driven. My husband hadn't driven a car, he had a motorbike in England. No it didn't appeal to me at all ... the lady interviewing us for the housing trust she said, 'Because of your situation we could give you a house straight away in Elizabeth or Salisbury'. She looked at me and she said, 'There's all these sports clubs and there's all the facilities social, education'. 'No', I said, 'no'. We came all this way to be in a different culture. My kids had changed school three times in a year and I didn't come all this way to live with a bunch of Poms. (DB, 30/10/14)

Finally, the influence of the attitudes of the broader community on migrants' settlement decisions also requires consideration. One of the reasons that ethnic residential clusters form is as a result of discrimination: migrants find it difficult to settle elsewhere because of problems in locating housing and work, or even feeling welcome in other areas (Poulsen, Johnston, and Forrest 2004). As noted previously, in the early years of the hostel system, assimilation was the key policy and much effort was put into educating the broader Australian public about the need for these (predominantly) European migrants and their contributions to Australia's future. While Southern Europeans were less well tolerated than Western Europeans, the prejudices against them were nowhere near as great as when the White Australia policy was abandoned and Asian and African migrants arrived in greater numbers. If we accept that ethnic residential clusters form in part as a reaction to discrimination, as well as serving as a 'base for action' in migrants' struggles against society in general (Boal 1978), it can be argued that they also form for community support. As Coughlan (2008, 12) concludes, Vietnamese refugees moved from the hostels to neighbouring suburbs and the networks and sense of community they formed enabled them to deal with the

> traumatic experiences of refugee flight and their perceived loss of their homeland. Over the long-term, these early enclaves [*sic*] provided psychological and social environments which would permit the successful integration of Vietnamese-Australians into mainstream Australian society.

Thus staying close to the hostel may have been particularly important for refugees, especially given community attitudes at that time.

Conclusion

Viewing hostels as the single biggest contributor to development of ethnic residential clusters is far too simplistic an argument, as we have demonstrated. An analysis of many underlying factors is important to develop an understanding of where, how, and why ethnic concentrations occur in communities. We have shown that the mere presence of a migrant hostel in a neighbourhood was often not correlated with the development of an ethnic concentration in that same neighbourhood. Consideration of the availability of low-skilled work opportunities and affordable housing was equally (or more) important in understanding why migrants moved to particular areas after leaving the hostel system. The changing nature of government policy over time (from assimilation towards multiculturalism), and shifts in the method of service delivery (to a decreasing number of more centralised hostels), provided environments more conducive to evolving ethnic residential concentrations, but without ever fostering the development of segregated enclaves or ghettos as seen in other parts of the world. It could even be contended that in many cases the presence of the hostels inhibited formation of ethnic enclaves, since the hostel system provided support and services often not present in countries lacking hostels, which likely caused migrants in those countries to seek support from people of the same ethnicity and thus generated residential clusters and even enclaves.

As shown throughout this paper, the departure records contained within hostel registers have considerable future potential for allowing researchers to trace migrants' settlement patterns in extensive detail. However, the process of transcribing this material is extremely time consuming. If we consider only the three largest centres in Adelaide, namely Finsbury/Pennington (55 volumes of names), Glenelg (24 volumes) and Woodside (2.25 m of individual cards arranged alphabetically by surname), the material is voluminous. Nevertheless, the complete transcription of registers for Gawler Hostel as part of the Hostel Stories project, together with samplings of other centres, has already produced some important information, and, perhaps more importantly, ongoing questions for consideration.[19]

What is particularly striking based on data from the hostel registers is that there are no obvious patterns of settlement. Consider, for instance, the diverse patterns described above during the period in which multiple ethnicities were represented in South Australia's hostels, especially in the early years (late 1940s through to the 1960s), and the varied patterns seen during the mass arrivals of specific ethnicities in the 1970s and 1980s. For example, Pennington received a number of groups of Yugoslavian refugees in 1973. Focusing just on one group of 64 people who arrived on the same day, it is interesting to note that some of these families left the hostel within days, while others were there for months. While some settled locally, others moved across the Adelaide metropolitan area and some into regional areas of South Australia. Ongoing data entry will allow for comparisons across hostels, across time, and within ethnic groups and will continue to provide answers—and, of course, raise further questions—about migrants' settlement patterns.

Methodologically, our work on the Hostel Stories project has highlighted the value of moving beyond aggregate census statistics to analysis of data on individual movements. These data, available in historical records (such as hostel arrival and departure registers) and oral testimony, allow us to better explore the range of human experiences associated with migrant transition into the community and the potential for development of ethnic or other social groupings. Furthermore, as we continue to accumulate richer, qualitative data, we will be able to examine migrant settlement over time and place in response to changing policies and a rapidly changing society. Through this study, we hope to contribute substantively and methodologically to ongoing discussions about the role of government policy and migrant accommodation provision in the development of a functional multicultural Australian society.

Notes

1. In addition to clusters discussed in the body of the paper, classic examples include Italian ethnic clusters in Leichhardt (NSW) and Carlton (VIC); see Reynolds (2000) and Jones (1964) respectively.
2. For example, the scholars of the Chicago School in the 1920s and 1930s such as Wirth (1928). For more recent studies, see, for instance, the various papers on ethnic settlements in cities such as London, Cologne, Berlin, Vienna, and Brussels contained in the special issue of *Urban Studies* 35 (10) (1998).
3. See, for example, Harney's work on Italians in Canada (Harney 1985) and Burnley's work (Burnley 1989) on the Vietnamese in Sydney.
4. For a good outline of the relevant international history, see Jupp, McRobbie, and York (1990).
5. However, some scholars have described these locales as 'ghettos' (e.g. Peel 2012, 96).
6. The term 'New Australian' was coined by Minister for Immigration Arthur Calwell in the late 1940s in an attempt to deter the use of the pejorative titles 'Balts' and 'Reffos' (short for refugees) which were being applied to new arrivals. It also fit with the wider rhetoric of the assimilationist policy of the day. The term 'New Australian' soon took on its own derogatory connotations.
7. Pauline Hanson was the co-founder (1997) and leader of One Nation, a populist political party with a highly conservative and anti-multiculturalism platform. She was expelled from One Nation in 2002, and rejoined the party in 2013, becoming its leader again in 2014. In 2016, she was elected to the Federal Senate representing Queensland.
8. The Cronulla riots (mid-December 2005) involved a series of race riots and outbreaks of mob violence, assault, and property damage in Cronulla, a beachside suburb of Sydney, and spread to surrounding communities. The riots stemmed from tensions between young people of Lebanese and Anglo-Celtic backgrounds.
9. Although the Australian Bureau of Statistics census website in numerous places stresses the need for continuity in the questions asked so that social changes can be measured over time,

they make 'new inclusions to meet emerging information needs'. So, for instance, ancestry data were included in the 1986 census but the results were inconsistent, especially for third- or fourth-generation immigrants who could not report their backgrounds accurately. Ancestry questions were removed in 1991 and 1996, and reinstated in 2001 with more specific instructions, but were coded using a standard classification system; see the ABS Fact Sheet at: http://www.abs.gov.au/AUSSTATS/abs@.nsf/ProductsbyReleaseDate/A7A0E94399353F1 DCA257148008018DC?OpenDocument. Because of these factors, the overall numbers and the picture provided vary significantly between census years.

10. At hostels such as Finsbury/Pennington, Glenelg, and Gawler (all SA), these records take the form of large, oversized register books that record all arrivals in black/blue and all departures, including first address, in red. Woodside Hostel (SA) used a system of individual registration cards noting names and dates of arrival as well as the departure dates and destination addresses. These records are available at the National Archives of Australia (NAA) office, Adelaide.

11. During and following both world wars, the Australian government required all 'aliens', that is foreign nationals or non-British subjects, living in Australia to register with local authorities. These forms include information on arrival, birthplace, occupation, marital status, and some-times physical descriptions or photographs. As all 'aliens' were required to notify change of address, these are useful records of places of abode. Registrations were required between 1916 and 1926, and again between 1939 and 1971, and records are held in the NAA. Information gathered varied over time but typically included name of ship, date of arrival, date and place of birth, occupation, marital status and current address. Later registrations also included physical description and even photographs. What is critical for our purposes is that every change of abode had to be registered, hence providing an excellent data source.

12. See, for example, NAA: J25, 1966/2798, Social Welfare—Migrant Accommodation—Problem Cases in Commonwealth Hostels, Queensland [Wacol and Colmslie Hostels], Brisbane. Length of stay was further limited in 1972 to 12 months.

13. Policy for the establishment of hostels stated that: 'Sites should be reasonably accessible from places of work; a total of approximately two hours travelling time per day is maximum. Use of established transport facilities is desirable.' See NAA, D618, IM4 PART 1, [Department of Immigration]—Gawler (SA) NA [New Australians] hostel accommodation, Sydney.

14. Citations to interviews provide an internal reference and the date of the interview from the oral histories which are part of the Hostel Stories project. Interviews will be lodged at the State Library of South Australia on completion of the project where permission from partici-pants was provided.

15. This sample is taken from vol. 4 in NAA, Series D2419, Adelaide.

16. The term 'Australian way of life', although never defined, was coined in the 1940s and came into regular usage in the 1950s in official, public and even advertising vernacular. It rep-resented a quintessential and idealised Australian lifestyle (Murphy 2000).

17. For example, the Indo-Chinese Refugee Association worked from an office situated in the Pennington Hostel itself, and social welfare workers no longer travelled between multiple hostels and work camps but also were located in the hostel itself.

18. For example, oral testimony has informed us about the formation of soccer and other sport-ing clubs in the hostels that later transferred to associated ethnic communities. Similarly, a Chilean migrant couple at Pennington Hostel explained how, seeking ethnic and cultural support, Chilean people met regularly under a particular eucalyptus tree and later went on to form the Chilean Club.

19. These transcriptions will be made publicly available at the conclusion of the Hostel Stories project.

Acknowledgements

We are grateful to our collaborators on the project for their input and contributions to this paper, and to the anonymous reviewers on the original version of this paper. We wish to dedicate this paper to the memory of our colleague Graeme Hugo, who served on the Advisory Board for the Hostel Stories project and was an enthusiastic supporter of it.

Disclosure statement

No potential conflict of interest was reported by the authors.

Funding

Research for this paper was supported by a Linkage Grant from the Australian Research Council [grant number LP120100553], 'Hostel Stories: Toward a Richer Narrative of the Lived Experiences of Migrants' to the University of Adelaide in collaboration with the Migration Museum and in partnership with the cities of Charles Sturt and Port Adelaide Enfield, State Records, and the Vietnamese Community in Australia (South Australian chapter).

References

Abrahamson, M. 1996. *Urban Enclaves: Identity and Place in America*. New York: St. Martin's Press.

Advocate. 1950. "Making Good Australians." Tasmania.

Ang, I., et al. 2002. *Living Diversity: Australia's Multicultural Future*. Artarmon, NSW: Special Broadcasting Service Corporation.

Ang, I., et al. 2006. *Connecting Diversity: Paradoxes of Multicultural Australia*. Artarmon, NSW: Special Broadcasting Service Corporation.

Birrell, B. 1993. "Ethnic Concentrations: The Vietnamese Experience." *People and Place* 1 (3): 26–32.

Blainey, G. 1988. "Australia Must Break Down the Walls of the Ghettos." *The Weekend Australian*, Sydney.

Boal, F.W. 1978. "Ethnic Residential Segregation." In *Social Areas in Cities: Processes, Patterns, and Problems. Vol. 1: Spatial Processes and Form*, edited by D. T. Herbert and R. J. Johnson, 57–95. Chichester: Wiley.

Burnley, I. H. 1989. "Settlement Dimensions of the Vietnam-Born Population in Metropolitan Sydney." *Australian Geographical Studies* 27 (2): 129–154.

Burnley, I. H. 1994. "Immigration, Ancestry and Residence in Sydney." *Australian Geographical Studies* 32 (1): 69–89.

Burnley, I. H. 1999. "Levels of Immigrant Residential Concentration in Sydney and Their Relationship with Disadvantage." *Urban Studies* 36 (8): 1295–1315.

City of Charles Sturt. 2011. "Pennington Community Profile." Accessed February 1, 2016. http://profile.id.com.au/charles-sturt/ancestry?WebID=150.

Collins, J. 1992. "Migrant Hands in a Distant Land." In *Images of Australia*, edited by G. Whitlock and D. Carter, 102–124. St Lucia, QLD: University of Queensland Press.

Coughlan, J. E. 2008. "The Changing Spatial Distribution of Australia's Vietnamese Communities." Paper presented at *The Re-imaging Sociology Conference of The Australian Sociological*

Association, Melbourne. Accessed January 3, 2016 https://www.tasa.org.au/wp-content/uploads/2011/01/Coughlan-James-Session-61-PDF.pdf.

Dunn, K. M. 1998. "Rethinking Ethnic Concentration: Cabramatta, The Case of Sydney." *Urban Studies* 35 (3): 503–527.

Grimes, S. 1993. "Residential Segregation in Australian Cities: A Literature Review." *International Migration Review* 27 (1): 103–120.

Hanson, P. 1996. "Maiden Speech." Accessed January 6, 2016. http://australianpolitics.com/1996/09/10/pauline-hanson-maiden-speech.html.

Harney, R. F. 1985. "Ethnicity and Neighbourhoods." In *Gathering Place: Peoples and Neighbourhoods of Toronto, 1834–1945*, edited by R. F. Harney, 1–24. Toronto: Multicultural History Society of Ontario.

Jones, F. L. 1964. "Italians in the Carlton Area: The Growth of an Ethnic Concentration." *The Australian Journal of Politics and History* X: 83–95.

Jupp, J., A. McRobbie, and B. York. 1990. *Metropolitan Ghettoes and Ethnic Concentrations: Volume 1*, Working Paper 1. Multicultural Studies, University of Wollongong.

Kempen, R., and Ş. A. Özüekren. 1998. "Ethnic Segregation in Cities: New Forms and Explanations in a Dynamic World." *Urban Studies* 35 (10): 1631–1656.

Marsden, S. 1977. *A History of Woodville*. Woodville: Corporation of the City of Woodville.

McKenzie, P. 1999. "Swimming in and out of focus: Second Contact, Vietnamese Migrant Others and Australian Selves." *The Australian Journal of Anthropology* 10 (3): 271–287.

Murphy, J. 2000. *Imagining the Fifties: Private Sentiment and Political Culture in Menzies' Australia*. Sydney: Pluto Press.

National Archives Australia—Glenelg North Hostel arrival and departure registers, D2419, Vol 4, 26 June 1959 to 1 February 1960, Adelaide.

National Archives Australia—Social Welfare—Migrant Accommodation—Problem Cases in Commonwealth Hostels, Queensland [Wacol and Colmslie Hostels], J25, 1966/2798, Brisbane.

National Archives Australia, D4878, Mrozek Henyrk, Adelaide.

National Archives Australia, D4881, Mrozek Henyrk, Adelaide.

National Archives Australia [Department of Immigration]—Gawler SA NA [New Australians] hostel accommodation, D618, IM4 PART 1, Sydney.

Peach, C. 1996. "Does Britain Have Ghettos?" *Transactions of the Institute of British Geographers* 21 (1): 216–235.

Peel, M. 2012. "A Place to Grow: Making a Future in Postwar South Australia." In *Turning Points: Chapters in South Australian History*, edited by R. Foster and P. Sendziuk, 88–102. Adelaide: Wakefield Press.

Poulsen, M., R. Johnston, and J. Forrest. 2002. "Plural Cities and Ethnic Enclaves: Introducing a Measurement Procedure for Comparative Study." *International Journal of Urban and Regional Research* 26 (2): 229–243.

Poulsen, M., R. Johnston, and J. Forrest. 2004. "Is Sydney a Divided City Ethnically?" *Australian Geographical Studies* 42 (3): 356–377.

Reynolds, A. 2000. "The Italian Heritage in Leichhardt: Sydney's 'Little Italy,'." In *In Search of the Italian Australian into the New Millennium*, edited by P. Genovesi and W. Musolino, 377–390. Melbourne: Gro-set.

Salt, B. 2010. "Our Cities Harbour a Lively Patchwork of Ghettos, Enclaves." Accessed February 4, 2016. http://www.theaustralian.com.au/business/opinion/our-cities-harbour-a-lively-patchwork-of-ghettos-enclaves/story-e6frg9jx-1225931946101.

Wirth, L. 1928. *The Ghetto*. Chicago: Phoenix.

Zang, X., and R. Hassan. 1996. "Residential Choices of Immigrants in Australia." *International Migration* 34 (4): 567–582.

Successful British Migration to Australia—what lies beneath the macro-level?

Janette Young, Lisel O'Dwyer and Richard McGrath

ABSTRACT

What is successful migration? At a macro-socio-political level migration by individuals may appear to be successful when it has met the objectives of governments, industries and domestic profit makers. However, delving beneath the surface can reveal contradictions and other measures of success at the individual, or micro-level. Within a broader critical historical ethnography, we interviewed 26 post-World War 2 (WW2) British migrants living in South Australia. All interviewees could be viewed as successful at the macro-level, having remained in Australia for many years and having established multi-generational Australian families. Their migration was a 'success' when measured against the priorities that were actively promoted by Australian governments in the post-WW2 period. At a micro-level, the migrants involved in this study reported mixed outcomes. While migration did result in self-identified aims of migration including employment, opportunities and adventure, some migrants reported high levels of distress and longing, linked to loss and dislocation from people and places in geographically distant locales. For some, these feelings extended into the present, raising questions over the 'success' of their migration experiences at a personal level. We argue that pro-active migration recruitment—such as that undertaken by Australian governments in the post-WW2 period—has the potential to pressure some persons into migration, creating ongoing and unresolvable tensions. Experiences of such disruptions merit further exploration to develop deeper critical understandings of migration success.

Introduction

Australia is an 'immigrant society'. In 2011, 25 per cent of the population were overseas-born (migrants) and close to a further 25 per cent of Australians were children of migrants (Australian Bureau of Statistics (ABS) 2012). Australians may be unaware that this national pattern is unusual. In comparison, in 2013 just over 12 per cent of the UK population and 13 per cent of the US population were overseas-born (OECD 2016). Given these patterns within the Australian population, it is important to understand and explore multiple facets and interfaces within the complex migration space (Hugo 2006; Massey et al.

1999). Migration includes not only migrants themselves but also the populations and people left behind in countries of origin, the interests of recruiting entities and the communities that migrants enter and in which they become resident (Morawska 2004). Migration literature tends to focus on migrant experiences; however, migration is inherently linked to international and domestic politics in at least two countries; and migrants leave and enter two different communities, impacting on both (Morawska 2004). This paper explores these multiple facets through the lens of post-World War 2 (WW2) British migrants living in South Australia. As space restrictions prevent us from exploring the broad contexts of our case study migrants in detail, we provide only a brief description, but recognise through a post-modern lens that their stories and experiences are embedded in a complex meta-story. We seek to point to these intersections.

Australian migration history has arguably been a largely macro-level political undertaking (Department of Immigration and Border Protection (DIBP) 2015), reliant upon recruiting individuals in response to powerbrokers' interests in long-term population and economic growth.[1] These same macro-level discourses continue at this time in contemporary Australia (Krockenberger 2015) with calls for economic growth underpinned by public policies that ensure migration-fed expansion, predominantly through skilled migration. This continued perspective can be found in various public statements and key policy documents. For example, Geordan Murray, a Housing Industry Association economist, argues that as a result of an increasing number of baby boomers exiting the workforce '[p]olicy makers must ensure that Australia remains an attractive place for skilled migrants to settle' (Clancy 2016, n.p.). In South Australia, the State government has identified the need to ensure migration continues (and is even increased) as a key aspect of its 30-year plan for Greater Adelaide (Department of Planning and Local Government 2010). Thus it is pertinent to consider the impacts on the lives of the previous generations of migrants to Australia of past government and industry-led initiatives that actively promoted migration. Such an approach can provide insights into the potential pitfalls of contemporary approaches to Australia's migration program.

In Australia, from a policy perspective the definition of migration 'success' is seen as long-term population and economic growth (Krockenberger 2015). We present data and analysis indicating that migration 'success' is far more complex and nuanced, with macro- and micro-level understandings at times paralleling or complementing each other. At other times, 'success' and 'non-success'—or indeed pain caused by migration —may co-exist in the lives of individuals. The paper is structured in two core sections. The first part considers macro-level definitions of migration success (amongst politicians, bureaucrats, capital interests in a migrant receiving/recruiting country) that can be perceived from Australian and South Australian migration history in the post-WW2 era, with a specific focus on the British migration program. The second half of the paper focuses on data from qualitative interviews undertaken with older British men and couples who were part of this migration flow, and elicits their micro-level (individual and familial) perspectives and experiences of migration 'success'. The data and discussion presented throughout this paper emerge from a multi-faceted critical historical ethnography that combined historical research into Australian migration patterns, archival research into migration records and interviews with older male British migrants to South Australia—which in some cases also involved their wives (see Young 2009, 2010). South Australia was used as a case study, as the post-WW2 assisted-migration program (described in a later section) was managed at the State level. We begin by defining in further detail how the terms 'macro' and 'micro' have been interpreted in this paper.

Definitions of migration 'success' at the macro- and micro-level

We define the 'macro-level' as encompassing entities and persons within the receiving country who are able to mobilise resources to drive, define and manage migrant recruitment. Various macro-level causal factors have been used to explain migration (Jennissen 2007). The flow of people across nations is often viewed through the macro-level lens of 'push–pull' causes (Lee 1966; Ravenstein 1885), located within economic, political, social and demographic contexts (Hagen-Zanker 2008; Kritz, Lim, and Zlotnik 1992). Migration may be defined as successful by macro-level entities—such as governments (political and bureaucratic) and profit-making industries (such as housing developers)—when it has met their objectives of numerical growth, political kudos and profit as exemplified in our case study.

In post-WW2 Australia, the key macro-level objectives of the migration program were to increase the size of the Australian population, to grow the workforce and to build infrastructure and the economy (Joynson 1995; Jupp 2002). Historically, this macro-level definition of migration success can be traced back to discourses of colonialism, and the need to recruit settlers to populate '*terra nullius*' (Elkins 2005). The Australian population did increase markedly and industrial development thrived through the post-WW2 era, progress seen as proof of the (macro-level) success of Australia's migrant recruitment policies (ABS 1998). That one quarter of Australians are the children of migrants can be seen as the fulfilment of post-WW2 Australian government migration aims to increase the long-term future population base of Australia (Jupp 2002). This pattern has shifted slightly in recent decades, with the proportion of temporary to permanent arrivals reversing (see Collins 2014; Hugo 2006; and Castles in this issue). While there have been some recent changes to the type of migration that is occurring in Australia, public migration discourses continue to position long-term (permanent) relocation of employable migrants able to produce the next generation as the desired option for Australia (Clancy 2016; Department of Planning and Local Government 2010; Krockenberger 2015).

While migration studies at the macro-level can provide insight into the collective numerical ebb and flow of peoples, individual agency in the motivation to migrate is often missing from these macro-level analyses (Hagen-Zanker 2008; Jennissen 2007). The definition of 'micro-level' migration experiences used here thus encompasses individual and family perspectives. The approach taken in this paper is to explore migration success predominantly from the (micro-level) perspective of a group of people who may be considered 'successful' migrants from a macro-level (or powerbrokers') perspective. We seek to intersect the stories and understandings of success held by a coalition of powerbrokers (political, bureaucratic and capitalist, e.g. housing developers) with the migration stories of some of these 'successful' migrants themselves. Some of our interviewees spoke of how powerbrokers impacted on their choices, for instance through targeted migration recruitment materials. Hints of other players and interests in their micro-level stories of migration emerge in our interviews as significant—for instance, families and loved ones left behind; and the impact of personal networks on choices to migrate (Massey et al. 1999). We argue that an awareness of these other players and their potential significance is largely absent from the macro-level powerbrokers' perspectives of migration success.

Micro-level Australian migration analyses have often problematised migration. The literature has become replete with books, government reports and academic papers portraying migration as a difficult and often negative undertaking and experience (see, for example, Jirojwong and Manderson 2001; Thomson 2005). Decades of research have actively worked to identify the needs and issues faced by various migrant groups. This was not unneeded work; however, it has largely been unaccompanied by explorations of unproblematic or even positive aspects of migration. This gives an unbalanced understanding of the migration experience. Our focus on the individual stories of ostensibly successful migrants presents insights into the nuanced and sometimes complicated realities—even of 'success'. We use data collected in interviews with British migrants to Australia to explore the experiences and perspectives that they, as migrants, have of their migration. From the interviews conducted it was clear that all interviewees fitted within the macro-level indicators of migration success. All are long-term residents of Australia, and most have raised multi-generational families in Australia; the Australian macro-level definition of migration success is embodied by these people. Yet their personal experiences of migration success were varied and, in some cases, raise questions over how successful migration ought to be understood. Findings from this study indicate that, at the micro-level, individual and familial experiences of migration can be complex and even contradictory, positing additional yardsticks of successful migration. From a post-modern perspective (Lyotard 1984), the contradictions that seem to emerge in their 'small' stories—as they intersect with larger stories of 'success'—are not a surprise. Rather, they illustrate the post-modern rejection of simplifying grand narratives (Agger 1991) in favour of heterogeneity and complexity.

Post-WW2 British migration to Australia: drivers and macro-level indicators of success

Migration from the UK has been the dominant source of non-fertility-based population growth in Australia since colonisation of Indigenous lands in 1788 (Jupp 2002). Between 1947 and 1971 British migrants formed 42 per cent of all migrants to Australia, contributing a net population gain of just over 1 million people (Burnley 1978). A range of publicly funded programs drove this inflow. Perhaps the best known was the 'ten pound Pom' scheme (Hammerton and Thomson 2005), whereby British citizens were able to emigrate to Australia for only £10. The major driver for this influx of persons from the Australian government's point of view was an interest in population growth, initially as a counter to fears of invasion post-WW2, then shifting to a more overt focus on industrial and economic development and growth (Jupp 2002).

Australian governments of the time (dominated by one federal government—the Menzies era from 1949 to 1966) developed migration agreements and assisted-passage schemes with many countries in the post-WW2 era, including Spain, Italy, Greece, Malta, Germany, Austria, Belgium and Holland (Jordens 2001; Joynson 1995). The agreements entitled Australia to select migrants according to its own criteria, and to share the costs of passage with the sending country's government. There was also a ready and accessible supply of non-British migrants to Australia immediately post-WW2 due to the vast numbers of European Displaced Persons (Jordens 2001). In the first few years immediately following WW2, these non-British migrants far outnumbered assisted British migrants to Australia. The numbers reduced as refugees were relocated internationally and the displaced persons scheme ceased in 1953–54 (Jordens 2001). The discussion below presents evidence based on archival research outlining the role of powerbrokers in promoting British migration to Australia, as well as the macro-level indicators of success around which the post-WW2 migration program was premised. Details of the archival research methods used can be found in Young (2009).

In the post-WW2 era, a macro-level understanding of successful migration (as that which contributed to long-term population and workforce growth, including via family formation in Australia) was discernable from media and recruitment posters. It was also present in archival files and related oral histories of the South Australian Immigration, Publicity and Tourist Bureau Department (IPTBD)—the State agency charged with managing the British immigration scheme. Newsreel footage of the Australian Minister for Immigration (Arthur Calwell) shows him welcoming the first group of post-WW2 British migrants and stating that the aim of the program is for these migrants to 'settle down in Australia, and … join our destiny … ' (National Film and Sound Archive n.d.). In South Australia, Thomas Keig (Chief Migration Officer 1959–77) noted that the State made a point of recruiting families as this would provide the next generation of workers (Keig 2000). Recruitment posters of the 1950s proclaim 'Australia—Land of Tomorrow' (Commonwealth Department of Information 1950) and 'Build your Children's future' (Powerhouse Museum 1955–60).

Historically, Australian migrant recruitment was underpinned by a strongly gendered narrative positioning men as workers and women as homemakers. This narrative is neatly illustrated by two posters proclaiming 'Men for the Land; Women for the Home' (Department of the Interior 1928) and 'There's a man's job for you in Australia' (Commonwealth Department of Information 1947). Australia's post-WW2 British migration program was framed around the male breadwinner role, the gendered masculine expectation that a man should undertake paid employment to provide economically and hence materially for his wife and children (Murphy 2000). The South Australian archival lists of British migrant arrivals generally only identified the employment field of male migrants, reflecting this gendered understanding of the labour market (Young 2009, 2010). Indeed, until the mid-1960s Australian Commonwealth and State governments required women who married to relinquish their employment (Strachan 2010).

South Australia appeared to have little interest in British migrants in the immediate post-WW2 era. Hugo (1996) mapped population growth combining migration and natural increase (fertility minus mortality) in the State in the post-WW2 era. Hugo's data indicated that migration into South Australia in the 15-year period immediately after WW2 was composed predominantly of non-British sources, mainly Displaced Persons from Europe (Jordens 2001). As this flow of Continental European migration subsided over time, British migration became pre-eminent. Figure 1 shows the net numbers of British-born migrants in Australia and South Australia and the relative percentages of British to non-British migrants between 1933 and 1991.

Figure 1 also shows the much greater emphasis on British migration to South Australia compared to Australia as a whole and the different timing of the period of peak flow. British migration to South Australia peaked in the mid-1970s (long dashed line

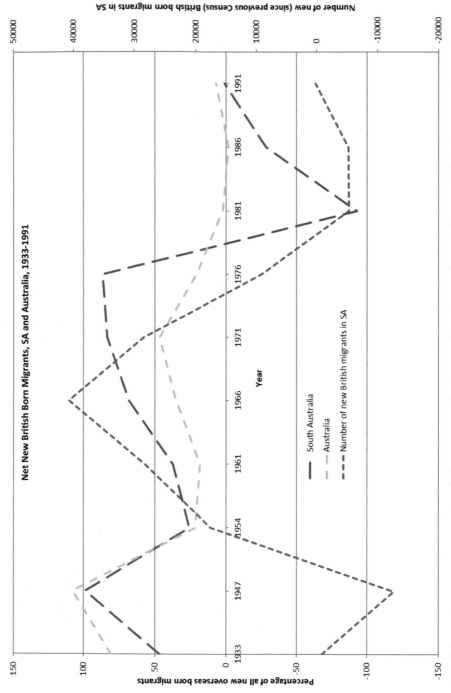

Figure 1. British-born migrants as a proportion of all migrants to South Australia and Australia, and the net number of new British migrants in South Australia (1933–91). Source: ABS (2014), Cat. No. 3105.0.65.001 and census (1933–91).

in Figure 1), whereas the peak for Australia as a whole was 5 years earlier (medium dashed line in Figure 1). Note that there was outmigration of non-British-born migrants from Australia in 1947 which accounts for the percentage of British-born migrants exceeding 100 per cent at that time (medium dashed line). The picture presented in Figure 1 can be seen to reflect the very active recruitment of British migrants by powerbrokers in South Australia through the 1960s, with its impact continuing into the early 1970s. The number of British migrants (short dashed line in Figure 1) shows the absolute numbers of new British-born migrants in South Australia, with the largest annual intake in 1966 and an exodus through the late 1970s and early 1980s.

The assisted British migration program was administered by States, but funding came from the federal government. Archival research by the authors showed that in South Australia (as in the rest of the country) the program became actively recognised as a tool for State-based economic growth and development from the early 1960s (Young 2009, 2010). In the archives of the IPTBD a coalition of public servants, political leaders and private housing developers can be seen to have enthusiastically sought to maximise the numbers of British migrants coming to South Australia (Young 2010). A more detailed description and exploration of this archival research can be found in other publications (Young 2009, 2010).

The agendas for these three powerful players were intertwined. A palpable bureaucratic pride in establishing a well-oiled recruitment system that connected government and private-sector players can be discerned in the archives of the IPTBD. This system can be defined as promoting 'successful' migration (as understood at the macro-level) as it led to disproportionately high numbers of British migrants coming to South Australia compared to other States. This bureaucratic system managed bids from private housing developers to become registered housing providers for migrants. Registration gave these developers a virtually guaranteed market of new home buyers. The State Bank was also enlisted in the system, offering home loans on special conditions to British migrants who were screened (by the IPTBD) into this system prior to their arrival in the State. The political player of the time was Premier Thomas Playford, who courtesy of a gerry-mander (Tilby-Stock 1996) was Premier of the State from 1938 to 1965.[2] The British migration recruitment and support scheme established by the IPTBD gave political kudos to Playford as Premier, enabling him to claim success on such things as the establishment of the city of Elizabeth—an industrial satellite city 25 km outside of the capital of Adelaide (Peel 1995). When existing local residents resisted relocation to this under-developed location, it was possible to direct newly arrived British migrants with no other accommodation options there instead (Peel 1995; Young 2009). The publicity material seen by prospective migrants in Britain did not match the reality of anticipated (but, at the time, undeveloped) infrastructure in Elizabeth, while existing residents knew the reality and could avoid it. In comparison, unknowing new arrivals were effectively trapped into moving to this satellite city as the alternative was long stays in substandard migrant hostels (Peel 1995; Young 2009).

Archival research undertaken by the authors showed that 'family men' were specific targets of South Australia's assisted-migration program in the post-WW2 era, as a means of progressing government agendas of economic and industrial growth and development (Young 2009, 2010). The gendered social positioning of men as workers and 'family breadwinners' in this era intersected with South Australian powerbrokers' labour supply aims (Young 2009, 2010). This economic and cultural context formed the rationale for specifically recruiting men as interviewees for this study, because they were the direct targets of Australian governments' migrant recruitment drives in the post-WW2 era. We now turn to the second substantive component of this paper in which the experiences of these British migrants are discussed.

Qualitative interviews with British migrants: methods

In 2007–08, 16 older British men were recruited via snowball sampling (Patton 2002) from informal networks for semi-structured interviews exploring the impact of migration on their lives. All chose to be interviewed in their homes. Consequently, roughly one-third of the interviews became couple interviews with another third having at least some input from wives, hence there were 26 research participants in total. All of the men came to Australia in the post-WW2 era (1945–70) and were retired at the time of interview. They represented the diversity of British ethnicities including English, Irish, Welsh, Geordie, and one Maltese British citizen. While some of the men had migrated while single, and some with different wives, the women who joined the conversations were very much co-participants in shared stories of migration. Interviewing husbands and wives together demonstrated that understandings of migration at a micro-level need to take account of the family/couple unit, not just individual migrants.

As part of an historical-ethnography approach the interviews were dominated by migration stories or oral histories. We were seeking understandings of how interviewees came to migrate (drivers, motivations) and how the migration process (from choosing to emigrate to their current settled scenario) happened for them. Hence the interviews complemented the archival research as interviewees spoke of the housing companies listed in the IPTBD files, the heavy public marketing presence in Britain (newspapers, Australia House, community information evenings) and the manner in which housing in the city of Elizabeth was positioned as an alternative to long-term stays in hostels. We were also interested in their reflections as older persons reflecting on the choices made by their younger selves. Interviewees were all asked 'would they do it [migrate] again?' This question encapsulates the focus of this paper on (micro-level) understandings of 'successful migration'.

Interviews were audio-recorded and then transcribed verbatim. Transcripts were analysed thematically using Dey's (1993) systematic framework with the aim being to identify themes that emerged from participants' stories of themselves. The research adopted an exploratory and emergent approach. It did not aim to prove or disprove theory; rather, as themes emerged extant theoretical perspectives were identified and meshed with the emerging data (Charmaz 2014). Our focus in the remainder of this paper is on the participants' perceptions of the success (or otherwise) of their own migration to Australia.

Macro-level migration success: employment and home ownership

One of the key macro-level interests that drove Australia's aggressive post-WW2 migration program was a perceived need to build the Australian labour force. Employment, or potential employment, was widely used as a 'pull' factor in promotional materials. Almost all of the men interviewed identified securing employment in South Australia as a rationale for leaving the UK, and all were able to secure employment. This outcome could easily be viewed as evidence of their 'successful' migration according to macro-level measures of migration success.

More nuanced analysis of the men's stories reveals how Australian recruitment tapped into populations of relatively vulnerable men who felt that their employment was under threat in Britain. This situation is consistent with Arendt's (1968) notion of 'superfluity'. Based on the Marxist notion of the need for capital to always have a surplus army of labour, Arendt (1968) argues that superfluous individuals (in this case, men) can be mobilised to become migrants. Almost all of the men who were interviewed reported some sense of market superfluity before emigrating, even though most were not unemployed. Poor employment prospects in the UK, and positive job prospects in Australia, respectively pushed and pulled British men to become migrants. Bill[3] noted: 'there wasn't a lot of work in England at the time ... you know, going down ... ', and Brian: 'things didn't seem very rosy in the UK. Like if you hadn't got a job it was hard to get one and if you had a job you had to hold onto it.'

It was common for both the men and women interviewed to identify the husband's employment vulnerability as a motivation for considering emigration to Australia. Here, Brian and Doreen provided a shared description of the labour market situation they identified as being part of their impetus to emigrate:

Doreen: ... Chrysler were ...

Brian: ... they were gearing down ... yeah ... I went to the tile works when there was a strike ... when the strike was over ... I went back to Chryslers. They were running well when I left.

Interviewer: Ok so employment ... the employment scene broadly was unstable?

Brian: Yes it was.

Doreen: It was going down ...

Brian: ... It was going down ...

Doreen: ... They were cutting down ...

Brian: ... they've shut down at Chrysler now. There's no longer a Chrysler in the UK.

Brian and Doreen described a labour market bedevilled by strikes and unstable demand, leading to cyclical worker retrenchment and recruitment. Advertisements promoting stable employment in Australia were a powerful force on these men and their families. Recruitment advertisements promising progress and prosperity tied emigration and industrial development to the needs of the 'Family Man' (Peel 1995). In sum, Australian migration recruiters tapped into British men's labour market superfluity and the gendered breadwinner role (Murphy 2000).

Once in South Australia, most of the men interviewed experienced relatively stable and long-term employment. At the time of interview, most were enjoying their retirement in self-owned homes. They had achieved long-term stability and so, in terms of employment, migration was successful at both the macro- and micro-level. Their home ownership dovetailed with the agendas of the South Australian migration recruitment interests of the post-WW2 era. These macro-level interests sought to see economic development, largely underpinned by migration-powered growth in the State's housing industry (Young 2009). However, a focus on micro-level migration narratives allowed evidence of personal agendas for migration to emerge, in addition to employment. Some of the men's narratives indicated a pre-existing emotive thirst for 'adventure' and most saw migration as offering opportunities beyond the employment prospects that were promoted so heavily in Australian governments' advertisements and recruitment materials. Yet the narratives of these men and women also provided insights into the tensions of migration—even for migrants who outwardly appear to have had a highly successful migration experience.

Micro-level migration success: adventure and opportunity

The notion of migration as 'adventure' is relatively novel and unexplored in recent Australian migration literature, yet 'adventure' was at the heart of many of the men's conversations. For example, Jim noted: 'I suppose I always had itchy feet and wanted to see a bit of the world … ', and Alf was ' … quite reasonably happy in England. But coming over here was an adventure'. Stuart, who came with his family when aged 18, shared his father's enthusiasm for whom it was 'an expedition almost, you know, going to see what's over there … I was all excited about coming over.'

Bill, who had little access to finances, was keen to emigrate somewhere, perhaps anywhere. He considered emigrating to other countries that were offering assisted passage including New Zealand and Canada. Bill resisted pressure to migrate to Australia via air. When asked why he choose to come to Australia by sea he replied: 'for a six week holiday'. When asked if he would 'do it all again', the following exchange unfolded:

> Bill: Oh most definitely—I wouldn't have any fears about it the second time though, I wouldn't have the adrenaline rush I had the first time.
>
> Interviewer: You'd take it a little bit more slowly you reckon, pace yourself?
>
> Bill: Oh yeah, I'd wait for a boat going the other way around.
>
> Interviewer: Get a longer trip?
>
> Bill: Yeah get a longer trip.

For some of the men, there was a sense that their whole history of being a migrant has been an adventure. Denis and Joseph shared an enduring sense of such:

> Denis: [T]he whole adventure has been quite good, and we've been here, what is it? Forty-five years this year?
>
> Joseph: Goodness gracious me. A country that has all the birds! All the animals! All the wide open spaces! Untamed wildernesses! My goodness, where do you start! Where do you finish! I haven't even seen it! Not even a quarter of it and I've been here now for forty-three years! … It's still an adventure!

'Adventure' was not just a masculine prerogative, as two of the interviews demonstrated. George's wife, Mary, who was not actively participating in the interview, joined in with heartfelt enthusiasm when George asked 'were you looking forward to migrating darling?' Mary responded, 'Ohhh yes! Something different!' Doreen, Brian's wife, stated: 'well it was … it was an adventure. It was a big adventure.' Both of these couples were still 'adventuring'. Doreen and Brian spoke of their travel to China, and George spoke of their planned 'grey nomad'[4] trip.

For many of the interviewees 'opportunity', as distinct from employment prospects, was the predominant motivation for migration. As indicated by Jean, the South Australian State's interests in housing development mirrored some of their individual interests as well. Maurice and Jean spoke of 'bettering themselves', something that could include housing.

> Jean: The major part was we were going to be able to come out here, we were going to be able to better ourselves ...
>
> Maurice: ... I think that was the main thing.
>
> Jean: ... a better life as far as a house. You know, somewhere to live you know, just that type of thing.

Alf saw migration as overarching opportunity:

> So I came for a month and they [daughter and son-in-law] took me around, you know, and met the grandchildren ... introduced me to their friends and I really liked it. I saw it as an opportunity for a new life.

This sense of 'opportunity' was often linked to parental concern for children's futures. Both Maurice and Alf above connected their migration decision to these relationships, but the more detailed nature of the connection between migration and opportunity for children varied. John, for example, saw migration as part of setting things up so that his children had the chance to achieve things he had not:

> John: ... Well ... it was always ... I always intended that my children would ... I mean ... do better than I did.
>
> Interviewer: Ok.
>
> John: And, you know, the expectation was there and of course they did.
>
> Interviewer: Yeah and so migration was part of that ...
>
> John: Yes.
>
> Interviewer: ... setting things up so that they could do better.
>
> John: Yes, yes.

While Howard had achieved upward social mobility in Britain, he felt that his position would not enable his sons to continue that trajectory. He spoke of 'giving his kids a chance'—chances that he perceived would be reduced for his children in the future in Britain. Migration was seen as being part of progress, if not for oneself then for one's off-spring. Howard noted, ' ... basically, we had our own house, of course, but that didn't seem to be any self-advancement for the children. And anyway, so we really came out [to South Australia] to give the kids a chance.'

Other research regarding fatherhood also notes the connection between personal choices and opportunity for children. Peel (1997) has written of how successful fatherhood in the post-WW2 generation of men (particularly working-class men) was not necessarily measured by their own success in employment or a career but rather by the achievements of their children. Most of the men interviewed for this study pointed to their children's achievements and 'opportunities' in Australia, as evidence of the 'success' of their migration decision.

Complicating micro-level migration success: longing and loss

Based on job prospects, fulfilment of opportunity and a desire for adventure, most of the men interviewed can be viewed as 'successful' migrants at both a macro- and micro-level. All gained employment, many experienced 'adventure' and they grasped 'opportunities' for themselves and their families. For those men who sought 'adventure', there was a sense at times that they had been 'won over' to emigrate to Australia (they had an interest in migrating *somewhere* and may have considered other countries). For other men and couples, Australia's active migration recruitment program planted an idea in their minds, for the first time.

Notwithstanding the widespread experiences of 'adventure' and 'opportunity' amongst the interviewees in the present study, the migration experience was nevertheless tainted for some. For many, the migration experience was fraught. It was both 'successful' and 'unsuccessful'. According to Hammerton and Thomson (2005), British migrants have often been positioned in Australian discourses as so similar to non-migrants that they could be ignored, even to the point of being eliminated from the very definition of the term 'migrant'. Past discussion on migration has often hinged upon language and major cultural differences (Colic-Peisker 2002 Jirojwong and Manderson 2001;), meaning that other significant issues—such as the leaving of loved ones, familiar spaces and places, and subtle cutural and linguistic differences—could be under-recognised. In more recent research, homesickness is increasingly being recognised as a part of relocation experiences—irrespective of the cultural or linguistic similarities between migrating and host populations (Stroebe, Schut, and Nauta 2015). Loss, particularly in relation to ties with people and places, was revealed during interviews by a number of men and couples. Both the present study's interviews and Thomson's work with returned British migrants (Thomson 2005) reveal that migration can become an ongoing source of distress often tied to the experiences of leaving and being physically distant from loved ones and familiarity.

For example, Maurice and Jean immigrated to Australia in 1968 to join Jean's sister and her husband. They may not have considered migrating had there not been a highly active process of migrant recruitment in Britain by Australia at the time. Jean pointed out that they were persuaded by the opportunities being advertised at the time:

> [y]ou can earn this, you can do this, you can do that, and you could set up your own business, you could build your house, you can ... everything was painted so rosy ... yeah, it really did lure you in.

The process of migration to Australia meant that Maurice and Jean's family ties were divided, resulting in close relatives located in both Britain and Australia. Maurice and Jean noted they arrived after some family (Jean's sister and her sister's in-laws) had already migrated to Australia. But they left a network of other relatives behind.

Interviewer: So when you came out here, who were you leaving behind?

Jean: Parents ...

Interviewer: ... Other siblings?

Jean: ... Yeah ...

Maurice: ... About four of five sisters.

Interviewer: ... Did they ever come out here?

Maurice: No.

Jean: ... too far ...

Maurice: ... It is a long way.

It was not just people who were missed. It was the lives and lifestyle that the interviewees had enjoyed with these people.

Jean: They were social things, you know, because we were very ... we did a lot of social things with my brother and people like that. I just missed it all ... the fact that you could hop on a bus and you would be somewhere, you know ...

Maurice: ... Social life in the UK at that particular time was far, far superior to here. There were more places you could go to.

The pain of dislocation and tension Maurice and Jean feel, caught between Australia and Britain, is now fading as the number of their loved ones in Britain declines. Maurice said:

... it's like cutting the thing from between here and the UK, the more that died—I can't say I'm not sorry—it's a strong word to say you're better off but eventually when they're all gone there will be no links.

Stuart experienced a similar sense of loss, but one related to wider social life rather than family. While initially migration was a huge 'adventure' for Stuart, he missed the 1960s London he had left:

... those English kids that had come out ... [from the] London area, where it was all discontent in the swinging sixties, they came to Adelaide where the pubs shut at six ... it was so backwards to us ... I mean there was no discotheque. It was 60/40 dance halls!

Stuart was part of a family group who had migrated. By the time he was able to act on his discomfort and return to Britain, he had his own family who went with him. When Stuart's Australian-born children decided to return to Australia, he followed them because he missed them terribly. Stuart continues to live in South Australia and has managed his sense of dislocation through the creation of a business acquiring and selling English antiques, an enterprise that requires him to spend several months of every year back in Britain.

Most of the interviewees presented themselves as originally loosely linked into extended family networks, where such links were not significant enough to the potential 'opportunities' that migration offered. At times, the families they were leaving did not share their perspective on the comparative value of hoped-for opportunities in Australia *vs* the value of familial ties in Britain. This difference could be extreme. Brian spoke of his father's response:

> When I said to him about it [migrating] he said, 'Oh well, I might as well say I've buried you', he says, 'because I shan't see you again' … And he wouldn't even answer the phone when I rang my mother when we got here like. I'd speak to her … and I said, 'Any chance of having a word with me Dad?' and she says, 'He doesn't want to speak to you.'

Other respondents reported a growing awareness of the significance of their migration decision in terms of relationships over time:

> Denis: If somebody said to me now, 'I'd like you to migrate', well, I couldn't do it. No. I talked to someone the other day. We left behind our mums and dads, brothers and sisters. Since I've been here I've lost all … most of me family except for one sister. It's sad. And Bet's lucky that she's only lost her mum and dad … yeah. It's a bit of a sad time in those circumstances.

> Betty: As you say you don't realise it until you get grandkids of your own that you think, 'Oh how did we do that to our parents?' you know. You think 'Oh … '

While many of the men were concerned about breaking familial ties via migration, they acknowledged that they placed more importance on their future. Maintaining everyday contacts with parents and grandparents was not seen, at the time, as important enough to limit 'opportunities' for the next generation (their children). What these excerpts hint at, however, is the little-explored area of the impacts of migration on those who are left behind (Morawska 2004). Concern for the potential devastation that may have been felt by individuals and communities in Britain from whom Australia (and South Australia specifically) eagerly recruited migrants was not evident in the South Australian-focused archival research. Apart from notions of building the future population (of Australia and South Australia respectively), there is no evidence in the archives that powerbrokers considered how a population of migrants might age, might regret or reflect painfully on migration, and perhaps feel that they had been manipulated (by aggressive recruitment tactics) into causing pain to others.

Discussion: successful migration or a 'honey trap'?

Previously published archival research, undertaken by the researchers, revealed a macro-level bloc of State bureaucrats, housing developers and politicians (powerbrokers) in South Australia in the post-WW2 era (Young 2010) who saw male British migrants as central to progressing an agenda of economic and industrial growth and development. The records reveal bureaucratic consternation when recruitment levels of British migrants decreased, and constant scheming and strategising to recruit further British migrants to the State. Migration 'success' (according to government objectives) was implicit in relatively high levels of migrants who purchased houses, found employment, remained in Australia over the long term, and formed their own families there.

From a macro-level perspective, all of the migrants interviewed in this study are highly successful. Simplistically, given that they are still in Australia and most are grandparents (and even great-grandparents) of Australians, they fit the macro-level definition of 'successful post-WW2 British migrants'. However, the interviews revealed a complex sense of both empowerment and disempowerment; success but at times painful entrapment. Maurice and his wife feel partially 'set up', for the pain they *still* experience, by the migration pull that occurred in South Australia in the era they migrated. Others had a cynical awareness of their depersonalised value to powerbrokers with a broad agenda of industrial and economic development. As Nich said:

> [w]ell it was … you could say, looked at cynically, you could say, it was a trap, you know … A honey trap, yeah. And … but … I mean, both parties benefited. The state benefited as did the people who were trapped …

The idea that migration could provide 'adventure' and 'opportunities' countered the interviewees' awareness of being pawns in more powerful players' agendas.

Theoretically, the picture that many of these migrants presented meshes with Giddens' (1984) notion of structuration. Individuals are born into and exist within socially constructed and constraining positions in society. Social position impacts on the choices and opportunities that individuals can make, but they are not completely constrained by it and can exercise agency or free choice. Working-class British men of the post-war era did not have the power to radically shift their (vulnerable) labour market position while in the UK. Their social position also meant they could be targeted by powerbrokers who had their own agendas on the other side of the globe. British men could thus be seen as pawns. However, as individuals, most of these men grasped 'opportunities' that enabled them to address their own and their families' interests. Essentially the men who were interviewed understood their position and role in society, understood that migration to Australia included playing into agendas of powerbrokers, but through their own agency (as well as their families') were able to recognise and seize 'opportunities' offered to them.

We believe there are cautions to be found in our research findings. Incentivised recruitment programs may initially encourage a particular population to relocate themselves and their families to another country. Individuals who naturally seek 'opportunities' and 'adventure' may be attracted to emigrate by macro-level recruitment programs tied to employment, housing and other future prospects. Achieving these tangible outcomes may indeed be used as a measure of success by both individuals and recruiters. However, as migration recruitment programs progress, and the initial market of adventurous opportunity seekers is drained, programs begin to draw on the less adventurous. These potential migrants may be motivated (pulled) by the growing population of relatives and friends who preceded them, and perhaps pushed by market superfluity in their place of origin. Some may keenly experience the tension of having close relatives in two countries and miss the lives and lifestyles that accompanied these relationships. The cost of what may be seen as 'successful migration' by governments and capital—long-term growth in markets and economies—may be personal trauma at the micro-level (Morawska 2004). The commodifying nature of economy-building migration recruitment programs has the potential to create lasting traumas and tensions in the lives of some migrants, and, as illustrated by Brian's family story, in the lives of those left behind.

Future directions in practice, theory and research

The experiences of people left behind are rarely considered in migration recruitment. It was not part of our research either. There is some transnational research exploring the micro-level connections between sending and receiving countries (see, for example, Waldinger 2014), and evidence of the nefarious effects of wealthy countries' recruitment strategies that target highly trained health professionals from lower income nations has led to the *WHO Global Code of Practice on the International Recruitment of Health Personnel* (World Health Organisation (WHO) 2010). Our participants' stories illustrate that macro- and micro-level analyses of migration must consider reactions in sender countries. Focusing simply on receiving countries' definitions of successful migration is insufficient and may overlook serious interpersonal costs between migrants and people left behind, and the trauma of feeling torn between two geographically distant places. Research on conceptualising 'successful' migration should consider its personal, extended familial and broader network implications, and needs to begin to research more carefully the implications of migration on those left behind.

Four themes emerged as core to the migration narratives of the individuals involved in this study: employment, adventure, opportunity and loss. The nuanced insights provided by the men and women who were interviewed stand in contrast to macro-level analyses suggesting that successful migration occurs when population growth and employment objectives are reached in the host society. The themes discussed by our interviewees are not necessarily fixed or separate. Migrants may seek 'adventure' and 'opportunity' in their younger years but come to realise 'loss' at a later stage in life.

The sense of adventure driving some individuals bears further exploration for policy implications. For example, although post-WW2 Australian migration policies had the motive of supplying British labour for manufacturing and mass production, the characteristics of migrants as 'adventurers' and 'opportunists' may be more suited to entrepreneurship and small business in the present Australian economic climate. Indeed, the tendency for migrants from other European countries and Asia to set up their own businesses in Australia is well known (Collins 2003; Collins et al. 1995; Wong 2003). Zgheib and Kowatly's (2011, 345) description of Lebanese expatriates also describes the characteristics of some of the British men in this paper: 'they are perseverant, innovative, risk-taking individuals who compete aggressively in the marketplace and are driven by a need for autonomy'. While Australia has had specific business migration policies for several decades, which have investment and experience requirements, we may be overlooking the broader aptitude and potential contributions of migrants who do not have these resources but do have the necessary psychosocial skills.

Our research suggests there is scope to explore and develop Australia's understanding of migration from English-speaking countries. Services such as language, counselling and employment support tend to be targeted at migrants from non-English-speaking backgrounds, but the disruption of social ties is common to all migrants. Awareness of more subtle impacts of migration could be beneficial. For example, in health settings where language and ethnic differences are routinely recognised as significant, migration experiences outside of these parameters are not. Our findings indicate that international migration, even in the absence of language and substantive cultural differences, can cause long-term distress; distress that having been relatively unrecognised to date is unexplored.

Macro-level drivers of migration and individual experiences of migration can be seen as the two ends of a continuum (De Jong and Gardner 2013). At the micro-end of the continuum, a step away from the individual, is the household (couple or family) which may often be the migration decision-making unit (King 2012). The focus of Australian post-WW2 migration recruitment was on British men, but the stories and perspectives of these migrants—as shown in this paper—involve familial circumstances. Migration is systemic, connecting macro-level international politics and policies with layers of communities, individuals and families. Understanding 'successful' migration must account for all of these layers, looking beneath broad macro-level layers and structures to the obscured individuals, households and families whose lives are changed by migration.

Notes

1. The understanding of power used throughout this paper is essentially Weberian (Weber 1978). Power is the ability to get what the individual (or a collective of individuals, i.e. power-brokers) wishes.
2. Following Playford's loss of leadership in 1965, the British migration scheme continued as it had become well established. However, the archives reveal that it did not survive the curtailing of federal funding in the early 1970s, alongside changes in domestic politics (Young 2010).
3. All interviewees are identified by a name that they chose to be known by. This was usually their first name but sometimes a nickname.
4. The term 'grey nomads" refers to Australian retirees who travel across the country for extended periods of time by caravan, motor-home, campervan, or converted bus (Onyx and Leonard 2007; Westh 2001).

Acknowledgements

Graeme Hugo was a generous colleague and friend to many people. We acknowledge his leadership in the field of migration studies. Lisel O'Dwyer would like to acknowledge the support, mentoring and friendship that Graeme offered her over many years. *Vale* good friend. We also thank and acknowledge the feedback from our reviewers and the editors of this special issue.

Disclosure statement

No potential conflict of interest was reported by the authors.

References

Agger, B. 1991. "Critical Theory, Poststructuralism, Postmodernism: Their Sociological Relevance." *Annual Review of Sociology* 17: 105–131.

Arendt, H. 1968. *The Origins of Totalitarianism*. San Diego: Harvest/Harcourt.

Australian Bureau of Statistics. 1998. *"Development of Manufacturing Industries in Australia" Year Book Australia 1988*. Canberra: AGPS.

Australian Bureau of Statistics. 2012. *2011 Census of Population and Housing*. Canberra: AGPS.

Australian Bureau of Statistics. 2014. Country of Birth, Australian Historical Population Statistics, Cat. No. 3105.0.65.001.

Burnley, I. 1978. "British Migration and Settlement in Australian Cities, 1947–1971." *International Migration Review* 12 (3): 341–358.

Charmaz, K. 2014. *Constructing Grounded Theory: A Practical Guide Through Qualitative Analysis*. 2nd ed. Thousand Oaks, USA: Sage.

Clancy, R. 2016. "Population Growth Slowed in Australia with Migration Growth Waning, AustraliaForum.com." http://www.australiaforum.com/information/australia/population-growth-slowed-in-australia-with-migration-growth-waning.html.

Colic-Peisker, V. 2002. "Croatians in Western Australia: Migration, Language and Class." *Journal of Sociology* 38 (2): 149–166.

Collins, J. 2003. "Cultural Diversity and Entrepreneurship: Policy Responses to Migrant Entrepreneurs in Australia." *Entrepreneurship & Regional Development* 15 (2): 137–149.

Collins, J. 2014. "Report marks Australia's Shift From Settler To Temporary Migrant Nation, *The Conversation*." http://theconversation.com/report-marks-australias-shift-from-settler-to-tempo rary-migrant-nation-34794.

Collins, J., K. Gibson, C. Alcorso, S. Castles, and D. Tait. 1995. *A Shop Full of Dreams: Ethnic Small Business in Australia*. Australia: Pluto Press.

Commonwealth Department of Information. 1947. "Poster, *There's a Man's Job for You in Australia*." National Library of Australia, PIC LOC Poster Drawer 94, digitized collection http://nla.gov.au/nla.obj-136976103.

Commonwealth Department of Information. 1950. "Poster, *Australia Land of Tomorrow*." National Library of Australia, PIC Poster Drawer 94, digitised collection http://trove.nla.gov.au/version/ 23237445.

De Jong, G., and R. Gardner. 2013. *Migration Decision Making: Multidisciplinary Approaches to Micro-Level Studies in Developed and Developing Countries*. New York: Elsevier.

Department of Migration and Border Protection. 2015. *A History of the Department of Migration: Managing Migration to Australia*. Belconnen, ACT: Government of Australia.

Department of Planning and Local Government. 2010. "*The 30 Year Plan for Greater Adelaide*." http://www.dpti.sa.gov.au/__data/assets/pdf_file/0006/132828/The_30-Year_Plan_for_Greater_ Adelaide_compressed.pdf.

Department of the Interior. 1928. Poster, *Australia—The Land of Opportunity*, National Archives of Australia, NAA: A434, 1949/3/21685, http://vrroom.naa.gov.au/records/?tab=group&ID=19038.

Dey, I. 1993. *Qualitative Data Analysis: A User Friendly Guide for Social Scientists*. London: Routledge.

Elkins, C. 2005. *Settler Colonialism in the Twentieth Century: Projects, Practices, Legacies*. New York: Routledge.

Giddens, A. 1984. *The Constitution of Society: Outline of the Theory of Structuration*. Cambridge: Polity Press.

Hagen-Zanker, J. 2008. "*Why do people migrate? A review of the theoretical literature*." Maastrcht Graduate School of Governance Working Paper No. MGSoG/2008/WP002.

Hammerton, A., and A. Thomson. 2005. *Ten Pound Poms: Australia's Invisible Migrants*. Manchester: Manchester University Press.

Hugo, G. 1996. "Playford's People: Population Change in South Australia." In *Playford's South Australia: Essays on the History of South Australia, 1933–1968*, edited by B. O'Neil, J. Raftery, and K. Round, 29–46. Adelaide: Association of Professional Historians.

Hugo, G. 2006. "Temporary Migration and the Labour Market in Australia." *Australian Geographer* 37 (2): 211–231.

Jennissen, R. 2007. "Causality Chains in the International Migration System Approach." *Population Research and Policy Review* 26 (4): 411–436.

Jirojwong, S., and L. Manderson. 2001. "Feelings of Sadness: Migration and Subjective Assessment of Mental Health among Thai Women in Brisbane, Australia." *Transcultural Psychiatry* 38 (2): 167–186.

Jordens, A-M. 2001. "Post-War Non-British Migration." In *The Australian People: An Encyclopedia of the Nation, It's People and Their Origins*, edited by J. Jupp, 69–70. Cambridge: Cambridge University Press.

Joynson, V. 1995. "Post-World War II British Migration to Australia 'The Most Pampered and Protected of the Intake'?" PhD dissertation, University of Melbourne, Melbourne.

Jupp, J. 2002. *From White Australia to Woomera: The Story of Australian Immigration*. Melbourne: Cambridge University Press.

Keig, T. 2000. "Sound Cassette and Transcript. Interview with Lizzie Russell." Commissioned by the Migration Museum Adelaide. State Library of South Australia OH580.

King, R. 2012. "Theories and Typologies of Migration: An Overview and a Primer." Willy Brandt Series of Working Papers in International Migration and Ethnic Relations (3/12), edited by E. Righard. Sweden: Malmo Institute for Studies of Migration, Diversity and Welfare.

Kritz, M., L. Lim, and H. Zlotnik. 1992. *International Migration Systems: A Global Approach*. England: Clarendon Press.

Krockenberger, M. 2015. *Population Growth in Australia*. Canberra: The Australia Institute. http://www.tai.org.au/content/population-growth-australia.

Lee, E. 1966. "A Theory of Migration." *Demography* 3 (1): 47–57.

Lyotard, J. 1984. *The Postmodern Condition: A Report on Knowledge*. Manchester: Manchester University Press.

Massey, D. S., J. Arango, G. Hugo, A. Kouaouci, and A. Pellegrino. 1999. *Worlds in Motion: Understanding International Migration at the End of the Millennium: Understanding International Migration at the End of the Millennium*. Oxford: Clarendon Press.

Morawska, E. T. 2004. "The Sociology and History of Immigration: Reflections of A Practitioner." In *International Migration Research: Constructions, Omissions, and the Promises of Interdisciplinarity Research in Migration and Ethnic Relations Series*, edited by M. Bommes and E. T. Morawska, 203–239. Aldershot: Ashgate.

Murphy, J. 2000. *Imagining the Fifties: Private Sentiment and Political Culture in Menzies' Australia*. Sydney: Pluto Press, UNSW Press.

NFSA. n.d. "Newsreel. *British Migrants welcomed to their new home*, Youtube." Accessed May 27, 2016. https://www.youtube.com/watch?v=3rMIqSw9kxA.

OECD. 2016. "*Foreign-born population.*" https://data.oecd.org/migration/foreign-born-population.htm.

Onyx, J., and R. Leonard. 2007. "The Grey Nomad Phenomenon: Changing the Script of Aging." *The International Journal of Aging and Human Development* 64 (4): 381–398.

Patton, M. 2002. *Qualitative Research and Evaluation Methods*. Thousand Oaks: Sage Publications Inc.

Peel, M. 1995. *Good Times, Hard Times: The Past and the Future in Elizabeth*. Melbourne: Melbourne University Press.

Peel, M. 1997. "A New Kind of Manhood: Remembering the 1950s." *Australian Journal of Historical Studies* 27 (109): 147–157.

Powerhouse Museum. 1955–60. "Poster, Advertising Migration to Australia, 1955–1960." Registration number 85/824, https://ma.as/52927.

Ravenstein, E. G. 1885. The Laws of Migration. *Journal of the Royal Statistical Society* 48: 167–235.

Strachan, G. 2010. "Still Working for the man?: Women's Employment Experiences in Australia Since 1950." *Australian Journal of Social Issues* 45 (1): 117–130.

Stroebe, M., H. Schut, and Maaike Nauta. 2015. "Homesickness: A Systematic Review of the Scientific Literature." *Review of General Psychology* 19 (2): 157–171.

Thomson, A. 2005. "My Wayward Heart: Homesickness, Longing and Return of British Post-War Migrants From Australia." In *Emigrant Homecomings; the Return Movement of Emigrants, 1600–2000*, edited by M. Harper, 105–130. Manchester: Manchester University Press.

Tilby Stock, J. (1996). "The 'Playmander' : It's Origins, Operation and Effect on South Australia." In *Playford's South Australia: Essays on the History of South Australia, 1933–1968.* edited by B. O'Neil, J. Raftery and K. Round, 73–90. Adelaide: Association of Professional Historians.

Waldinger, R. 2014. "Emigrants and Emigration in Historical Perspective." UCLA International Institute. http://escholarship.org/uc/item/0j2996nz.

Weber, M. 1978. *Economy and Society: An Outline of Interpretive Sociology*, edited by G. Roth and C. Wittich. Berkeley, CA: University of California Press.

Westh, S. 2001. "Grey Nomads and Grey Voyagers." *Australasian Journal on Ageing* 20 (3, Supp 2): 77–81.

Wong, L. 2003. "Chinese Business Migration to Australia, Canada and the United States: State Policy and the Global Migration Marketplace." *Asian and Pacific Migration Journal* 12 (3): 301–336.

World Health Organisation. 2010. *WHO Global Code of Practice on the International Recruitment of Health Personnel, Sixty-Third World Health Assembly—WHA63.16*. Geneva: WHO.

Young, J. 2009. "Migrating Australian Migration: Thinking Beyond Ethnicity." PhD dissertation, University of South Australia, Adelaide.

Young, J. 2010. "Migration, Ethnicity and Privilege: An Exploration of Representation and Accountability." *Systemic Practise and Action Research* 23: 101–113.

Zgheib, P., and A. Kowatly. 2011. "Autonomy, Locus of Control, and Entrepreneurial Orientation of Lebanese Expatriates Worldwide." *Journal of Small Business & Entrepreneurship* 24 (3): 345–360.

Australian Immigration Policy in Practice: a case study of skill recognition and qualification transferability amongst Irish 457 visa holders

Fidelma Breen

ABSTRACT

Immigration quotas in Australia are guided primarily by economic policy—the needs of the nation are quite rightly the principal concern of policymakers. Using data from a mixed-methods study, this article engages in a dialogue between labour geography and population and migration studies through an examination of the lived experiences of migrating workers and their families. The paper examines a number of cases where policy and practice have detrimentally affected the migration experience of Irish migrants who came to Australia under the Temporary Work (Skilled) (subclass 457) visa. The case study focuses on interviews with secondary 457 visa applicants, namely the wives of former 457 workers. These interviews highlight the problems faced by newcomers to Australia who arrive under this visa. The experiences relayed by these women demonstrate how ill-considered policy relating to qualification transferability makes entering the workplace and, therefore, transition to life in Australia more difficult than it needs to be. While the cases that underpin this paper ended in migration failure (the families involved returned to Ireland), the core issues of qualification transferability and skill comparison were replicated in the wider study dataset (of 1022 survey responses and 80 qualitative interviews). While acknowledging that citizens generally have more freedom and more civil rights than non-citizens, the findings of this study indicate that further review of Australia's 457 visa is needed, particularly with respect to the limitations placed on the agency of migrant workers. Greater awareness amongst prospective migrants as to the purpose and limits of the 457 visa is also necessary, in order to avoid misinterpretation and the extreme personal repercussions outlined in this paper.

Introduction

The data in this article derive from a PhD project on the experiences of Irish migrants to Australia from 2000 to 2015, which commenced under the generous supervision of Professor Graeme Hugo AO, as part of a larger diaspora project that he was involved in at the time of his passing. The specific focus of this paper is on Irish migrants who came to Australia as temporary skilled workers, but who had returned to Ireland at the time of the research due to their difficult migration experiences. Their accounts of the convergence of events which resulted in dramatic migration failure are unique amongst the study participants. However, the core issues of skill recognition and qualification transferability are replicated across the author's broader dataset of 1022 survey responses and 80 qualitative interviews with Irish migrants to Australia.

This article's primary concern is with the effects of immigration policy on individual migrants. It builds on Hugo's work on theories of international migration and his concern that 'policies and programs, both formal and informal' are developed to 'maximize the benefits and minimize the costs of the greatly enhanced mobility' that the world is now experiencing (Hugo 2002, 13). Its concentration on the ability of migrants to fully use their training and qualifications provides a specific illustration of what Hugo deemed a key issue in the skills focus of Australia's immigration policy (Hugo 2014b). However, the case studies presented in this paper also highlight the emotional cost of some migrations. The article opens a dialogue between labour geography and population and migration studies in its examination of the lived experiences of migrating workers and their families and in its contemplation of the individual agency of the workers in the case study (Castree 2007). The concerns articulated by Hugo and other leading scholars (Howe 2013; Howe and Reilly 2014) about Australia's temporary labour schemes are mirrored in this study of the Temporary Work (Skilled) (subclass 457) visa, hereafter 457. This visa allows skilled workers to work in Australia on a long-term but temporary basis. The relevance of the article, then, is in its capacity to relay something substantive about the experiences of labour migrants, and their families, under the most popular temporary skilled visa Australia currently offers (DIBP 2015a).

It is nothing new for government policy to be mechanistic and impersonal. Entry systems in immigrant nations such as Australia are complex, immense and are supported by complicated legal and regulatory frameworks. Immigration quotas in Australia are guided primarily by economic policy—the needs of the nation are, quite rightly, the principal concern of policymakers. What has arisen as a matter of concern from this research project, regarding recent emigration from Ireland to Australia, is that Australia's skilled visa system has seemingly little regard for the emotional well-being of the migrants involved or their personal circumstances. Setting aside the contention that Australia's policy provides an opportunity to skilled workers, they are viewed primarily as economic entities and 'care' is given in so far as it is consistent with the main aim of enabling them to provide an economic benefit. This article first provides a broad overview of the 457 visa. It then briefly describes the development of Australia's immigration policy from White Australia to the revision of the points system for skilled migration in 1999 in order to demonstrate the shift in focus from population augmentation to targeted, skilled migration that benefits the Australian labour market. The empirical portion of the paper considers the experiences of three highly qualified linesmen from Ireland who were employer sponsored under the 457 visa. In the examples presented—relayed through the workers' female partners—the lack of full disclosure and misrepresentations made by employers caused migration failure and the return of the 457 workers and their families to Ireland. Stress, alcoholism, domestic violence, house repossession and relationship breakdown also resulted. While the examples presented are arguably extreme cases, they raise important questions about qualification transferability and skill comparison that were replicated in the wider study dataset. They show that even highly skilled migrants can struggle with the migration experience if labour expectations are not transparent. The paper argues that there is a need for more transparency in the pre-migration stages on the part of employers, for bilateral qualification recognition between Australia and its source countries and, where retraining is needed, utilisation of online delivery methods whenever possible.

The Temporary Work (Skilled) (subclass 457) visa: an overview

Introduced in 1996 under the Howard administration and initially called the Business Long Stay (subclass 457) Visa, the visa was renamed the Temporary Work (Skilled) (subclass 457) visa in November 2012. The 457 visa allows

> skilled persons to come to Australia to work for an approved employer, accompanied by their immediate family members for a period of between one day and four years ... The program involves a three-stage process whereby an employer applies to become an approved sponsor and then nominates a skilled overseas worker to fill a specific position. The skilled overseas worker completes the process by lodging a linked temporary work skilled visa application. (Parliament of Australia 2013)

The Department of Immigration and Border Protection (DIBP 2016b) states that 'the Temporary Work (Skilled) (subclass 457) visa is designed to enable employers to address labour shortages by bringing in genuinely skilled workers where they cannot find an appropriately skilled Australian'. The intake is uncapped and driven by employer demand (Cully 2011).

Empirical evidence shows that some 457 visa holders are particularly exposed to the vagaries of their employers. This reflects the precarious labour market position of temporary workers, their disenfranchisement and the coercive relationships that often exist between these workers and their employers (Berg 2015; Boese et al. 2013; Campbell and Tham 2013; Costello and Freedland 2014; Howe 2013; Velayutham 2013). While the 457 visa allows long-term temporary residence, it is also a primary source of transfer to permanent residency—indeed, almost 40 per cent ($n = 61$) of the 457 respondents in the current study indicated that they did not consider their move to Australia to be temporary at the time of departure from Ireland. Irish people comprised 5.3 per cent of total subclass 457 visa recipients in the year to 30 September 2015 (DIBP 2015b).

The difficulty the 457 visa presents in a comparative exercise lies in the fact that, while temporary, it is a means of attracting highly-skilled workers (Hugo 2014b) rather than (so-called) low-skilled workers such as those used in California's *bracero* scheme or Australia's seasonal employment programme (Connell 2010; Mitchell 2013). Nevertheless, some striking similarities remain between this newest skills shortage reduction scheme and California's importation of an agricultural workforce from the 1930s until the 1960s. California's scheme was described and analysed by Mitchell (2013, 224) as 'a large-scale program of importation of a highly controlled—indeed indentured—force of temporary labourers, one that required unprecedented levels of state involvement in labour procurement, housing, and regulation—in essence state command over the labour market'. This is not to imply that Australia's 457 visa is a form of indentured labour *per se*[1]—although the fact that migrants are tied to an employer sponsor[2] and are unable to move jobs with ease or progress their careers[3] is morally questionable. The lives of migrant workers and their families are highly regulated and restricted outside the employment sphere by their visa conditions, which inhibit access to the welfare support network and subsidised health-care and education enjoyed by Australian residents. It is a requirement that 457 visa holders take out and maintain high-level healthcare insurance, as prescribed by the DIBP under Condition 8501, to insure against the cost of all medical treatments not covered by a Reciprocal Health Care Agreement. Welfare payments cannot be accessed for a period of 104 weeks after arrival (DHS 2015). The issue of high public school fees for the children of 457 families is discussed in some detail in the empirical portion of this paper.

There appears to be little *quid pro quo* in this visa arrangement: it is an unequal trans-action if one considers the cost and risk undertaken by the skilled migrant in moving to Australia to fill a gap in the labour market, with little or no support. Concerns about the 457 visa have also been vociferously expressed by union bodies, focused on the protection of individual Australian workers and the state of industry in Australia. The Australian Council of Trade Unions (ACTU) made a comprehensive submission to the Governmen-tal Review of skilled migration and 400 series visa programmes[4] which highlighted their concerns over the growing reliance on 'demand driven' temporary skilled immigration and the preoccupation with deregulation rather than the integrity of the programme (ACTU 2014). The following section of this paper turns to review the shift in the nature of Australia's immigration programme over recent decades. Particular attention is paid to the shift towards temporary skilled migration driven by economic imperatives.

Development of Australia's immigration policy: from population augmentation to skilled migration

As one of the world's traditional immigration countries (Castles 2014), international migration has been core to the development of Australia since white settlement. A notable feature of its post-war development was an expansive, planned immigration pro-gramme profoundly shaped by government policy (Hugo 2014a). From Federation in 1901 to the close of the Second World War, a protectionist attitude towards market regu-lation and trade fostered a tight, race-restricted immigration control, known as the White Australia policy, which continued the largely Anglo-Saxon immigration trend that had characterised the country's population expansion since European settlement (Tavan 2005). The debunking of race theory (Barkan 1992), and the ascendancy of assimilation in the post-war period, aided a change in immigration policy priority which determined that the country must 'populate or perish'. The government aimed to increase the popu-lation through immigration by 1 per cent per annum. Various changes (from the revision of the *Migration Act* in 1958, which abolished the dictation test, to the reforms under Harold Holt and Gough Whitlam in 1966 and 1973) respectively dismantled the White Australia policy (Meaney 1995; Oakman 2002). It took almost 30 years and many timid, piecemeal steps to remove the concept of White Australia from the lexicon of Aus-tralian government.

Population augmentation was the main focus of Australian immigration policy until the late 1970s. More targeted skilled migration became the focus with the introduction of the Numerical Multi-Factor Assessment System (NUMAS) in 1979 (Davis, McAllister, and Manning 1980). This assessed migrants on occupation, skills, English-language competence and other factors such as family ties. Non-English-speaking applicants with low skills had little chance of passing. In 1985, the then Labor government increased immigration and introduced a new selection system that favoured the most skilled of sponsored relatives. Opposition to increasing Asian immigration and multiculturalism in general found a voice in a wide-ranging public inquiry held in 1987 and chaired by Dr Stephen Fitzgerald. The Committee to Advise on Australia's Immigration Policies (CAAIP) was assigned the task of investigating public opinion on immigration and charged with making recommendations to guide future immigration policy. The CAAIP used Hugo's research, which had shown that family reunion pathways were generating a relatively low-skilled intake, to support their new rationale for increased immigration on economic grounds (Birrell and Betts 1988; CAAIP 1988; Hugo 1988).[5] The CAAIP appealed to Australians for support by emphasising the fiscal benefits that higher skilled immigration could bring in terms of international competitiveness and strengthening the economy (CAAIP 1988). Selection processes then turned away from extended family sponsorship towards applicants with business and labour skills.

A points system—the Structured Selection Assessment System (SSAS)[6] (Mackellar 1979)—had been in place since 1973 but it was the introduction of a Canadian-style 'Points Test' system (Shachar and Hirschl 2013) in 1989 that consolidated the expectation that migrants be self-reliant upon arrival. Going forward, applicants were allocated visas on the attainment of points related to their age, health and character criteria, English-language proficiency, skill level and the possession of credentials recognised in Australia (Boucher 2013; Miller 1999). The rule-based system made rational selection easier but the end result was a minefield of visa classes and permit types (Jupp 2007). While Labor laid the foundations for some of the harshest financial aspects of the new economically rationalist immigration policy—introducing a 6-month waiting period for social welfare and charging for visa applications, appeals and English tuition—these were all to become lengthier and more expensive under the Liberal administration of John Howard. Social welfare assistance, for example, was denied to new arrivals for a period of 2 years, despite research showing that this period was the most crucial in terms of securing stable employment and housing (Jupp 2007).

Immigration policymakers were necessarily circumspect of the domestic situation. Immigration policy was balanced, after all, on the combined support of social and organised business interests and presented an electoral vulnerability. When Howard's Liberal–National Coalition government came to power in 1996, deliberate moves were made to rebalance the immigration system in favour of targeted, skilled immigration while the value of family ties was concurrently reduced. Changes to the Points Test in 1999 demonstrated this new focus: the pass mark was raised by 10 points to 120; the required International English Language Testing System (IELTS) scores were raised; points were given for pre-arranged employment and removed for familial links; the age for which maximum points could be attained was halved to 30 years; and the 80 points previously given for general occupational skills was replaced by a maximum 60 points for targeted skills which were listed in the Migrant on Demand List (MODL). These elements remain the basic structure of the entry system to Australia to the present day.

As a result of this changed focus, the composition of Australia's immigration intake has changed dramatically over recent decades. In particular, the proportion of migrants entering Australia through the 'skill' and 'family' components has diverged significantly since the tightening of the points system (see Figure 1). The permanent migration programme now comprises 68 per cent skill and 32 per cent family visa grants (DIBP 2015a). Migrants now constitute 27.7 per cent of Australia's overall population of 24 million people.

To demonstrate the importance of the temporary migration programme against permanent intake: in the year 2014–15 year 51 125 visas were granted under the 457 programme, alongside 214 830 Working Holiday Maker visas, against the 189 097 places in the permanent programme (DIBP 2015a).

Figure 2 illustrates the flexible and uncapped nature of the 457 visa against the family and skill components of the permanent migration programme, while Figure 3 charts the 457 visa programme intake, part of the temporary programme, since its inception. In the 5 years following December 2010, the number of 457 visa holders in Australia rose dramatically, largely because Australia was viewed as having escaped the worst effects of the Global Financial Crisis (DIBP Economic Analysis Unit 2013). In this 5-year period to 2015, temporary 457s from India increased by 155.5 per cent, those from China (excluding Special Administrative Regions of China) increased by 115.4 per cent, and from Ireland by 59.1 per cent (DIBP 2016a, 13).

Given that Australia's skilled migration programme aims to assist Australian employers by allowing long-term work rights to people with the skills they need to conduct their business, it is in everyone's interest that the migration process—including the 457 visa programme—is a success (DIBP 2016b). Despite the high 457 intake figures outlined above,

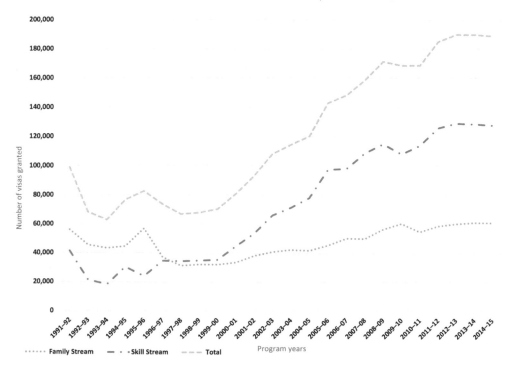

Figure 1. Infographic depicting the divergence between the family and skill entry streams, 1991–2015. *Source*: adapted from DIBP Migration Programme Statistics (DIBP 2015).

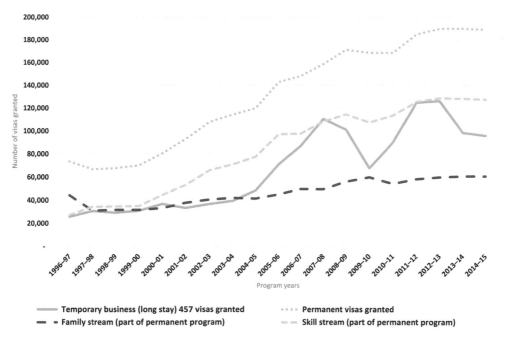

Figure 2. 457 visa intake shown against the permanent programme intake, 1996–2015. *Source*: 'Migration to Australia: A Quick Guide to the Statistics' (Janet and Simon-Davies 2016).

Australian policymakers should be wary of complacency as the strong economic focus of Australia's immigration policy, application costs and generally high entry requirements may have contributed to the increasing importance of alternative destinations for the Irish in recent years. In 2015 the UK returned to its position as the lead destination for Irish emigrants, while Australia attracted less than half the number of Irish migrants it did in 2013 (CSO 2015). Australia attracts only around 5 per cent of the world's total skilled migrants (UN-DESA & OECD 2013).

As shown throughout this section, Australian immigration policy has undergone profound changes over recent decades in implementing a stronger focus on economic-based

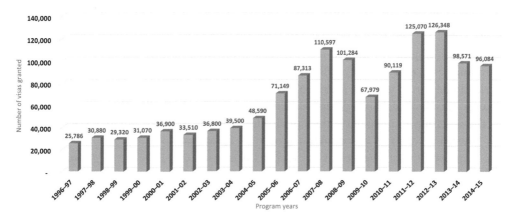

Figure 3. 457 visas granted since 1996. *Source*: 'Migration to Australia: A Quick Guide to the Statistics' (Janet and Simon-Davies 2016).

policy. Elements of these policy changes include 'a shift away from a human capital focus toward more targeted selection based on labour market demand for specific skills [and] increased emphasis on temporary foreign worker programs' (Akbari and Macdonald 2014, 801). Hugo (2005, 205) noted an overwhelming focus in Australia on skills in migrant selection, which he described as 'searching for talent'. This shift has occurred despite the fact that demographic and economic changes in Australia, like in other OECD nations, have meant that there is growing demand for both unskilled and skilled workers. Hugo (2005, 207) expressed concern about: 'a growing mismatch between immigration policies focused on skill and a tightening labour market with demand for labour across a broader skill spectrum'. He also voiced concern that the focus of immigration research attention in Australia 'remains predominantly on permanent settlement [but] there is an increasing realization that temporary migration is of significance when considering the effect of migration on Australia's economy and society' (Hugo 2006, 113; see also Castles, in this issue). He added that 'temporary migration [is] dealt with quite separately from permanent-settlement migration [in both policy and academic writings] although there is considerable overlap' in practice (Hugo 2006, 113). The case studies presented in this paper demonstrate that the 457 visa is particularly important when looking at the pathway to permanent residency, given the expectation voiced by many of the families interviewed that the 457 was 'a foot in the door'. The DIBP (2014b) has recognised this trend in transition to permanency stating that 'the proportion of the Migration Programme places filled by people on a temporary visa in Australia has increased from around 30 per cent in 2004–05 to 50 per cent in 2013–14'. This trend has been 'particularly pronounced in the Skill stream where the proportion of onshore grants has increased from 35 per cent in 2004–05 to 57 per cent in 2013–14' (DIBP 2014b). The paper turns now to case studies of recent Irish 457 emigrants in an effort to highlight some of the existing difficulties with this visa.

Methods

The cases presented in this paper are part of a broader mixed-methods study of the experiences of Irish migrants to Australia from 2000 to 2015. The study dataset comprises two sections: Irish migrants living in Australia and Irish migrants who have departed Australia. The total study dataset contains 1022 usable survey responses and 80 qualitative interviews. The three case studies used in this article are taken from 15 interviews recorded with former Irish migrants who had left Australia at the time of interview.

In an example of the snowball method of recruitment common to ethnographic fieldwork and best used to reach sensitive cohorts (Brace-Govan 2004), contact with the interviewees was gained through an administrator of the Irish Families Living in Perth Facebook group. Snowballing then proceeded through the first interviewee's network. The interview schedule was chronological in its investigation of the early life, education and employment experiences of participants before delving into the migration decision and journey and post-arrival experiences in depth.

Telephone interviews were conducted by the author in January 2016 as the participants were in various parts of Ireland and the researcher was based in Australia. The three vignettes featured in this paper are based on interviews with the female partners of men who had migrated to Australia as 457 visa holders sponsored by an energy subcontractor in Western

Australia. The women interviewed were thus secondary 457 applicants and their migration experiences no doubt differ somewhat from their partners' experiences. The recruitment of female partners as interviewees came about because the initial contact in Perth knew one of these women (Julie)[7] personally and so Julie became the natural point of contact. Julie, in turn, introduced the researcher to Natasha and Emily. It was not possible to speak to Julie and Emily's partners (Michael and Tom) due to their work commitments and Peter was no longer in a relationship with Natasha at the time of interview.

The women's voices are used here as a bridge to feminist labour geography—it was their experiences which combined with the employment experiences of their partners to produce migration failure. As home workers, Julie and Natasha are viewed in this paper as equal economic migrating partners prepared to undertake the provision of care. Their intended role as workers in the social economy, or the non-market sector of the diverse economy (Cameron and Gibson-Graham 2003), was confounded by the failure of employment of their partners in the capitalist economy (under the 457 visa). The experiences of the women and their partners, as a group of migrating workers, deserve holistic attention. Too often the employment aspects of labour migration are the sole or paramount focus of research, when in reality the personal and family circumstances of a migrating worker can affect the prospect of migration success as much as the conditions of employment (Castree 2007).

The case studies presented in this paper were chosen because each highlights the dramatic effect the programme's restrictions can have on a migrant, and in this case, his family, through the conjuncture of events and the divergence of expectations and realities. In each case the worker was shown to be at the mercy of the employer and the company's market success or failure which was, in turn, affected by the general economic climate. In each instance the worker was bound by the terms of the 457 visa which restricted both his (and his family's) agency and access to social support infrastructure. In two of the cases, lower than expected earnings severely compromised the families' ability to survive financially.

Case studies—the experiences of three Irish families under Australia's 457 visa programme

As noted above, the three women whose interviews form the basis of the remainder of this paper had partners who were all employed (under the 457 visa scheme) by the same energy subcontractor in Western Australia. Michael and Peter came across the positions offered by the energy company through an internet search for job opportunities in Australia while Tom received information from a former work colleague who had relocated to Perth. Visa applications for all three linesmen and their families were handled by an Australian-based migration agent. The interviewees mentioned a couple of instances where the migration agent had given them incorrect information regarding entitlements. However, the option of progressing to permanent residency was promised to all three workers by the employer and is relayed by Natasha below. Overall, there was confusion about the pathway to permanency open to them as 457 visa holders and little distinction made between the information-giving entities they encountered. The employer, the migration agent and the host government were seen as one and the same and the veracity of the information provided by them was never doubted.

The company paid relocation costs and an administrative worker within the energy company was, by all accounts, most helpful to the newcomers. A relocation agency retained by the energy company for the purposes of assisting settlement did little for the families in terms of finding long-term accommodation or suitable schools, but there was little complaint as the migrants started their new lives. According to the women, their partners (the 457 employees) were initially happy at work. However, after induction, they were informed that 'upskilling' was required to enable them to meet Australian safety standards. The company offered to pay the cost of retraining, which was approximately $5000 each, but only if the men agreed to sign a document extending their contract time with the company by 1 year. There had been no mention by the employer of the need for training during the 457 visa assessment or application process. Given that the workers and their families had arrived at different times—Peter arrived in February 2012 and Tom and Michael arrived in February 2013—it is unlikely that this was a new development that the company was unaware of prior to offering sponsorship. Julie (married to Michael) said:

> He'd signed into the extra year's contract—that was hanging over us all the time. If he'd left we'd have had to repay the relocation. He was told they [the new 457 employees] weren't to Australian standards and they wouldn't be taken on by another company if they didn't do the training course. Even though they'd [the company] come to Ireland to take skilled labour over there? So how can they tell you then you're not up to standard? It's totally immoral. You're enslaved ... we'd be up for all that [repayment of relocation] money. I think it was well into $20,000 ... that was hanging over us all the time—that we'd have to pay the money back. We were in contract but they [the employer] never thought to tell us that if they broke the contract they'd have to pay to send us all home.

After retraining, the men assumed work would begin in earnest and that the overtime the company had promised would also start. Julie continued:

> The deal was that he [Michael] would be away three weeks out of the month so we said grand because I'd be able to be home with the kids—[in Ireland] we were passing each other by on shift work and the kids were spending a lot of time in childcare—so we thought it would be a better life there [in Australia].[8]

However, the basic employment hours (40 hours per week at the base pay rate) continued, with the company saying that there was no overtime available despite having claimed that the workers would have a substantial wage top-up with overtime earnings:

> Then they [the men] went for their first bout of 'away' work and were there three days and they had to come back—whoever had planned it had planned it for the wrong time of the year and the landowners wouldn't let them in because it was crop season. So they spent three days in the back of beyond and then they had to come home ... it actually cost them [the workers] money because they had to pay for the digs [accommodation] and they only got paid the basic pay. (Julie)

An outcome of this failure, by the company, to fulfil the earnings promise was that Natasha and Julie each had to find a job to supplement the household income:

> They [the company] promised us everything. Everything was grand at the start ... It was kind of holiday mode because he was doing courses and coming home early. He was promised overtime and he never got it. Money started getting tight then because I couldn't get work. It was just impossible. Julie minded my wee lad and I got a job cleaning—it was for

> buttons. I was away from 7am until 7pm and got maybe $50 for it because they only give you the money for the hours and I was travelling 1.5 hours there and the same back so it didn't pay me to work by the time I paid Julie. (Natasha)

> ... they [the men] didn't have the work that they thought they'd have ... the great news was that they could come home every evening but it was a quarter of the wages [that they had been promised] so that meant Mike was getting about $1200 a week and our rent was $560 a week so I then had to go back to work ... it meant we were back to square one—he was working, I was working and the kids were being minded. (Julie)

Michael and Julie's family of seven could not survive on the income they had and Julie obtained a job at a play café. Paid childcare was undertaken by local Irish mums or university students and, while much cheaper than the registered centres it was still a strain on the families' finances because 457 families cannot access the Child Care Rebate.[9] Social fallout followed:

> I was great at the start because I was able to go to all the GAA [Gaelic Athletic Association] matches, the hurling and all is brilliant over there—the standard was fantastic but once I started to work ... like I used to go to an Irish mothers and toddler group—we had all the catch-ups, coffee mornings and everything and then once I had to work, I worked weekends and I missed out on all that and totally lost the connection. (Julie)

Relationship breakdowns also ensued. Julie recalls that the company often miscalculated the wages and on one particular occasion, the wages were late when the rent was due. This was her breaking point. She asked the company for flights home for her and the five children and left after 15 months in Western Australia. Michael stayed, and for a while she was unsure if he would return to Ireland. The implications for their marriage have been long-lasting:

> My marriage is totally different. There's a lot of resentment built up and a lot of defensiveness and it might sort itself out and then again it might not, but it definitely changed us totally ... we had to start from scratch coming back and we lost a lot, the house and all and we're going to be in debt forever and we might never get a loan again. (Julie)

Six months after the interview Julie revealed: 'My marriage has not survived. I'm on my own now with the kids. I firmly believe that the disaster that was Oz played a big part in the breakdown.' Natasha and Peter also separated:

> Peter ended up turning violent because of the stress from work. He was coming in and everything was getting to him—he wasn't getting what he was told he'd get and he was taking it out on me. The last straw was when he hit me ... When we first went [to Australia] he was grand. It started to turn bad about 8 months in and he ended up hitting me one night. I had a nervous breakdown and I had just had enough and I had to come home with the child ... I just had to go and I had to leave him [Peter] behind and I had to leave everything behind. I just had to.

Only Emily and Tom returned from Perth unscathed by the experience. Tom was made redundant and there seemed little point in pursuing another 457 sponsor as Emily fell pregnant and the couple decided to have the baby at home in Ireland. However, Emily explained that at the time he was made redundant Tom had questioned the extended contract that had been presented in exchange for the company paying for the required retraining. The migrant workers viewed signing an extension to their employment contract as additional job security rather than a burden of servitude. But the burden of fulfilling the 1-year contract extension was entirely on the workers. The company, as it emerged, was not required to keep the workers on for the additional year:

> ... he [Tom] thought at the time when he signed for the course that he did sign something and then when we questioned it when he was being made redundant, because he'd only the two years served, they came back and said no, there wasn't any obligation on them to give him work for another year. (Emily)

The experience seems to have had less of a negative impact on Emily and Tom than on Natasha and Peter and Julie and Michael. Emily said that the lesser hours did not affect them as severely as the other couples because they did not have children and were able to save money each week. When they returned to Ireland, they did so to the home they had built previously:

> We were both working [in Australia]—like poor Julie and that, they'd an awful experience even money-wise with the five children whereas myself and Tom, we were never short of money—we were so lucky: both working, no kids so we got to save a bit of money there whereas Julie had the opposite end of the stick—they had an awful time of it.

Overall, Emily and Tom's journey differed substantially from that of the other couples. They didn't see themselves completing the 4 years of the 457 visa and didn't intend to stay in Australia:

> I knew we weren't staying forever and the whole thing about Australia was the two of us were going out there, something different and to make money, you know ... We'd only ever planned on staying maybe 3 of the 4 years of the visa. (Emily)

While Julie, Natasha and Emily's spouses all worked for the same company, the problems they faced in terms of retraining and a lack of information about qualification transferability was not specific to either their employer or the industry. Other respondents from the larger dataset who worked in childcare, social work and teaching all faced similar hurdles to career employment in Australia. Teachers reported registration taking several months; a childcare worker had the option of paying $6000 to get the Australian equivalent of the Diploma she already held, or work as an assistant rather than room leader; and an educational psychologist had to pay astronomical costs for qualification assessment and supervision for registration.[10]

As highlighted by Julie's and Natasha's migration experiences (in comparison to Emily's), family composition played a part in the migration failures related here, since the lack of anticipated earnings severely affected the ability of families with children to settle in Australia. Fewer work hours and consequently lower wages have been a feature of previous temporary labour schemes and it may be that labour requirements are difficult for companies to estimate (Connell 2010). However, the migrant families interviewed as part of this study had placed their trust in a sponsoring employer. The information they were given with regard to expected income and prospective permanency, as well as their access to schooling and medical care, was provided by the employer and a registered MARA migration agent.[11] Their feelings that the employer had been duplicitous when promised enticements did not materialise are somewhat justified. Lower earnings impacted a myriad of factors—housing affordability, general lifestyle expectations, a need for costly childcare providers—all of which affected the happiness of the migrants, particularly the mothers who had to seek paid employment. Furthermore, changes in state school fees had a detrimental effect on these families, particularly on Julie and Michael who had four children of school age. When the West Australian State government introduced a $4000 per year charge for each 457 child in its public school system, a group of 457 families, mostly Irish, challenged the State. Through protest and negotiation, the charge was reduced to $4000 per family per year. During the debate the point was made by the protestors that changed policies, whether federal and state-sponsored, should, in fairness, only be applicable to new applicants, not to those already *in situ*.[12] The retrospective application of ministerial decisions and ensuing changed policies is inherently unfair.

The changed school fees were introduced by the State government under whose remit education policy falls. The 457 families protested against this as a new condition of their visa since it was their visa status which made the new State education policy applicable to them. They did not consider the two to be separate. It is important to consider at this point who the migrants saw as responsible for controlling their experiences in Australia; who could affect their well-being. From the interviews, it is clear that this was the 'Australian government' with no distinctions made between separate departments or even State and federal levels of government. The visa was the control mechanism for access to all employment, financial and social benefits and it was administered by the federal government so all ensuing benefits or discriminations were also seen to be coming from this source. Consideration should also be given to the 'sales' method of Australian visas by Australian companies, and indeed migration agents, given Julie's statement about their understanding of what the 457 meant in the long term. During the school fees campaign Julie rang the local MP's office:

> … it was actually a British man I spoke to and I asked what did they think of the whole thing and he said 'What are you on about? What are you complaining about? You're only here to fill a gap or to train up Australians and then you're supposed to go home. It's a temporary visa'. I said that we were of the understanding that once we were in we could apply for residency and he was like, 'No, no, that's not the idea of it'. And I said 'Well that's the way you're selling it [the 457 visa] in Ireland—'c'mon and get your foot in the door' and he said 'No that's not the idea of it. You must understand you're just here to fill a gap' and I said 'Do you really think that people would be uprooting their families to come over here and fill a gap? Come on now'. But they just don't care.

A route to permanent residency for 457 visa holders exists in a further visa process through the Employer Nomination Scheme (ENS). The ENS allows for a permanent residency application after two years for eligible 457 visa-holders under the Temporary Residence Transition stream. The families here were somewhat confused by this distinction. They thought that, by being in Australia on a 457 visa, they were automatically entitled to apply for a permanent visa. Confusion on the part of the migrating families is understandable when the possibility of permanent residency under the 457 is introduced. Promises made by the energy subcontractor to Natasha's partner, Peter, that 'they would put him through PR at some stage' only strengthened the migrants' impression that permanent residency was a matter of course. While there may have been an element of naivety or trustfulness on the part of these families, it is not unreasonable for them to have expected to have been treated fairly by the employer and migration agent, and in the larger sense by the Australian government. Neither is it reasonable to expect them to independently verify all the information they are given considering the various health, education and other social structures a new arrival in Australia has to contend with after the visa process. Various interviewees reported that even some Medicare staff were unsure of the Reciprocal Health Care Agreement and the differences in rights between migrants from Northern Ireland and those from the Republic of Ireland. The multitude and very separateness of the various departments one needs to deal with upon arrival in Australia prompts the question of the feasibility of introducing a whole-of-government, interdepartmental source for incoming migrants which could help fill a perceived responsibility of the Department of Immigration and Border Protection to extend its remit of care beyond the simple issuance of entry visas.

Conclusion

The participants involved in this study are not alone in raising concerns about issues of skill recognition and qualification transferability under the 457 visa programme. In a submission to the Productivity Commission, the ACTU (2015) noted that the Australian government should seek opportunities to improve the recognition of overseas qualifications obtained at high-quality institutions, including through bridging courses. Union support comes 'provided the qualifications are formally assessed by an independent body against the Australian Qualifications Framework (AQF) occupational requirements and other relevant Australian endorsed standards' (ACTU 2015, 6). The argument here is not that there should be any undue concessions made for skilled migrants, but that the qualification matching process should be clear from the outset and that it should be as efficient as possible in order to facilitate an expeditious entry to the labour market. Similarly, in its response to the Discussion Paper *Reviewing the Skilled Migration and 400 Series Visa Programmes* (DIBP 2014a) in September 2014, the Australian Psychological Society acknowledged the difficulty that migrants faced in obtaining a general registration and that they were not considered 'work ready' until this was granted since employers did not want psychologists with a provisional registration who required supervision. One of its recommendations was that 'the Department of Immigration work with the national assessment authority and the regulator to move forward on the issue using "comparability" as the basis for decisions about migration' (Hammond 2014, 5). The Law Society of Australia's submission to the above-mentioned Discussion Paper stated that 'poor legislative drafting ... and Reg. 5.19 of the Migration Regulations 1994 has resulted in a range of confusion and (presumably) unintended consequences'. The submission went on to cite inflexibility as a barrier to career development and progression to the permanent subclass 186/187 visa (Law Society of Australia 2014).

This article has examined a number of cases where policy and practice have detrimentally affected the migration experiences of Irish migrants. The stories told by the partners of three former 457 visa holders highlight a number of problems faced by newcomers to Australia under this visa. Specifically, they demonstrate how ill-considered policy relating to qualification transferability makes entering the workplace, and, therefore, transition to life in Australia, more difficult than it needs to be. It also supports previous research which illustrates the particular vulnerability of this worker group because of their hope for continuing employer sponsorship so that they can achieve permanent residency (Howe and Reilly 2014). Based on these data, this article concludes that to ensure that the actual performance of policy is both practicable and fair, a bilateral commitment to national industry and trade recognition of qualifications between Australia and the origin countries of its migrants is required, most especially for occupations on the Skilled Occupation List. This is not to suggest that industry bodies should set the skills requirements for prospective migrants (Howe and Reilly 2014); rather, they should work with a government agency to streamline qualification assessment *before* migration occurs. Pre-migration assessment should be done in a timely fashion so that any necessary training gaps can be addressed as soon as possible, thereby maximising the possibility of a migrant being work ready upon arrival. The provision of a clear and accessible information portal for this process would assist prospective migrants in understanding the intricacies of skilled migration, and reduce the likelihood of migration failure—as experienced by the three families whose stories are discussed in this paper.

Qualitative data from this research project suggest that elements of the Temporary Work (Skilled) (subclass 457) visa are illogical and outdated. Migrants are not made aware of the cost and time required for qualification assessment, and skilled workers hired for the very skills they possess are left uninformed about necessary retraining. There appears to be no stated obligation in the 457 sponsorship requirements on an employer to inform prospective sponsored employees of Australian industry standards. Whilst acknowledging the need for strict regulation of health and safety procedures, especially in a highly dangerous sector such as electricity and energy, time and technology could also be utilised more effectively to assess overseas qualifications and provide online training to assist newcomers to meet Australian industry standards in qualifications and safety prior to migration. While there is currently no evidence of the use of technology to close skill gaps in trade work, a body of research outlines the progress of online training in the vocational education and training sector (Curtain 2002; McKavanagh et al. 2002; Salmon 2004; Webster, Walker, and Barrett 2005). Telemedicine or telehealth is another area where e-learning is evident. This could be similarly successful in filling competency gaps in skilled trades (Foroudi et al. 2013; Mather, Marlow, and Cummings 2013). Online workplace health and safety training is also the norm for many workplaces across different sectors in Australia currently.[13] With reference to the case study explored herein, the receipt of necessary training prior to migration would have equipped the migrant workers to be ready to start their jobs in Australia immediately. The uncertainty regarding hours of employment and expected income would likely have been greatly reduced. Whilst acknowledging that the decline in the Australian resource sector in general, and mining in particular, played its part in the difficulty that the power company experienced obtaining tenders (and thus its inability to provide the promised overtime work hours for Michael, Tom and Peter), there is a possibility it may have met with more success with a ready workforce.

In referencing Hugo's (2005) call for the migration experience to be a 'win' for all, including the migrant, the practical and personal effects of the 457 visa deserve scrutiny. Despite knowing, as we do, that 'neo-liberalist governments which dominate in OECD nations like Australia [are not] sensitive to policy advice which is based in part on ethical considerations, altruism, social justice and [which] flies in the face of what are perceived as "market forces"' (Hugo 2005, 201), an onus remains on researchers to highlight aspects of policy practice where there is little consideration given to ethical behaviour and aspects of social justice. In short, policy is rarely based on justice, morals or ethics yet the case studies presented in this paper show the disturbing effects of this—the very real, human effects. Union bodies, of course, do this in a systematic way by dealing with individual worker protection issues and breaches of employment legislation on a day-to-day basis. The focus here on the moral geography (Castree 2007) of this migrating cohort adds weight to the suggestion that policy and practice should be amended to improve the prospects of immediate and fulfilling participation in the labour market for skilled temporary migrants. Critical evaluation of the case study leads not just to policy prescription but to a bridge between migration geography and labour geography in considering the 'normative issues at the level of both policy and principle' (Castree 2007, 860).

The likely cost of streamlining qualification recognition across industries should not deter the governments of Australia and Ireland and other migrant source countries from a commitment to this activity since faster and more efficient entry to the workplace is a highly probable outcome. Given that 457 migrants earn middle to high incomes and pay tax on those incomes, while presenting no burden to the state in terms of health care or social assistance, they deserve better than the scenarios outlined above.

Notes

1. Betzien (2008) argued that the proposed Pacific guest worker scheme would 'create a semi-indentured subservient class of workers, *as has occurred under the 457 visa program*' (emphasis added).
2. Visa holders have 90 days in which to secure a new employer sponsor if they have a valid reason for terminating the contract with their current employer.
3. DIBP officials stated recently (April 2016) that promotion could jeopardise the continuation of a 457 visa and/or hamper a subsequent permanent residency application since discontinuing employment in the position specified in the employment contract meant the visa holder did not comply with the conditions of the visa granted.
4. The 400 series visa programme allows for the temporary entry of people to Australia for economic, social or cultural purposes (DIBP 2014a).
5. It should be noted that Hugo's report stated that

   In 1986/87 the highest workforce participation rates (in excess of 50 per cent) were in the independent, skilled labour, concessional and 'other' categories although those of the Family 1C and 1A and refugees were also high (between 40 and 49 per cent). On the other hand, the rates for the Family 1B (18 per cent), Special Eligibility (29.2) and Business (31.9) categories are relatively low. (Hugo 1988, 17)

6. The SSAS was the first step towards a structured migrant assessment protocol. Immigration officers had to complete a two-part interview report—Part A related to economic factors and Part B required the interviewing officer to make an 'Assessment of Personal and Social Factors' regarding the applicant (Hawkins 1991). 'NUMAS was an amalgamation of the Canadian points system and the SSAS which preserved the two-part assessment format but added numerical weightings to a total of 100' (Hawkins 1991, 142). Opposition to it lay in the widely held perception that it indicated a return to the White Australia policy. The Canadian system did not require family migrants to be assessed under the points system and for other migrants, 90 of the 100 points focused on economic factors with only 10 points awarded for 'personal suitability'.
7. Pseudonyms are used to de-identify participants.
8. Julie's assertion that everything would be 'grand' reflected her ability to stay at home with the children rather than a sense of pleasure at Michael being away 3 weeks out of 4. In her eyes, the reduction of the time the children would need to spend outside the home at childminding facilities was the primary benefit of the move. For Julie, having to work outside the home was an economic necessity in Ireland, rather than a choice.

9. The Department of Human Services pays a Child Care Rebate of 50 per cent of out-of-pocket childcare expenses for approved childcare, up to an annual limit of $7500 per child, for eligible carers (usually permanent residents and citizens) who meet the Work, Training, Study test for the rebate.
10. $1500 to have her transcripts assessed by the Australian Psychological Society, $5400 for supervision and around $3000 on registration.
11. Having fulfilled the requirements to be registered as a migration agent with the Office of the Migration Agents Registration Authority.
12. Lobbying group, Interview 45.
13. Some examples: Flinders University: http://www.flinders.edu.au/ppmanual/health-safety/ohs-training.cfm; University of Sydney: http://sydney.edu.au/whs/activities/training.shtml; Coles Group contractors: http://contractor.colesgroup.com.au/content150.asp; DFAT: http://dfat.gov.au/about-us/publications/corporate/annual-reports/annual-report-2014-2015/dfat-annual-report-2014-15.pdf.

Acknowledgements

The author gratefully acknowledges the kind and generous supervision provided by Professor Graeme Hugo AO under which the research project that provided the data for this article commenced.

The author wishes to expressly thank the anonymous reviewers of this work for their thorough and knowledgeable feedback.

Disclosure statement

No potential conflict of interest was reported by the author.

References

ACTU. 2014. Review of skilled migration and 400 series visa programs.

ACTU. 2015. *ACTU Submission Productivity Commission Draft Report into the Migrant Intake in Australia.*

Akbari, A. H., and M. Macdonald. 2014. "Immigration Policy in Australia, Canada, New Zealand, and the United States: An Overview of Recent Trends." *International Migration Review* 48 (3): 801–822. doi:10.1111/imre.12128

Barkan, E. 1992. *The Retreat of Scientific Racism: Changing Concepts of Race in Britain and the United States Between the World Wars.* Cambridge: Cambridge University Press.

Berg, L. 2015. *Migrant Rights at Work: Law's Precariousness at the Intersection of Immigration and Labour.* Abingdon: Routledge.

Betzien, J. 2008. *Bonded labour? Pacific 'guest worker' scheme.* Green Left Weekly 764. http://www.greenleft.org.au/node/40144.

Birrell, R., and K. Betts. 1988. "The FitzGerald Report on Immigration Policy: Origins and Implications." *The Australian Quarterly* 60: 261–274.

Boese, M., I. Campbell, W. Roberts, and J.-C. Tham. 2013. Temporary Migrant Nurses in Australia: Sites and Sources of Precariousness. *The Economic and Labour Relations Review* 24 (3): 316–339.

Boucher, A. 2013. "Bureaucratic Control and Policy Change: A Comparative Venue Shopping Approach to Skilled Immigration Policies in Australia and Canada." *Journal of Comparative Policy Analysis: Research and Practice* 15 (4): 349–367. doi:10.1080/13876988.2012.749099

Brace-Govan, J. 2004. "Issues in Snowball Sampling: The Lawyer, the Model and Ethics." *Qualitative Research Journal* 4 (1): 52–60.

CAAIP, C. t. A. o. A. s. I. P. 1988. *Immigration, A Commitment to Australia*. Canberra: Australian Government Publishing Service.

Cameron, J., and J. K. Gibson-Graham. 2003. Feminising the Economy: Metaphors, Strategies, Politics. *Gender, Place and Culture: A Journal of Feminist Geography* 10 (2): 145–157.

Campbell, I., and J.-C. Tham. 2013. "Labour Market Deregulation and Temporary Migrant Labour Schemes: An Analysis of the 457 Visa Program." *Australian Journal of Labour Law* 25 (3): 239–272.

Castles, S. a. 2014. *The age of Migration: International Population Movements in the Modern World*. 5th ed., edited by Stephen Castles, Hein de Haas and Mark J. Miller. New York: Guilford Press.

Castree, N. 2007. "Labour Geography: A Work in Progress." *International Journal of Urban and Regional Research* 31 (4): 853–862. doi:10.1111/j.1468-2427.2007.00761.x

Connell, J. 2010. "From Blackbirds to Guestworkers in the South Pacific. Plus ça Change …?" *The Economic and Labour Relations Review* 20 (2): 111–121. doi:10.1177/103530461002000208

Costello, C. e., and M. R. e. Freedland. 2014. *Migrants at Work: Immigration and Vulnerability in Labour law*, 1st ed., edited by Cathryn Costello and Mark Freedland. Oxford: Oxford University Press.

CSO. 2015. *Population and Migration Estimates 2015*. http://www.cso.ie/en/releasesandpublicati ons/er/pme/populationandmigrationestimatesapril2015/.

Cully, M. 2011. "Skilled Migration Selection Policies: Recent Australian Reforms." *Migration Policy Practice* 1 (1): 4–7.

Curtain, R. 2002. Online Delivery in the Vocational Education and Training Sector: Improving Cost Effectiveness [and] Case Studies: ERIC.

Davis, M., M. McAllister, and I. Manning. 1980. *Diary of Social Legislation and Policy 1980*.

DHS. 2015. Availability of Welfare Payments. https://www.humanservices.gov.au/customer/ enablers/newly-arrived-residents-waiting-period.

DIBP. 2014a. *Discussion Paper: Reviewing the Skilled Migration and 400 Series Visa Programmes*: C. o. Australia. Retrieved from https://www.border.gov.au/ReportsandPublications/Documents /discussion-papers/skilled-migration-400-series.pdf.

DIBP. 2014b. *Setting the Migration Programme for 2015–16 Discussion Paper* Canberra: Commonwealth of Australia. http://gofastvisa.com/wp-content/uploads/2015/09/Migration-Programme-2015-16-Discussion-Paper.pdf.

DIBP. 2015. *Migration Programme Statistics*. https://www.border.gov.au/about/reports-publications/research-statistics/statistics/live-in-australia/migration-programme.

DIBP. 2015a. *Annual Report 2014-2015*. ACT: Commonwealth of Australia.

DIBP. 2015b. *Subclass 457 Quarterly Report Quarter Ending at 30 September 2015*. Canberra: C. o. Australia.

DIBP. 2016a. *Temporary Entrants and New Zealand citizens in Australia*. Canberra. http://www. border.gov.au/ReportsandPublications/Documents/statistics/temp-entrants-newzealand-dec31. pdf.

DIBP. 2016b. *Temporary Work (Skilled) (subclass 457) visa*. Canberra: Commonwealth of Australia. http://www.border.gov.au/Forms/Documents/1154.pdf.

DIBP Economic Analysis Unit. 2013. *Country Profile—Ireland*. Belconnen, ACT: C. o. Australia.

Foroudi, F., D. Pham, M. Bressel, D. Tongs, A. Rolfo, C. Styles, … T. Kron. 2013. "The Utility of E-Learning to Support Training for a Multicentre Bladder Online Adaptive Radiotherapy Trial (TROG 10.01-BOLART)." *Radiotherapy and Oncology* 109 (1): 165–169.

Hammond, S. 2014. APS Submission to the Australian Government Department of Immigration and Border Protection On Discussion Paper (September 2014): Reviewing the Skilled Migration and 400 Series Visa Programmes Melbourne.

Hawkins, F. 1991. *Critical Years in Immigration: Canada and Australia Compared / Freda Hawkins*. 2nd ed. Montreal: McGill-Queen's University Press.

Howe, J. 2013. "Is the Net Cast Too Wide-An Assessment of Whether the Regulatory Design of the 457 Visa Meets Australia's Skill Needs." *Fed. L. Rev.* 41: 443–469.

Howe, J., and A. Reilly. 2014. *Submission to the Review of Skilled Migration and 400 Series Visa Programmes.* http://www.border.gov.au/ReportsandPublications/Documents/submissions/joanna -howe-alexander-reilly.pdf.

Hugo, G. 1988. Outputs and Effects of Immigration in Australia.

Hugo, G. 2002. "Effects of International Migration on the Family in Indonesia." *Asian and Pacific Migration Journal* 11 (1): 13–46.

Hugo, G. 2005. Migration Policies in Australia and their Impact on Development in Countries of Origin. *International Migration and the Millennium Development Goals*, 199.

Hugo, G. 2006. "Globalization and changes in Australian international migration." *Journal of Population Research* 23 (2): 107–134. doi:10.1007/BF03031812.

Hugo, G. 2014a. "Change and Continuity in Australian International Migration Policy." *International Migration Review* 48 (3): 868–890. doi:10.1111/imre.12120

Hugo, G. 2014b. "Skilled Migration in Australia: Policy and Practice." *Asian and Pacific Migration Journal* 23 (4): 375–396. doi:10.1177/011719681402300404

Janet, P., and J. Simon-Davies. 2016. *Migration to Australia: A Quick Guide to the Statistics.* Canberra: Australian Capital Territory, Parliament of Australia.

Jupp, J. 2007. *From White Australia to Woomera: The Story of Australian Immigration.* Cambridge: Cambridge University Press.

Law Society of Australia. 2014. *Skilled Migration and 400 Series Visa Programmes Discussion Paper.* https://www.lawcouncil.asn.au/lawcouncil/images/LCA-PDF/docs-2800-2899/2896_-_Skilled_ Migration_and_400_Series_Visa_Programmes_Discussion_Paper.pdf.

Mackellar, M. 1979. "The 1978 Immigration Decisions—A Reply." *The Australian Quarterly* 51 (2): 93–103. doi:10.2307/20635010

Mather, C., A. Marlow, and E. Cummings. 2013. "Digital Communication to Support Clinical Supervision: Considering the Human Factors." *Stud Health Technol Inform* 194: 160–165.

McKavanagh, C., C. Kanes, F. Beven, A. Cunningham, and S. Choy. 2002. *Evaluation of Web-Based Flexible Learning.* Leabrook: ERIC.

Meaney, N. 1995. "The end of ' White Australia' and Australia's Changing Perceptions of Asia, 1945– 1990." *Australian Journal of International Affairs* 49 (2): 171–189. doi:10.1080/ 10357719508445155

Miller, P. W. 1999. "Immigration Policy and Immigrant Quality: The Australian Points System." *The American Economic Review* 89 (2): 192–197.

Mitchell, D. 2013. "Labour's Geography and Geography's Labour: California as an (ANTI-) Revolutionary Landscape." *Geografiska Annaler: Series B, Human Geography* 95 (3): 219–233. doi:10.1111/geob.12022

Oakman, D. 2002. "'Young Asians in our Homes': Colombo Plan Students and White Australia." *Journal of Australian Studies* 26 (72): 89–98. doi:10.1080/14443050209387741

Parliament of Australia. 2013. *The Subclass 457 Visa: A Quick Guide.* http://parlinfo.aph.gov.au/ parlInfo/download/library/prspub/2840657/upload_binary/2840657.pdf;fileType=application/pdf.

Salmon, G. 2004. *E-moderating: The key to Teaching and Learning Online.* London: Psychology Press.

Shachar, A., and R. Hirschl. 2013. "Recruiting "Super Talent": The new World of Selective Migration Regimes." *Indiana Journal of Global Legal Studies* 20 (1): 71–107.

Tavan, G. 2005. "Long, Slow Death of White Australia [online]." *The Sydney Papers* 17 (3/4): 135–139.

UN-DESA, and OECD. 2013. *World Migration in Figures.* http://www.oecd.org/els/mig/World-Migration-in-Figures.pdf.

Velayutham, S. 2013. "Precarious Experiences of Indians in Australia on 457 Temporary Work Visas." *The Economic and Labour Relations Review* 24 (3): 340–361. doi:10.1177/ 1035304613495268

Webster, B., E. Walker, and R. Barrett. 2005. "Small Business and Online Training in Australia: Who is Willing to Participate?" *New Technology, Work and Employment* 20 (3): 248–258.

Visualising 30 Years of Population Density Change in Australia's Major Capital Cities

Neil T. Coffee, Jarrod Lange and Emma Baker

ABSTRACT

This study uses a novel spatial approach to compare population density change across cities and over time. It examines spatio-temporal change in Australia's five most populated capital cities from 1981 to 2011, and documents the established and emerging patterns of population distribution. The settlement patterns of Australian cities have changed substantially in the last 30 years. From the doughnut cities of the 1980s, programs of consolidation, renewal and densification have changed and concentrated population in our cities. Australian cities in the 1980s were characterised by sparsely populated, low density centres with growth concentrated to the suburban fringes. 'Smart Growth' and the 'New Urbanism' movements in the 1990s advocated higher dwelling density living and the inner cities re-emerged, inner areas were redeveloped, and the population distribution shifted towards increased inner city population densities. Policies aimed at re-populating the inner city dominated and the resultant changes are now visible in Australia's five most populated capital cities. While this pattern has been reported in a number of studies, questions remain regarding the extent of these changes and how to analyse and visualise them across urban space. This paper reports on a spatial method which addresses the limitations of changing statistical boundaries to identify the changing patterns in Australian cities over time and space.

Introduction

The morphology of Australian cities has undergone several key changes over the past 30 years. At the beginning of the 1980s, capital city development was outward focused with new land sub-divisions on the suburban fringe dominant and inner city areas in decline. This was termed the 'doughnut city' (Croxford 1977; Department of Infrastructure 1998). During the 1990s, Commonwealth policy targeted inner city decline in all of Australia's capital and regional cities and promoted urban consolidation and dwelling densification through programs such as 'Greenstreet' and 'Building Better Cities' (Hall 2010). Concerned individuals formed groups such as 'Smart Growth' (http://www.smartgrowth.org/)

or 'New Urbanism' (http://www.newurbanism.org/), to promote city developments that increased dwelling densities and development around major transport hubs (transit-oriented developments), and called for suburbanisation to be halted. Inner cities' dwelling densities have increased, as observed in the five most populated Australian capital cities (Adelaide, Brisbane, Melbourne, Perth and Sydney). Whether this is attributable to these programs is not clear. This paper reports on these changes, comparing spatial patterns over time and between these five capital cities, with particular attention to changing spatio-temporal patterns of population density.

An earlier study (Baker, Coffee, and Hugo 2000) examined the spatial patterns of population density change in Australian cities in the 15 years from 1981 to 1996 and provided a simple, yet novel method for comparing change across time and space. Utilising the Baker, Coffee, and Hugo (2000) methodology, this paper revisits the spatial pattern of population density change in Australia's five most populated capital cities for the 1981–2011 period. The earlier paper supported the hypothesis that a process of suburbanisation and re-urbanisation was occurring in the 1981–96 period (Baker, Coffee, and Hugo 2000). This paper extends the examination to a 30-year period to investigate whether the re-urbanisation process continued after 1996, and whether urban consolidation policies resulted in the predicted increased population densities beyond inner and middle city areas, and towards the outer suburban fringe.

Analysing Australian Bureau of Statistics (ABS) Census of Population and Housing data, this paper utilises Geographical Information Systems (GIS) to monitor spatial patterns of population growth and population density change over the 1981–2011 period. The method addresses the data limitations of changing statistical geography to provide a means for comparing change between major urban areas over time. This paper does not focus on the policy environment as such; rather, the focus is on the application of a spatial method to allow the identification of changing patterns in Australian cities over time and space.

Monitoring urban change

Compared to similar post-industrial nations, Australia is a sparsely populated wide brown land (Mackellar 1990), where most of the population (85 per cent) lives in urban areas (ABS 2014). Australian cities have been shaped by history, infrastructure, natural landscapes, and, importantly, policy. Across all of the major mainland cities, the development of accessible and cheap private transport after the Second World War enabled the population to spread into relatively cheap land on the expanding urban fringes. As Paris (1994) has noted, the dispersed urban form of our cities was further reinforced by low levels of development of public transport and high levels of home ownership. By the beginning of the study period (1981), Australia's five most populated capital cities were characterised as 'doughnut cities', with inner area population declines encircled by outer suburban growth (Forster 2006). In the late 1980s, potential environmental considerations, housing affordability (Yates 2001), and social consequences of sprawling cities led the Commonwealth government to focus a new policy agenda on urban consolidation (McLoughlin 1991) and dwelling densification (Bunker et al. 2002). Despite opposition (Stretton 1991; Troy 1992, 1995), interventions (e.g. Building Better Cities) and city planning strategies (such as Sydney's 1988 Metropolitan Strategy) attempted to concentrate people within existing metropolitan areas and reduce allotment sizes, especially at the fringes of the cities (Forster 1999). Sub-division, dual occupancy, infill development, smaller block sizes and the repurposing of non-residential buildings were used to encourage more compact and less sprawled metropolitan areas.

Urban consolidation policy in Australia focused on re-populating the inner and middle areas through the late 1980s to early 1990s, but subsequently the focus evolved towards increased dwelling density across whole metropolitan areas (Thompson and Maginn 2012). Increasingly, State and Territory strategic plans promoted urban consolidation, and a concentration on new housing development within defined zones, with transport accessibility or urban renewal potential. The current South Australian 30 Year Plan for Greater Adelaide targets growth in 'current urban lands' (Government of South Australia 2010, 3), especially along major transport corridors and hubs. Similarly, the Plan Melbourne, Metropolitan Planning Strategy (2014) aims to contain new housing development within existing urban boundaries and focuses new development on urban renewal precincts (The State of Victoria 2014).

Our earlier study (Baker, Coffee, and Hugo 2000) showed convincing patterns of density growth in the middle and inner areas of Australia's five most populated capital cities, concluding that there was strong evidence of re-urbanisation and continuing suburbanisation having occurred during the 1981–96 period. In looking forward, the study predicted that while the urban peripheries would continue to grow, a sustained increase in growth in the inner and middle zones of these major urban areas might occur from 2001 onwards. In recent years, there has been a renewal of research interest in understanding the spatial patterning of Australian cities. Though much of this research focused on examining patterns in just one or two urban areas (Searle and Filion 2010; Chhetri et al. 2013; Randolph and Tice 2014), rather than generalised patterns across urban areas, these studies support the general patterns predicted in our 2000 study (Baker, Coffee, and Hugo 2000). In a subsequent study, Hugo (2002) suggested that both suburbanisation and re-urbanisation were occurring in Australian cities in the 1990s. Importantly, this work highlighted that it was not population size per se, but population growth and the changing composition of the population that were implicated as drivers for shifts in population distribution. That is, the changes driving both suburban and inner urban growth stemmed not just from population growth but also from the changing composition of the population, including migration, ethnic mix, ageing and the lifestyle choices of baby boomers, Generation X and Generation Y (Hugo 2002). These changes, as predicted in the Hugo (2002) paper, have indeed continued to shape major Australian cities.

Although in recent years the generalised focus of policy has been towards consolidation and the compact city, research evidence suggests that patterns of densification do not directly conform to policy intention. Chhetri et al. (2013), in their analysis of evolving urban residential densities in Melbourne, provide a good example. Examining changes to dwelling counts between 2001 and 2006, they found a notable 'hollowing' of densities in the inner city areas, and a corresponding 'rapid' densification along the outer city areas. Searle and Filion's (2010) comparison of urban intensification in Sydney and Toronto indicates a quite complex emerging spatial patterning of residential density in Sydney, where transport linkages and infill development may lead to higher density hot spots. Similarly, Grosvenor and O'Neill (2014) highlight the complexity of analysing urban density in their examination focused on Sydney.

As a background to current and future policy planning and evaluation, this paper contributes a means to monitor patterns of urban density change—between and within cities—over time. Building on our previous work, the paper details and applies a spatial method for monitoring and comparing urban density change.

Measuring urban change

Australia has one of the best housing and population censuses in the world, in terms of frequency and content. The 5-yearly census of the Australian population provides a rich source of data for research on population trends, and how people and housing are distributed across space. To assist with these analyses, the ABS provides spatial geography through the Australian Standard Geographic Classification (ASGC 1981–2006) and the Australian Statistical Geography Standard (ASGS 2011 onwards). Any analysis of the Australian census is based upon one of the spatial units defined in the ASGC or ASGS. However, the changing size, structure and distribution of the Australian population have resulted in ongoing changes to the spatial geography standards that confound temporal and spatial research. As suggested by Hugo (2002, 15):

> It is necessary to be able to fit population data to meaningful socially, environmentally and economically defined areas in order for them to be optimally useful in planning, and relevant to social, economic and environmental research.

The smallest spatial unit for most of the study period was the Collection District (CD; 1981–2006) and for the most recent census the Statistical Area 1 (SA1), which replaced the CD. A CD was traditionally designed to represent an area that one census collector could cover, delivering and collecting census forms over a 10-day period (approximately 200–300 households) (ABS 2006a). SA1s were designed specifically for the release of census data and comprised 200–800 persons (ABS 2011). As the time period for this research is based primarily upon the CD, this section is focused on issues associated with the CD.

Spatial boundaries for a census are considered current only at the census time and, as a consequence, CD boundaries have been subject to change at each census. Boundaries are fixed for the release of census data, but as the population changes from one census to the next so do many of the spatial boundaries (such as CDs, Statistical Local Areas (SLAs), census-derived postcode areas or local government areas), especially in regions experiencing rapid growth or decline. The lack of a spatially and temporally consistent framework for analysis is an important issue for researchers and policymakers concerned with how housing and population characteristics vary over time and across space. In Australia, as with other similar countries, variation in the definition of areal units limits the development of consistent temporal and spatial data (Hugo et al. 1997; Blake, Bell, and Rees 2000; Walford and Hayles 2012). Specifically, the changes to Australian statistical geography that occur every 5 years confound understanding of how variations are occurring at the more detailed spatial scale.

This is a problem that has a few dimensions. First, many of the administrative spatial units of Australian cities were never designed to be units of analysis but were based upon existing historic boundaries (such as suburbs, postcodes and local government areas). Over the years, these have been changed for political, social (local government areas and suburbs) or workload (postcodes, census CDs) reasons. Second, beyond these changes, spatial units have been modified to manage the changing demography over time. As areas developed, CDs were split or merged (in response to population growth or decline) to maintain workload numbers for census collectors. They were also altered to concord with larger spatial unit changes, especially changes to local government area boundaries. After each census, the ABS provided a comparability code for CDs to highlight changes and alert users to CDs that could not be compared from one census to the next due to the changed boundaries. The percentage of CDs that can no longer be compared differs from census to census, hampering temporally consistent comparisons, especially over extended time periods. Areas undergoing substantial population change are usually the focus of comparisons, to manage the provision of services and infrastructure and to provide sound planning outcomes. Problematically, these are the areas that are least able to be compared due to changing spatial boundaries.

For larger spatial units, changing boundaries also impact on comparisons over time (suburb, postcode area, SLA, local government area and statistical division (SD)) both in terms of comparability over time and diminished capacity to identify where change is occurring. Very few papers report on changes over time for smaller geographical areas as changes to spatial boundaries compromise reliability of time series data (Blake, Bell, and Rees 2000; Adamo 2011). Four approaches to managing temporal consistency were reported by Blake, Bell, and Rees (2000). These were to: (1) freeze history; (2) use the latest spatial units; (3) construct designer units; and (4) geocode data (create a geographic reference for data, usually based upon street address). Of these, geocoding provides a solution, but due to cost and issues of confidentiality has not yet formed the basis of census data collections. The other options do not provide a simple means of comparison over time and space and are often complex to construct and data hungry. In light of these challenges, the methodology presented in this paper offers a response to the problem of how to represent change in Australian cities over time and space. The method proposed in this paper is novel, simple and flexible. As presented, this methodology can be used to compare population density change across cities and over time. The featured methodology can also be modified to: (1) compare one or more suburban areas within a capital city region; (2) be segmented along transport corridors within a capital city area or large metropolitan region (e.g. comparing density change between the northern and southern urban areas of Sydney, Australia); or (3) be adjusted in size to accommodate metropolitan areas that are more or less densely populated.

Urban policy in Australia is increasingly concerned with infrastructure costs, environmental issues related to fringe development/urban sprawl, healthy cities, and the socio-spatial polarisation of our population. There is a clear need for a simple and accessible means for policymakers and planners to understand current population patterns and to extrapolate these as future policy. This research provides elements of both of these requirements by sectioning our cities into smaller bands that allow analysis over time and within cities, and also across cities. Australian census data are a tremendous resource that is currently under-utilised for urban and housing policy development. This paper offers a methodology for better utilising this information, and for making the results accessible to a wider audience base to inform policy development and future decision-making processes.

Methodology

This study analysed changes in population density in Australia's five most populated capital cities (Adelaide, Brisbane, Melbourne, Perth and Sydney) from 1981 to 2011. This was achieved by deriving the geometric centre (centroid) of each CD/SA1 for each census from 1981 to 2011. Census population data were then allocated to all centroids. The Central Business District (CBD) for each capital city was represented as the geometric centre of the ABS SLA that contained the CBD. ESRI ArcGIS 10.3.1 software was utilised to construct 5 km concentric rings around each city's CBD, with the centroid of the capital cities' SLAs used as the start point for concentric ring construction. The CD/SA1 centroids were spatially joined with the concentric ring and the population summed for each concentric ring. The concentric ring area and associated population counts were determined by assigning each CD/SA1 to a concentric ring based on each spatial unit's geometric centre (centroid). Once all CD/SA1 areas within the study region were assigned to a concentric ring, population density within each concentric ring was calculated by dividing the sum of assigned population by the sum of assigned area (sq. km) (Figure 1).

The boundary of analysis for each of these five capital cities was defined as the ABS Capital City Statistical Division (ABS 2006b). As in any study of this nature, the choice of boundary can impact on the results and the use of the 2006 SD boundary does impact the outcomes. It would be remiss in a spatially based analysis not to make reference to the modifiable areal unit problem (MAUP) (Openshaw 1984) and the potential for this to impact on the patterns described in this paper. Changing the ring bandwidths could provide different population density patterns, as could reducing or enlarging the overall city boundary. The 2006 SD boundary was considered appropriate for all of the cities in this analysis, as the SDs were predominantly urban in character and designed to represent the current and anticipated urban development of a capital city (ABS 2006b) compared with the ABS GCCSA boundaries which include large peri-urban populations unsuitable for this analysis.

The choice of 5 km concentric ring distances was tested by analysing the 'goodness of fit' of CD/SA1 concentric ring membership consistency over time. Each CD/SA1 for the 1981, 1996 and 2011 periods was allocated to a ring and to test 'goodness of fit', the CD/

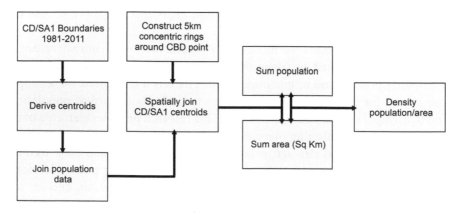

Figure 1. Methodology flow diagram.

SA1 boundaries were overlaid on the rings, and the area within and outside of the allocated ring calculated. Overall, concentric rings up to 30 km from the CBD point for all five capital cities had, on average, greater than 90 per cent of all CD/SA1s within their allocated concentric ring. For the more densely populated cities of Melbourne and Sydney, an average of greater than 90 per cent was achieved up to 45 km and 55 km respectively from the CBD.

Results

Population density outcomes in Australia's five most populated capital cities across the study period are shown in Figure 2. Notably, population density increased in every period across every city and for the distances between 0 and 35 km. Figure 2 provides evidence that these five mainland cities have increased population density and that the increases were particularly noticeable within 0–10 km of the CBD, deteriorating as distance from the CBD becomes greater. It is of particular interest to examine the extent to which the population density within the 10 km ring in each city has increased, especially for Sydney, Melbourne and Brisbane. Is this evidence that consolidation policies have succeeded or does it reflect population growth pressure? It is difficult to extricate cause and effect, but the policy changes in the early 1990s were a reaction to expanding suburban areas and inner city declines. There is little doubt that this paper highlights that the population density declines of 25 years ago have been halted and inner cities are now the preferred residential choice of a larger percentage of the population.

To better understand the changes for each city, Figures 3–7 demonstrate substantial changes to population density in Australia's five most populated cities between 1981 and 2011. These five capital cities are different in terms of their land area, population size, and rates of population growth, but this comparison highlights similar patterns of population density change across all five capital cities. What is evident is that all five capital cities increased population densities across this period, especially Sydney, Melbourne and Brisbane, and this was particularly pronounced in the 10 km band.

As was evident from Figure 2, Sydney experienced the most population density growth in the 0–5 km ring over this period, increasing from more than 4000 persons to

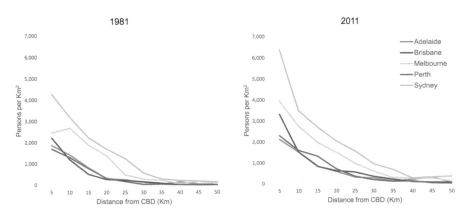

Figure 2. Population density by distance from CBD for 1981 and 2011. *Data source*: ABS (1981–2011).

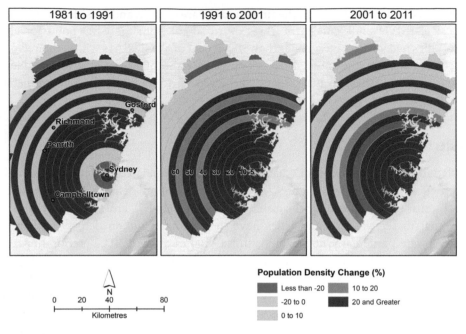

Figure 3. Population density change by 5 km concentric ring distance from CBD—Sydney. *Data sources*: ABS (1981–2011); ESRI et al. (2014); Geoscience Australia (2006).

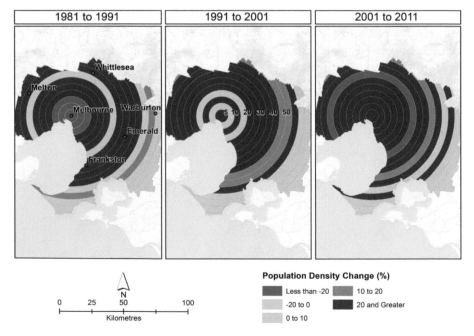

Figure 4. Population density change by 5 km concentric ring distance from CBD—Melbourne. *Data sources*: ABS (1981–2011); ESRI et al. (2014); Geoscience Australia (2006).

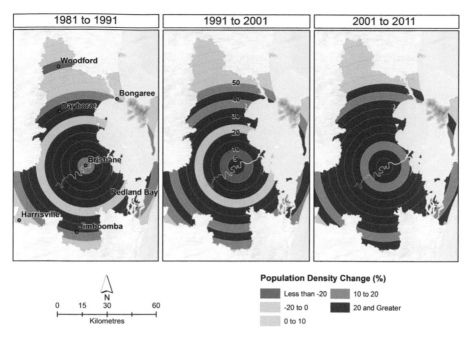

Figure 5. Population density change by 5 km concentric ring distance from CBD—Brisbane. *Data sources*: ABS (1981–2011); ESRI et al. (2014); Geoscience Australia (2006).

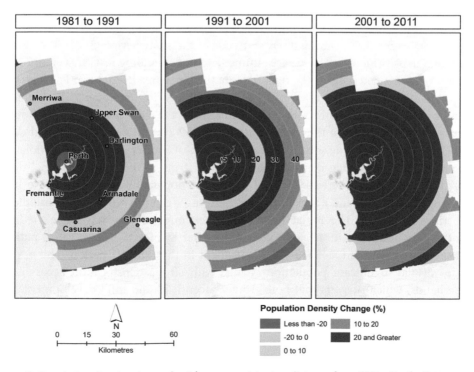

Figure 6. Population density change by 5 km concentric ring distance from CBD—Perth. *Data sources*: ABS (1981–2011); ESRI et al. (2014); Geoscience Australia (2006).

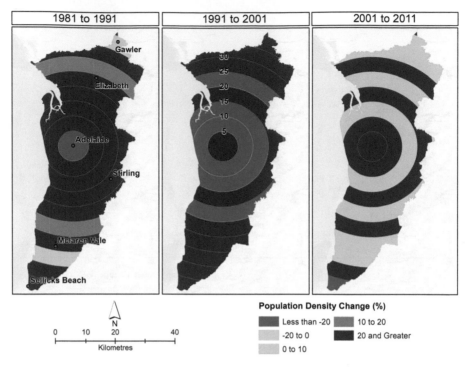

Figure 7. Population density change by 5 km concentric ring distance from CBD—Adelaide. *Data sources*: ABS (1981–2011); ESRI et al. (2014); Geoscience Australia (2006).

approximately 6500/km². During the first decade of the study period (1981–91), Sydney's population density increased in the area immediately surrounding its CBD (Figure 3), but decreased or remained stable in the remainder of the inner area (5–20 km). The increase in population density between 1981 and 1991 was highest in what was at that time the expanding urban fringe of Sydney (20–50 km from the CBD), and less pronounced towards the outer extent of the metropolitan area (50 km+). During the second decade (1991–2001), population density increased across all of Sydney's inner and middle rings (0–65 km), with decreasing densities on the urban fringes (65 km+). The final decade (2001–11) can be characterised by solid continued population density increase in the inner rings (0–35 km), and a less consistent pattern of changes through the middle and outer rings (35 km+). The initial premise of the doughnut city has some credibility from this visualisation, with negative change in inner Sydney (5–10 km band) and sub-stantial suburbanisation (20–50 km bands) evident in the 1981–91 period. The next two decades display significant population change with strong densification across almost all the bands, which suggests that both fringe (greenfield) and infill developments were contributing to these changes.

Melbourne (Figure 4) exemplifies what was termed the 'doughnut city' in the first decade (1981–91), with noticeable population density decline throughout the inner city areas (0–15 km), substantial and largely uniform density increases through the middle rings (15–30 km), and generally small density increases in the outer rings (30 km+). This pattern was starting to reverse during the second decade (1991–2001) in the inner

5 km ring, but fringe growth was still the clear pattern at that time, with increased population density in the 20–50 km rings. The final decade of the study (2001–11) shows continued population density gains extending beyond the inner 0–30 km rings. Melbourne was clearly a doughnut city over the 1981–91 period, more so than any other Australian capital city. As was evident for Sydney, the changes over the next two decades suggest that strong population growth in greenfields and infill have both contributed to the higher densities in almost all of the bands.

The population density distance patterns for Brisbane (Figure 5) are interesting. They show an almost continuous population density increase throughout the inner and middle rings (0–25 km) and smaller density changes in the outer rings (25 km +) between 1981 and 1991. Between 1991 and 2001 the rate of population density increase appears to have slowed, and there was a net decrease in parts of the inner rings (5–10 km and 20–25 km). Between 2001 and 2011 all of Brisbane's inner, middle and the edge of the outer suburban area rings (0–50 km) increased their population densities, and there was a levelling or even reduction of density in the outermost rings. What this method provides is a basis for analysing population density change over time and space to highlight both the similarities and differences between Australia's five major capital cities. Brisbane has a different pattern of change to both Sydney and Melbourne, both of which displayed doughnut city patterns in the early decades. Brisbane has been increasing density in most distance bands across all of the decades considered, and this would indicate different drivers perhaps or even that differing topography played some part in how the city developed. This visualisation provides a starting point for understanding why Brisbane exhibited these patterns of change over this period.

The patterns of density change for Perth (Figure 6) are perhaps the clearest example of the process of population density consolidation. In the first decade (1981–91), Perth's inner city population density declined (0–5 km), there was solid density growth throughout the middle rings (5–30 km), and small density increases through to the outer metropolitan rings (30 km+). By the second decade of the study, population density increased throughout the inner and towards the middle rings of Perth (0–20 km). By the 2001–11 period, all of Perth's inner and middle rings (to a distance of 40 km from the CBD) greatly increased in population density.

Of the five mainland capitals, Adelaide (Figure 7) has the smallest population, economy and lower levels of absolute population growth. Though conforming to the generalised patterns of the other five capitals, the pattern seen in Adelaide during the study period has some variations, and reveals notable declines in density, especially in the 1991–2001 period. Nevertheless, between 1981 and 1991, inner city (0–5 km) decline is evident alongside population density growth throughout the remainder of the city. During the second decade (1991–2001), while the inner and outer rings (0–5 km and 25 km+) experienced substantial increases in population density, these changes were accompanied by density decreases across all of the middle rings (5–15 km) and the 20–25 km ring. The third decade (2001–11) shows a different pattern of population density change, with increasing densities in the inner rings (0–10 km), and a relative decrease at the edges of the metropolitan area (30–40 km).

Discussion and conclusion

This paper offers a novel spatial approach to analyse the population densification pattern in Australia's five most populated capital cities across three decades, and strongly supports the hypothesis of suburbanisation (1981–91), re-urbanisation (1991–2001) and city-wide population densification (2001–11). Overall, it appears that densification and urban consolidation policies have resulted in increased population densities beyond the inner and middle rings of these cities, and towards the outer suburban fringe. Despite the similarities over time, there are some key differences across the five cities that are likely to reflect jurisdictional planning strategies and policies, State political and economic conditions, variations in land availability and geography, and different starting points on the urbanisation trajectory. Perth, for example, experienced a period of unprecedented mining-related economic growth and a housing boom through the latter years of the study, and this is reflected in the solid pattern of population density increase to 2011. The pattern seen for Adelaide, on the other hand, shows staggered growth in population density over the period, where the uneven banding pattern probably reflects smaller areas of regeneration-led growth.

The development of sound urban policy requires quality data analysis and visualisation for as wide an audience as possible. The concentric ring methodology outlined in this paper offers an effective geographic visualisation means to understand and monitor the outcomes of previous policy decisions including the capacity to inform future strategic planning and development of our capital cities. Importantly, the methodology is immune to spatial data definition changes, allowing cities to be compared using high-quality national data over time. The approach is also able to be applied to other datasets and variables of interest. While in this paper we have focused on cities, we note that the methodology could be reliably applied to smaller jurisdictions such as local government areas, and for each inter-censal period.

One perceived limitation of the methodology may be its capacity to analyse population density change *within* capital city areas/regions. The current methodology has a monocentric underpinning (where the city centre is the focus of the region), which may be viewed as irrelevant to the analysis from a polycentric perspective (where two or more major cities/suburbs exist within a capital city area/region) (Angel and Blei 2016; Vasanen 2012). However, this methodology has the potential to be applied to address both a monocentric and polycentric urban setting by varying the concentric ring distances and using more than one central point within a capital city area/region. In addition, the rings can be applied to just urban areas of the capital city, divided into segments, to reflect direction from the city (north, south, etc.), and the rings can be varied in width. Further, as highlighted by Mees (2009), there is a need for spatial measures of urban density that are independent of administrative boundaries. Provided the choice of concentric ring size in conjunction with the spatial unit being used is validated to ensure: (1) each spatial unit assigned has consistent ring membership over time; and (2) each spatial unit area is located predominantly within one concentric ring, comparisons within and across capital cities remain possible, using population counts and/or other census variables of interest (e.g. household income, socio-economic variables). Although the methodology has focused on 5 km concentric ring distances to measure population density change over time, adaptation of the methodology to suit different census variables and census spatial units in other countries remains an area for future research.

Overall, the findings of this analysis stand alongside a growing body of work (Searle and Filion 2010; Randolph and Freestone 2012; Grosvenor and O'Neill 2014; Randolph and Tice 2014) indicating that Australian cities are increasingly complex in their form. Our analysis shows that population density change patterns in Australia's largest capital cities have perhaps become less predictable over the last 15 years. The model of inner and middle city decline and growth in the outer rings of urban development has been replaced by a more complex pattern in which the positive correlation between population density growth rate and distance from the city centre has been eroded. Undoubtedly this is partly due to demographic factors. It has been shown that in the early post-war years of rapid population growth and lateral expansion of Australian cities, whole suburbs tended to be initially settled by people in the young family formation ages and their children (Hugo 1986). As these groups have aged in place and their children have left, the population has declined. As these original settlers of suburbia, migrate, die or move into aged care accommodation, their houses will come onto the market and be purchased by younger people. Thus one occupant of a home is replaced by two or three occupants, causing a new cycle of population growth. Hugo (2002) implicated not just absolute population change but the components of population change as driving inner, middle and outer population density changes. However, it is clear that other elements are also at work. Local and State governments are encouraging urban consolidation and some cohorts of the population are exhibiting a preference for inner city 'café society' lifestyles. This has undoubtedly caused increased population density growth in inner and middle rings. Growth on the periphery has certainly not ceased, especially in the fastest growing cities, with the 2001–11 period experiencing continued density increases in the inner area and a less consistent pattern of increase through the middle and outer rings. To this end, it is important that the forces encouraging growth in the different ecological contexts of the major cities be identified if accurate anticipation of future changes in population growth is going to be achieved as a basis for better metropolitan planning.

Acknowledgements

When Professor Graeme Hugo led the successful bid to establish a National Key Centre for Social Applications of Geographical Information Systems (GIS) in 1995, one of the aims was to 'make public and private sector planning in Australia "smarter"'. The establishment of the Key Centre was a seminal moment in the use of GIS for social science in Australia and was in many respects an idea before its time. When we fast forward 20 years, GIS and social science are now interlinked and this is in no small part due to the Key Centre and Graeme's vision. The authors wish to acknowledge the valued contributions of Professor Graeme Hugo. This paper has, as its genesis, a paper prepared for the 2000 Australian Environment report by Emma Baker, Neil Coffee and Graeme Hugo. Since the original paper was written there have been three subsequent censuses, and Graeme Hugo, along with the authors, intended to update the analysis. Unfortunately, Graeme did not get to work on this final paper, but his words are still incorporated from other works and internal documents we wrote along the way. It would be fair to say that this paper was a long time in the writing and it is a little sad that Graeme did not get to see the final analysis, but this paper was a marriage of two areas that Graeme was a major supporter of over the years— population and GIS. Thanks Graeme for your support of this idea all those years ago.

Disclosure statement

No potential conflict of interest was reported by the authors.

References

Adamo, S. 2011. Specialist Meeting—Future Directions in Spatial Demography.

Angel, S., and A. M. Blei. 2016. "The Spatial Structure of American Cities: The Great Majority of Workplaces are No Longer in CBDs, Employment Sub-centers, or Live-work Communities." *Cities* 51: 21–35.

Australian Bureau Statistics. 1981; 1986; 1991; 1996; 2001; 2006 and 2011. *Australian Census of Population and Housing.* Canberra: Australian Capital Territory.

Australian Bureau Statistics. 2006a. *Statistical Geography Volume 1—Australian Standard Geographical Classification. Cat No. 1216.0.* Canberra: Australian Capital Territory.

Australian Bureau Statistics. 2006b. *Census Dictionary, 2006. Capital City Statistical Division (Capital City SD). Cat No. 2901.0.* Canberra: Australian Capital Territory.

Australian Bureau Statistics. 2011. *Australian Statistical Geography Standard: Volume 1 Main Structure and Greater Capital City Statistical Areas. Cat No. 1270.0.55.001.* Canberra: Australian Capital Territory.

Australian Bureau Statistics. 2014. *Australian Historical Population Statistics. Cat. no. 3105.0.65.001.* Canberra: Australian Capital Territory.

Baker, E., N. Coffee, and G. Hugo. 2000. *Suburbanisation Vs Re-Urbanisation: Population Changes in Australian Cities, Report for Environment Australia.* Canberra: Department of the Environment and Heritage.

Blake, M., M. Bell, and P. Rees. 2000. "Creating A Temporally Consistent Spatial Framework for the Analysis of Interregional Migration in Australia." *International Journal of Population Geography* 6 (2): 155–174.

Bunker, R, B. Gleeson, D. Holloway, and B. Randolph. 2002. "The Local Impacts of Urban Consolidation in Sydney." *Urban Policy and Research* 20 (2): 143–167. doi:10.1080/08111140220144461.

Chhetri, P., J. H. Han, S. Chandra, and J. Corcoran. 2013. "Mapping Urban Residential Density Patterns: Compact City Model in Melbourne, Australia." *City, Culture and Society* 4 (2): 77–85.

Croxford, A. 1977. *Editorial, Living City, Spring/Summer.* Melbourne: Melbourne Metropolitan Board of Works.

Department of Infrastructure. 1998. *From Doughnut City to Café Society.* Melbourne: Victorian Government.

ESRI, DeLorme, GEBCO, NOAA NGDC, and other contributors. 2014. Ocean/World Ocean Base Map. [Online]. Accessed 7 March 2016. http://www.arcgis.com.

Forster, C. 1999. *Australian Cities: Continuity and Change.* Melbourne: Oxford University Press.

Forster, C. 2006. "The Challenge of Change: Australian Cities and Urban Planning in the new Millennium." *Geographical Research* 44 (2): 173–182.

Geoscience Australia. 2006. *Populated Places and Place Names.* Geodata Topo 250 K Series 3 Dataset. Canberra: Australian Capital Territory.

Government of South Australia Planning, S. A. 2010. *The 30-Year Plan for Greater Adelaide.* Adelaide: Government of SA.

Grosvenor, M., and P. O'Neill. 2014. "The Density Debate in Urban Research: An Alternative Approach to Representing Urban Structure and Form." *Geographical Research* 52 (4): 442–458.

Hall, T. 2010. *Life and Death of the Australian Backyard*. Collingwood: CSIRO Publishing.

Hugo, G. J. 1986. *Australia's Changing Population: Trends and Implications*. Melbourne: Oxford University Press.

Hugo, G. 2002. "Changing Patterns of Population Distribution in Australia." *Journal of Population Research* Special ed. 2002, Sept 2002: 1–21.

Hugo, G., D. Griffith, P. Rees, P. Smailes, B. Badcock, and R. Stimson. 1997. *Rethinking the ASGC: Some Conceptual and Practical Issues*, Monograph Series 3. Adelaide: National Key Centre for Social Applications of GIS, University of Adelaide.

Mackellar, D. 1990. *I Love A Sunburnt Country: The Diaries of Dorothea Mackellar*. Edited by J. Brunsdon. North Ryde: Angus & Robertson.

McLoughlin, B. 1991. "Urban Consolidation and Urban Sprawl: A Question of Density." *Urban Policy and Research* 9 (3): 148–156. doi:10.1080/08111149108551499.

Mees, P. 2009. "How Dense Are We?: Another Look at Urban Density and Transport Patterns in Australia." *Canada and the USA, Road & Transport Research: A Journal of Australian and New Zealand Research and Practice* 18 (4): Dec 2009: 58–67.

Openshaw, S. 1984. *The Modifiable Areal Unit Problem, Concepts and Techniques in Modern Geography # 38*. Norwich: Geo Books.

Paris, C. 1994. "New Patterns of Urban and Regional Development in Australia: Demographic Restructuring and Economic Change." *International Journal of Urban and Regional Research* 18 (4): 555–572.

Randolph, B., and R. Freestone. 2012. "Housing Differentiation and Renewal in Middle-Ring Suburbs: The Experience of Sydney, Australia." *Urban Studies* 49 (12): 2557–2575.

Randolph, B., and A. Tice. 2014. "Suburbanizing Disadvantage in Australian Cities: Sociospatial Change in an era of Neoliberalism." *Journal of Urban Affairs* 36 (s1): 384–399.

Searle, G., and P. Filion. 2010. "Planning Context and Urban Intensification Outcomes: Sydney Versus Toronto." *Urban Studies* 0042098010375995.

Stretton, H. 1991. "The Consolidation Problem." *Architecture Australia* 80 (2, March): 27–29.

The State of Victoria. 2014. Plan Melbourne, Metropolitan Planning Strategy.

Thompson, S., and P. Maginn. 2012. *Planning Australia: an Overview of Urban and Regional Planning*. Cambridge: Cambridge University Press.

Troy, P. 1992. "The New Feudalism." *Urban Features* 2 (2): 36–44.

Troy, P. N., ed. 1995. *Australian Cities: Issues, Strategies and Policies for Urban Australia in the 1990s, Reshaping Australian Institutions*. Cambridge: Cambridge University Press.

Vasanen, A. 2012. "Functional Polycentricity: Examining Metropolitan Spatial Structure Through the Connectivity of Urban sub-Centres." *Urban Studies* 49 (16): 3627–3644.

Walford, N. S., and K. N. Hayles. 2012. "Thirty Years of Geographical (In)Consistency in the British Population Census: Steps Towards the Harmonisation of Small-Area Census Geography." *Population, Space and Place* 18 (3): 295–313.

Yates, J. 2001. "The Rhetoric and Reality of Housing Choice: the Role of Urban Consolidation." *Urban Policy and Research* 19 (4): 491–527.

Spatial Concentration in Australian Regional Development, Exogenous Shocks and Regional Demographic Outcomes: a South Australian case study

Peter Smailes, Trevor Griffin and Neil Argent

ABSTRACT

As a tribute to the massive contribution of our friend and colleague Graeme Hugo to the population and settlement geography of Australian rural areas, this paper presents a longitudinal study from his home State. It forms part of a wider study of the long-term demographic relationships between Australia's rapidly growing regional cities and their surrounding functional regions. Of particular interest is the question of what effect the accelerating concentration of population and economic activity into a given regional city will have for the longer term demographic sustainability of its functional region as a whole. Taking the case of Port Lincoln, regional capital of most of South Australia's Eyre Peninsula, it examines the nature of change in the functional region over the period 1947–2011, and investigates the forces feeding, and partly counteracting, the population concentration process, informed by concepts of evolutionary economic geography. In particular it traces the demographic impact (particularly differential migration and ageing trends) of exogenous shocks to the region's essentially primary productive economic base during the period of major change from 1981 to 2011.

Introduction

Graeme Hugo was a colleague, co-researcher and teacher to us severally, and a friend and ongoing inspiration to us all. His prodigious output includes some 40 refereed papers or book chapters devoted exclusively to rural and regional Australia. He was a pioneer in the study of rural newspaper circulation analysis, counter-urbanisation, accessibility/remoteness measurement, immigration to rural areas, welfare-led migration and welfare of the elderly in small rural towns. He was part of a previous extended study of two small rural communities conducted over 40 years (Smailes and Hugo 2003), and we dedicate the present long-term demographic study to him. The paper is concerned with demographic outcomes likely to arise from the rapidly increasing concentration of Australian non-metropolitan growth into a limited number of regional cities, each of which dominates an extensive functional region occupied by smaller communities experiencing decline or, at best, much more limited growth. A wider study in progress investigates six such functional regions, of which the present paper focuses on just one: the region dominated by Port Lincoln, undisputed regional capital of most of South Australia's Eyre Peninsula. It presents a longitudinal study of demographic trends over the period 1947–2011, tracing changes in the dominant central city, the minor surrounding communities, and the functional region as a whole—in each case differentiating between the urban (clustered) and rural (dispersed) population components. In what follows, the term 'Eyre Peninsula' refers to the whole landmass within the Ceduna/Port Lincoln/Port Augusta triangle. 'Functional region' refers to the more restricted southern and western parts of the Peninsula interacting dominantly with Port Lincoln. It includes seven Local Government Areas (LGAs), which collectively encompass eight distinctive local communities spatially defined as social catchments (see Figure 1).

In this outlying, extractive resource-based economy, then, is the increasing concentration of population growth into a single centre—Port Lincoln—likely to promote or impair the long-term demographic sustainability of the functional region as a whole? This umbrella inquiry incorporates four subordinate questions.

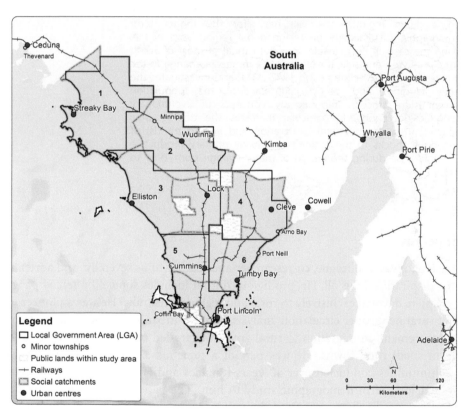

Figure 1. Port Lincoln's functional region in its spatial setting. Local Government Areas are: 1, Streaky Bay; 2, Wudinna; 3, Elliston; 4, Cleve; 5, Lower Eyre Peninsula; 6, Tumby Bay; 7, Port Lincoln. Social catchments are defined for eight towns: Streaky Bay, Wudinna, Elliston, Lock, Cleve, Cummins, Tumby Bay and Port Lincoln.

(1) Did the rural crisis years of 1982–94 cause a fundamental change in a former intra-regional demographic equilibrium?
(2) Following the crisis, has the functional region become locked in to a downward demographic spiral, starting from the periphery?
(3) How far have resistance and resilience modified the effects of the crisis and the region's future demographic prospects?
(4) What potential relevance does the Port Lincoln case study have for other Australian functional regions?

South Australia's Eyre Peninsula in many ways forms a microcosm of an extractive resource-based economy, providing an ideal test bed for the above research questions. Its development impinges on a wide range of interlocking geographical discourses, including staples theory, the 'sponge city' hypothesis, the notion of 'uncoupling' of country towns from their agricultural service base, the 'spread' and 'backwash' concepts of Myrdal and Hirschman, and concepts drawn from evolutionary economic geography— particularly selection, path dependence, lock-in, and the impact of exogenous shocks (Tonts, Argent, and Plummer 2013). At the risk of over-simplifying the complex, multi-scalar interactions between these discourses, we interpret demographic change since first colonial settlement primarily through the lenses of Innisian staples theory and evolutionary economic geography (Innis 1930, 1933) recently applied elsewhere in Australia by Tonts, Martinus, and Plummer (2013).[1] We concentrate here on the demographic responses and outcomes in Port Lincoln's functional region. Taking account of the initial spatio-temporal contingencies of the developing economic landscape, we seek to expose the nature as well as the results of demographic change and the mechanisms that produced it, and emphasise continuity and resilience as well as change.

Historical antecedents

In the nineteenth and much of the first half of the twentieth century, the large triangular mass of Eyre Peninsula, known to Adelaide residents as 'the West Coast' was to all intents and purposes an island, separated from Adelaide ('the Mainland') by Spencer Gulf and Gulf St Vincent, and normally accessed by sea. The passage from Adelaide to Port Lincoln by sea was 280 km, against a land distance of some 650 km, including a forbidding semi-arid stretch around the top of Spencer Gulf (Figure 1). Port Lincoln itself lies close to the southern extremity of the Peninsula. Its magnificent natural harbour made it a contender in 1836 for the site of the new Province's capital, though finally rejected in favour of Adelaide due to its lack of fresh water. The first settlers arrived and the town was surveyed only 3 years later, in 1839, and its first jetty was built in 1857. This apparently 'peripherally located centre' gained such a dominant initial advantage that for 25 years it was the only surveyed town on the Peninsula, whose sparse settlement was then mainly coastal and dominated by wool production, shipped out on shallow-draft coastal vessels. By sea, Port Lincoln was indeed centrally located between the Peninsula's west and east coast pastoral settlements, throughout the nineteenth century. East- and west-coastal travelling stock routes and telegraph lines also converged on Port Lincoln (Sumerling 1987).

In the twentieth century the extractive resource economy was in effect turned 'outside-in' from coastal wool and fishing to inland agriculture, through the construction of Eyre

Peninsula's isolated narrow gauge railway system—commencing in 1905 and completed by 1926. This system, along with the northward reticulation of water from Port Lincoln's Tod River reservoir not only facilitated agricultural expansion through clearance of huge areas of mallee scrub (Heathcote 1996) but also focused the land as well as sea routes and indeed the entire peninsula's settlement system firmly on Port Lincoln. However, the rise in the 1930s, 1940s and 1950s of Whyalla and (to a much lesser extent) Ceduna/Thevenard later reduced the city's influence in the northeast and far northwest of the peninsula (Figure 1). In 1840 Port Lincoln had just 270 settlers; by the first Commonwealth census of 1911 the population of Eyre Peninsula as a whole had risen to 12 788 persons. The Peninsula's central and northeastern areas, then untapped by railways, remained almost unpopulated, with iron ore mining still in its infancy. By the conclusion of the Second World War, boosted by the rise of Whyalla as a steel-making and shipbuilding centre, the whole peninsula population had more than doubled, to 28 000 in round figures. Of this total, Port Lincoln's defined functional region, on which the paper now focuses, had 16 000 or some 57 per cent.

Port Lincoln's region, 1947–96: from prosperity to crisis

By 1947 Port Lincoln's population was just under 4000. Its dominance was undoubted, but the demographic balance between the city and its functional region had been under constant change through ongoing land clearance, droughts, rabbit plagues, the Depression years and the impact of two world wars (Heathcote 1996). No clear period of demographic or economic equilibrium between city and functional region can be discerned in the first 100+ years. From the end of the Second World War until around 1973, however, Eyre Peninsula (and rural Australia in general) experienced almost three decades of relative prosperity (the 'Long Boom') under a strongly productivist national agricultural policy (Hefford 1985). From 1947 to 1976 the population of Port Lincoln itself grew rapidly from 4000 to 10 000 persons, while primary industry employment in the functional region grew by almost 30 per cent to its 1976 peak of 4050 workers. The loss of around 1500 of these mostly farm-based workers by 1991 was to be by far the most important driver of demographic change during the crisis years extending from the early 1980s to the mid-1990s, described below. Partial recovery of around 500 workers by 2006 was due mainly to the rapid development of fishing and aquaculture rather than farm employment.

Beyond Port Lincoln's own immediate social catchment, the functional region's settlement system includes seven smaller communities, each centred on a country town with its surrounding social catchment incorporating dispersed farms and (in some cases) small clusters around former outposts/fishing harbours or over-optimistically surveyed railway townships. Spatially this total of eight social catchments collectively fits closely within the outer perimeter of a bloc of seven LGAs,[2] though within the functional region the spontaneously evolved catchments depict local town/country interaction patterns that cut across the rectilinear administrative boundaries (Figure 1). These catchments are frequently referred to here as 'communities', a term inclusive of both their town and rural population components. Methods used in their definition are described in detail elsewhere (Smailes et al. 2002).

Prior to the economic crisis the regional economy was supported by an infrastructure almost entirely publicly or co-operatively owned and managed—including ports, jetties, railways, roads, water supplies, wheat, wool and barley marketing, electricity, postal services, telephone and schools. At the 1981 census, employment in federal, State and Local government provided over 2500 jobs, or 22 per cent of the entire functional region's workforce (rising to 33 per cent in Port Lincoln city). The drought-proof incomes of these centrally-funded jobs, along with pension and retirement funds, provided an important buffer for country businesses in bad farming years. Most (80 per cent) of the public-sector jobs were State employees, soon to be decimated by cuts, 'regionalisation' and privatisation following the near-collapse of the State Bank in 1991. Eyre Peninsula in the post-war years experienced closer integration with the State space economy and a gradual shrinkage of its island-like status, though it remains remote in terms of the water and electricity supply networks.

The economic and social impact of exogenous shocks visited on the study region in the crisis years from 1982 to around 1995 have been discussed in detail for Australia as a whole by many authors (e.g. Pritchard and McManus 2000; Gray and Lawrence 2001) and for Eyre Peninsula in particular (Smailes 1996b, 1998). They are summarised here as a brief timeline of main events (Table 1).

The Eyre Peninsula's problems began as a drought- and credit-induced crisis, and were exacerbated from about 1990 by the nation-wide recession and the initiation of ongoing neoliberal policies that saw the gradual dismantling of national marketing boards and privatisation of practically all of the Peninsula's extractive industries infrastructure, apart from roads. The severity of the crisis, the misery and privation it caused, and the far-reaching consequences of the new politico-economic ground-rules can hardly be over-emphasised. By 2011 farming families had faced not only these crisis years but also the 'millennium drought' which from 2004 to 2010 caused serious to severe rainfall deficiencies in eastern Eyre Peninsula, with 'lowest on record' rainfall experienced around Cleve (Australian Bureau of Meteorology 2012).

Change in total population, 1947–2011

Figure 2 provides an overview of the demographic response to change in the functional region over the whole period 1947–2011 (for comparison including Whyalla's rise and

Table 1. Summary of events during Eyre Peninsula's crisis years

Year	Event(s)
1982	Severe drought across Australia
1983	Australian dollar floated
1984	Banking sector deregulated. Loans readily available, farmers over-commit to major purchases
1985	Farm land values peak as farmers compete to enlarge holdings
1986	Interest rates on farm loans peak at 20 per cent
1987	Collapse of farm equity as land values down around 50 per cent on 1985
1988	Severe local drought and wind erosion in Eyre Peninsula after 4 years' diminishing rainfall
1989	A good farming year gives some respite. Farmers' Action Groups picket banks, disrupt farm foreclosure auctions
1990	National recession and credit squeeze. Collapse of wool and wheat prices
1991	USA/EEC subsidised grain export war impacts Australian wheat markets. Collapse of State Bank of SA. AWC wool stockpile reaches 4.8 million bales, wool floor price scheme ended
1992	Heavy rain throughout grain harvest season, much of sprouting Eyre Peninsula crop reduced to feed grade
1993	Severe mouse plague follows wet year. Serious entrenched farm debt across Eyre Peninsula
1994	Widespread drought returns. Rural debt in SA reaches $1.4 billion
1995	Rural poverty, distress, suicides. Eyre Peninsula Regional Task Force set up to address problems

Source: Smailes (2006, Appendix IV: 'Chronology of the Rural Crisis, 1984–1994').

decline). Not surprisingly, the immediate impact of the crisis years is reflected most strongly in the 'rural remainder' population component which dropped substantially from 1981, as a lagged response to the farm labour force's decline after its 1976 peak. This fall, however, is further lagged and only weakly reflected in the total population, which by 1996 had already resumed its upward trend. The seemingly muted response of the 'rural remainder' to such a severe crisis results partly from its heterogeneity—including both farm and non-farm dispersed settlement plus minor clusters, some with fishing, recreational and retirement housing. Of particular interest is the relationship between the growth of Port Lincoln (1947 population 3972) and the aggregate of its seven outlying minor towns (2418). With a still growing and prosperous rural service population (9542 in 1947) there was clearly scope for both Port Lincoln city and its seven minor towns to expand their service functions, and all of them grew up to about 1966. However, in these early post-war years rapidly increasing scale economies and restructuring in retailing entrenched the supremacy of regional cities like Port Lincoln and Whyalla, since they were the only non-metropolitan centres able to support national or State supermarket and other retail chains entering rural South Australia (Smailes 1979).

Further emphasising Port Lincoln's status as regional capital were the introduction in the 1950s of the deep-sea tuna fishery and of the bulk handling of grain, with the first major silos built at Port Lincoln in 1958, followed in 1969 by the decision to upgrade the port to take ships of up to 100 000 GRT (South Australia: The Civic Record 1986). Between 1947 and 1966 the ratio of Port Lincoln's population to that of the combined minor towns in its functional region rose from 1.6:1 to 2.3:1. For the next 20 years, though, this ratio remained almost constant as the regional city and its minor towns grew at practically the same rate, providing the regional population's nearest approach

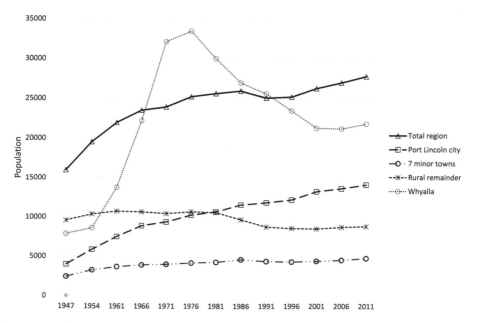

Figure 2. Total population change in Port Lincoln's functional region, by settlement components, 1947–2011.

Source: ABS data by LGA and urban centre/locality (UCL), 1947–2011 censuses.

to a dynamic growth equilibrium. After 1986 Port Lincoln's advantage grew slowly to 3:1, as the smaller towns, providing low-order goods and services with a generally low-income elasticity of demand, were disproportionately hard hit by falling farm numbers and work-force, and by a rapid decrease in the friction of distance on road travel (Smailes 1996a).

Fertility and natural increase, 1947–2011

Traditionally the relatively high fertility of Australian rural populations has provided a sufficient natural increase to at least partially offset net out-migration. With annual birth/death statistics available at LGA level only from 1986, as a surrogate the child/woman ratio (children aged 0–4 per 100 women aged 15–44) indicates that the functional region's fertility was well above the Australian average throughout the study period. The disparity in the child/woman ratio was at a maximum in 1961 (Australia 53, Eyre Penin-sula 72), narrowing gradually to a minimum by 2001 (Australia 30, Eyre Peninsula 36) after which the gap again increased. The region's fertility thus mirrored the national trend, but at a consistently higher level. Throughout the period Port Lincoln city's child/woman ratio remained below that of the rest of the region, while still exceeding the national average. Total Fertility Rates (TFRs) are available at the LGA level for certain years, including 1992–94 at the height of the crisis. The TFR for the whole func-tional region and for all except one of the seven constituent LGAs (Tumby Bay) then com-fortably exceeded replacement level, averaging 2.42. By 1997–99 this average remained almost unchanged at 2.39.

Not surprisingly, natural increase remained high from 1947 to the onset of the crisis years. The slow decline in natural increase rate and numbers from 1986 to 2006, with a slight rise in the most recent intercensal period mirrors the general fertility trend described above. All seven LGAs experienced this general decline, with some fluctuations (Table 2). However, despite its lower child/woman ratio, from 2001 onwards Port Lincoln itself accounted for over 60 per cent of the region's numerical increase. It also maintained its natural increase rate better than the hinterland LGAs—due to the impacts of migration and age structure, discussed below. Declining natural increase was particularly severe in Cleve and Elliston, while Tumby Bay, an attractive coastal retirement venue, experienced actual natural *decrease* from 1996 onwards.

The spectacular, though steadily diminishing, growth of agricultural prosperity in the immediate post-war period dominates the region's intercensal population change up to 1981 (Figure 3). The emergence in the 1960s of a cost-price squeeze and falling real farm incomes (Hefford 1985) is reflected in the deficits or very low gains in the 'remainder' (rural) population in the two decades leading up to the 1982–94 crisis years, while the increasing importance of Port Lincoln city's contribution to the total regional growth is clearly evident right up to 2001.

Migration currents, 1981–2011

The population changes depicted in Figure 3 for the period 1981–2011 are the net out-comes of an enormous web of inward, outward and local individual movements interact-ing with the region's natural increase. With fertility relatively high and declining only slowly throughout the period, fluctuating migration flows dominate the net outcomes.

Table 2. Natural increase/decrease by LGA, Port Lincoln functional region, 1986–2011: **numerical** and (percentage) change

	1986–91	1991–96	1996–2001	2001–06	2006–11
1 Streaky Bay (DC)	**87** (4.0)	**100** (5.3)	**57** (3.0)	**17** (0.9)	**60** (2.9)
2 Wudinna[a] (DC)	**89** (4.5)	**106** (6.3)	**74** (5.0)	**55** (3.9)	**48** (3.5)
3 Elliston (DC)	**100** (7.1)	**86** (6.5)	**69** (5.7)	**44** (3.7)	**39** (3.3)
4 Cleve (DC)	**115** (5.2)	**68** (3.4)	**51** (2.7)	**77** (4.2)	**27** (1.1)
5 Lower Eyre Peninsula[b] (DC)	**179** (4.7)	**138** (3.7)	**127** (3.3)	**94** (2.3)	**118** (2.7)
6 Tumby Bay (DC)	**33** (1.2)	**22** (0.9)	**−13** (−0.5)	**−26** (−1.1)	**−5** (−0.2)
7 Port Lincoln (C)	**589** (4.9)	**527** (4.5)	**500** (4.1)	**396** (3.0)	**482** (3.6)
Total, less Port Lincoln	**602** (4.2)	**519** (4.0)	**365** (2.8)	**261** (2.0)	**287** (2.1)
Total, whole region	**1190** (4.5)	**1045** (4.2)	**865** (3.5)	**657** (2.5)	769 (2.8)

Notes:
[a]Formerly Le Hunte (DC).
[b]Formerly Lincoln (DC).
Source: Australian Bureau of Statistics (1989, 2001, 2007, 2011, 2012, 2013).

Between 1947 and 1954 the region had an estimated total net in-migration of over 1500 people; census by census, this inflow shrank to become a small deficit by 1981, followed by a sharp plunge to a net loss of 1000 by 1986, as the rural crisis set in. From 1954 onwards net migration gains accrued only to the city, increasingly offset by net outflows from the hinterland.

For a closer examination of migration during and after the rural crisis years we employ a database for which the Census Collection District (and, for 2011, the Statistical Area 1) data covering the functional region have been reallocated to the eight communities defined in Figure 1, differentiating between the central towns and their respective rural catchments. By 1986 the severe drought of 1982 and the effects of financial over-commitment

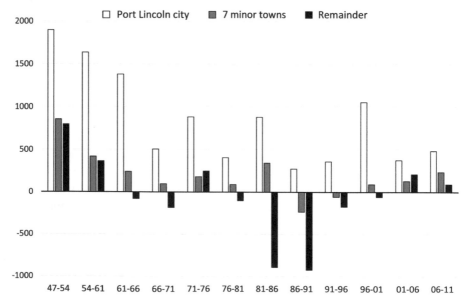

Figure 3. Net intercensal population increments by settlement type, Port Lincoln functional region, 1947–2011.

Source: ABS data by LGA and urban centre/locality (UCL), 1947–2011 censuses.

had already launched out-migration, but the strongest impact of the crisis occurred in the 1986–91 period when estimated net migration losses in the region as a whole amounted to −11 per cent of the 1986 population, and in the seven smaller communities ranged between −9 per cent (Tumby Bay) and −26 per cent (Lock). Not surprisingly, the losses were much greater in the rural than in the urban elements of these seven communities. In aggregate their estimated 1986–91 net migration losses were −23 per cent (rural) and −6 per cent (urban) of their 1986 population.

Most severely affected by out-migration were the 15–19 and 20–24 school leaver and young adult age groups, though fortunately the key reproductive age groups (aged 25–44) were somewhat better balanced by inflows, giving a more modest net loss. Port Lincoln community itself (including both the city and its social catchment) also experienced heavy losses in the 15–24 age groups, but retained a far larger proportion of the key reproductive 25–44 age groups than the smaller outlying communities. After 1991 the total volume of outflows moderated a little, but the general pattern described above was maintained. The effects of net outflows varied greatly between individual local communities, but Figure 4 shows their aggregate impact.[3]

Clearly, the total region's relatively favourable migration balance in the key reproductive age groups of 25–39 years, and also in the later working-age groups up to 64 years, is almost entirely due to the role of the regional city. The hinterland communities have a small favourable balance in just the one, albeit important, 25–29 age group. Figure 5 demonstrates the strong contrast between these small communities' urban and rural population elements. Collectively the towns themselves are still attracting an appreciable inflow of people in the older age groups, especially the pre- and early retirement age groups of 55–64 years, and even register a small net gain in the 25–34 family formation age brackets. The dispersed rural population of their social catchments, however, has net outflows practically throughout the age structure but most prominently in the highly mobile 15–24 age groups.

Figure 5 differentiates the migration streams by age group, but gives no idea of their spatial pattern. However, Port Lincoln's net migration gain, coupled with the hinterland communities' heavy net losses, suggests that the regional city may progressively be

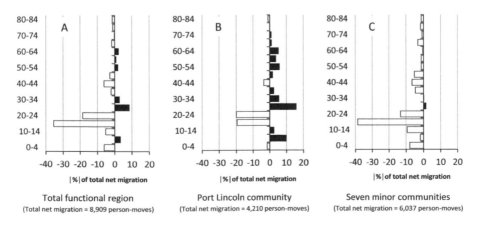

Figure 4. Estimated net migration by quinquennial age group, Port Lincoln functional region, 1981–2011.

Source: authors' estimates from ABS census data, using the cohort survival method.

draining the hinterland of its people. Full ABS inter-SLA migration matrices, available for the four most recent intercensal periods, allow us to construct detailed in- and out-migration fields for the city of Port Lincoln, summarised below (Table 3). The data show a vigorous two-way exchange between the city and its total migration field, with over 5000 person-moves in most intercensals giving a net change of just a few hundred people, and very low migration effectiveness[4] over the two decades.

The largest and most effective flow, from which Port Lincoln consistently *loses* people, is that between the city and metropolitan Adelaide. Second in importance, and crucial to this paper's central research question, is the exchange between the city and its functional region, from which the city consistently *gains* people—particularly strongly in the 2001–06 period, when effectiveness reached 20 per cent. However Port Lincoln's 1991–2011 net gain of 475 people from its hinterland amounts to only 20 per cent of its total population growth over the same period, and to only 4 per cent of the 1991 hinterland population. If we include the whole Peninsula extending to Ceduna in the far northwest and Cowell (but not Whyalla) in the northeast, the equivalent figures rise only to 31 per cent and 5 per cent, respectively—a useful contribution, but not enough to account for Port Lincoln's growth, or to offset its outflow to Adelaide. The population exchange between Port Lincoln and the nearby South Australian regional capitals of Whyalla, Port Pirie and Port Augusta (Figure 1) is very limited, though Port Lincoln has small net gains from all of them.

Ageing: town and country

We turn next to the vital and closely related question of ageing. The distribution of the net migration flows suggests that, particularly in the seven smaller communities of Port Lincoln's hinterland, the relatively small losses of 1986–96 in total population in the minor towns, and their small gains since then (Figures 2 and 5), may conceal a seriously deteriorating age structure.

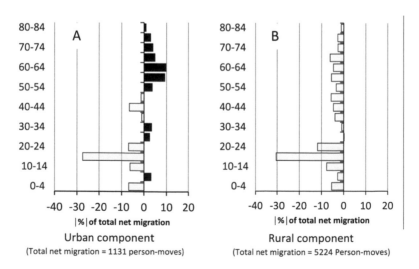

Figure 5. Estimated net migration by quinquennial age group, 1981–2011, urban and rural population components of seven minor communities within Port Lincoln's functional region.

Source: authors' estimates from ABS census data, using the cohort survival method.

Table 3. Spatial distribution and effectiveness of net migration flows to/from Port Lincoln city, 1991–2011

	1991–96	1996–2001	2001–06	2006–11	1991–2011
Other SLAs in functional region	70	141	243	21	475
Migration effectiveness	**+6.5**	**+13.0**	**+20.8**	**+1.9**	**+10.7**
Rest of Eyre Peninsula (incl. Whyalla)	53	71	94	81	299
Adelaide Statistical Division	−337	−343	−140	−212	−1032
Migration effectiveness	**−17.4**	**−19.2**	**−9.4**	**−14.1**	**−15.4**
Rest of South Australia	83	−140	79	362	384
Other Australia	−243	134	−7	−77	−193
Total migrants (in + out)	5650	5295	4915	5323	21 183
Migration effectiveness	**−1.46**	**−2.59**	**+5.47**	**−0.18**	**+0.22**

Source: Australian Bureau of Statistics (2008, 2015).

Figure 6 provides a long-term view of ageing in the study region, relating it to Jackson's (2014) study of sub-national ageing in New Zealand, which shows how an apparently robust national age structure may be dependent on strong growth in just one or two core central regions, masking impending population decline in many peripheral regions. Her (roughly sequential) suggested series of 13 early warning indicators of impending decline begins with a youth deficit (age groups 15–24 < 15 per cent) and ends with natural decrease (Jackson 2014, 22).

Although the time sequence is different, seven of Jackson's indicators occur in the 1947–2011 demographic trajectory of Eyre Peninsula (total region). Figure 6 shows the timing of their appearance, superimposed on trends in one of the most significant indicators, the workforce entry/exit ratio. The figure provides evidence of a fundamental change in the trajectory after 1986, as the rural crisis began to take its toll. The period of parallel development of the city and its subsidiary communities came to an end.

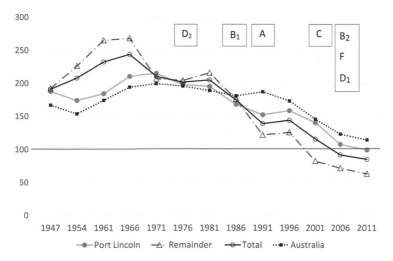

Figure 6. Port Lincoln functional region: workforce exit/entry ratio, 1947–2011, in relation to Jackson's (2014) indicators of ageing. A = youth deficit (15–25 years > 15 per cent); B_1 = workforce entry/exit ratio below national average; B_2 = workforce entry/exit ratio > 100; C = ages 25–39 below national average; D_1 = ages 65+ above national average; D_2 = ages 65+ > 20 per cent; F = ages 65+ > ages 0–14.

Source: ABS data by LGA and urban centre/locality (UCL), 1947–2011 censuses.

Whereas the hinterland had consistently provided the region's main supply of young workforce entrants, after 1986 this role was taken over by the city itself, the hinterland's exit/entry ratio drying up to fall below parity by 2001. In the same year the hinterland's ratio of persons aged 65+ to children 0–14 rose sharply to 70:100, from 42:100 in 1996 —exceeding the national average for the first time. Space does not permit discussion of Jackson's early warning indicators for individual communities, but their clustering at the total regional level in the early 2000s confirms a serious, spatially differentiated impending ageing problem.

The main measure of ageing used here (Table 4) is the Relative Ageing Index (RAI), introduced and fully described elsewhere (Smailes, Griffin, and Argent 2014). Essentially it measures the net percentage points by which a regional population has aged faster or slower than that of the nation as a whole.[5] Negative values express ageing, positives express (re)juvenation. Technically the RAI could range from +200 to −200, but in real populations it rarely extends beyond ±50 and is highly unlikely to reach ±100. In the 33 years between 1947 and 1981 (in effect entirely since 1966), Port Lincoln's functional region had already experienced a very slight ageing relative to the national average (RAI −4.7). However, our focus is on the following 30 years from 1981 to 2011, in which three times as much relative ageing occurred (RAI −15.6). Because of the age selectivity of net migration, intercensal change in the RAI mirrored the volume of net migration—highest in 1986–91 due to out-migration of the young, and again in 2001–06 due to in-migration of the old. Table 4 shows the continuous severe ageing of the seven minor communities (in aggregate) compared to Port Lincoln's much lesser, and interrupted, ageing trajectory. Recall that the RAI measures ageing *over and above* the already considerable ageing of the entire national population.

The collectively high RAI index for the minor communities conceals a large range. At the extremes are Cummins (RAI −4.9, buoyed by some rejuvenation in the last two inter-censals) and the small coastal community of Elliston (RAI −46.1, with very severe losses in the 1991–96 period). The incidence of ageing varies in inverse relation to population size, occupational diversity, and remoteness. Most significant, though, is the contrast between the rural and the urban components of the hinterland communities and between coast and inland.

Four of the seven communities centre on inland towns, two of them (Lock and Wudinna) having no coastline at all. In general, the inland communities have aged dominantly by out-migration of the potentially reproductive age groups. Ageing in the more environmentally attractive coastal communities is exacerbated by retirement and pre-retirement in-migration. Hence the coastal communities collectively have the greater RAI of −29.0, against −21.1 collectively in the inland communities. However, the

Table 4. Port Lincoln functional region: movement of the Relative Ageing Index (RAI) in relation to net migration flows, 1981–2011

	1981–86	1986–91	1991–96	1996–2001	2001–06	2006–11	1981–2011
RAI: Port Lincoln	−0.94	−2.97	0.63	1.61	−3.83	−2.87	−8.37
RAI: seven minor communities	−3.47	−9.47	−2.27	−4.10	−5.14	−1.28	−25.74
RAI: total region	−2.35	−5.92	−0.54	−0.48	−4.50	−1.86	−15.64
Total region: net migration	−5.7	−10.6	−4.5	0.2	1.4	−3.4	−23.0

Source: calculated by the authors from Australian Bureau of Statistics census data, 1981–2011.

consequences of ageing are more serious in the inland communities (Figure 7) due to their greater shrinkage in absolute numbers, and lesser alternative employment potential.

Figure 7A is an ageing profile showing how the RAI net ageing of −21.1 percentage points relative to the national population is distributed through the sub-population's age structure. A population ageing at exactly the national rate would have an RAI of zero and a very narrow range, or none, on either side of the Y axis. The bars on Figure 7 add up to −21.1, and all quinquennial groups that have aged more than the national average appear on the left (negative) side of the Y axis, while just one rejuvenating age group appears on the right (positive) side. The most serious ageing has occurred through *shrinkage* in the vital young adult early career and family formation age groups, together with great *expansion* in the oldest age group. The one case of rejuvenation is at the upper extreme of potential female childbearing.

Figure 7B speaks for itself. It shows the end result that the demographic events set in motion by the rural crisis have produced in just 30 years for these communities' very youthful, almost pre-transition, 1981 age pyramid.

Discussion

Conceptual linkages

A central concern of evolutionary economic geography (EEG) is the precise charting and explication of regional economic dynamics, drawing on Darwinian concepts of variety, selection and competition across space (Essletzbichler and Rigby 2007). A particular interest lies in identifying and explaining 'shocks' and their impact on the long-run developmental trajectory of regions. Viewed through the conceptual lenses of EEG and Innisian staples theory, the late nineteenth- and early twentieth-century-development of the case study region via its integration into the broader South Australian economy can be seen at least in part as a path- and place-dependent process. Established as essentially

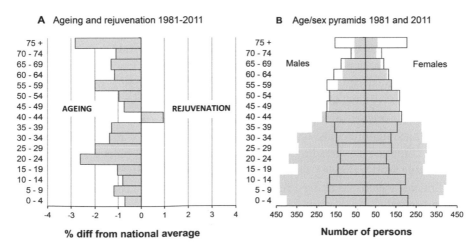

Figure 7. Change in four inland communities (Cummins, Lock, Wudinna and Cleve), 1981–2011. A: relative ageing profile; B: age/sex structures 1981 (shaded) and 2011 (blank).

Source: authors' calculations based on ABS 1981 and 2011 censuses.

a livestock production region, with later cereal production boosted by the inland rail system, the Eyre Peninsula sent growing volumes of largely untransformed farm products, via the local port, primarily to export markets. For three decades after the Second World War the substantial financial returns from this form of production—plus growing exploitation and a degree of local processing of marine resources—helped attract more population and wealth in a circular and cumulative causation process. Consistent with a staples theory perspective on the relative strengths and weaknesses of export-based natural resource dependence, Port Lincoln and its hinterlands experienced both prosperity and, more recently, economic, social and demographic restructuring in response to shocks arising from droughts, corrupt international markets and the adoption of neoliberal national farm, rural and trade policies.

Continuity and resilience

Recently, the notion of resilience has been incorporated into the EEG research agenda as a means of measuring the adaptive capacity of places in response to the inevitable crises they are subject to (Simmie and Martin 2010). Notions of resistance (i.e. the magnitude of the initial impact of a 'shock') and robustness (i.e. the nature of the region's adjustment to the 'shock') are centrally important to regional resilience. As shown above, the exogenous shocks brought to Eyre Peninsula's resource-based economy by the rural crisis set off a demographic chain reaction, still far from over. However, resistance, continuity and resilience also require recognition. Unlike, say, Western Australia's Pilbara region, or the current (2016) woes of the steel city of Whyalla, Eyre Peninsula's extractive economy is based on many individually small, relatively labour-intensive and locally owned and/or managed capital units, giving both incentive and capacity to withstand shocks. Importantly, the local community structure with its sporting, social and service clubs, church affiliations, meeting places and businesses, though greatly thinned out, remains in place. Also, due largely to the very low population densities, the 1947 spatial local government structure has survived several rounds of cost-cutting amalgamations essentially unchanged (apart from Port Lincoln's extended city boundaries). For a rural community under pressure the guaranteed local voice, identity, perceived self-determination and permanence conferred by Council status is more conducive to survival than absorption into a distant unit. At the same time, the constant and long-standing population exchange between the regional city and its hinterland helps build bridging social capital and a feeling of broader regional identity. Long distances and low density still afford the surviving small local businesses some protection. Also, despite privations and amalgamations the family farming system remains structurally changed but basically intact, through technical innovation, climatic risk-spreading through purchase of distant blocks, and many stratagems to escape from crippling debt, aided by the government-funded rural counsellor system (Smailes 1996b). Moreover, fertility levels in all but one community remain high enough to generate a small natural increase.

Coast and inland

The changing workforce balance from farming towards fishing, aquaculture, recreation and retirement is clearly differentiating Eyre Peninsula's coastal and inland communities.

At its maximum in 1976, the region's farm workforce reached 4047 against just 203 fishers, a ratio of 20:1. In 2011, only 1435 were employed in farming against 726 in fishing and aquaculture—a ratio of just 2:1. Though concentrated heavily in Port Lincoln (74 per cent) smaller coastal settlements, especially Elliston, Streaky Bay, Coffin Bay and Arno Bay, have also benefited strongly, while the inland communities have borne the brunt of the drastic decline in farm employment. The role of local leadership, technological advances in fish and oyster farming, processing and direct international marketing of both tuna and shellfish (Kroehn, Maude, and Beer 2010) give future promise, and may be considered a spread effect from Port Lincoln initiatives.

The inland communities of Cummins, Lock, Wudinna and (outside the study area) Kimba, on the narrow-gauge rail net down the centre of the Peninsula, are faced not only with migration and ageing problems but with the uncertain future of their aged parent, the rail net itself. By the early 2000s an average of 1.2 million tonnes of bulk grain reached Port Lincoln by rail, against about 0.54 million tonnes by road (Penfold 2006). However, with the system sold to private ownership in 1997 and the Australian Wheat Board, Ausbulk (farmers' co-operative bulk handler) and the Eyre Peninsula ports also corporatised, demutualised or privatised, the rail line which revolutionised inland settlement is now hampered by low axle load capacity, many stretches with maximum speed limits of 20 or 30 kmph, and new silos built off the rail net altogether (Rail, Tram and Bus Union 2002). Lock, the smallest and most central rail town of the Peninsula, had by 2011 already shrunk below minimum size for census recognition as a clustered settlement. These towns are not without hope of alternative industries such as a potential low-level nuclear waste storage facility, or mining if the market for iron ore, copper, graphite and other bulk mineral products improves sufficiently to warrant development of significant deposits around Lock, Cummins and Wudinna. Large-scale infrastructure developments would be needed, including possible new rail lines or underground slurry pipelines to carry bulk ore between the central Peninsula and either Whyalla or a 'greenfield' deep-water port north of Tumby Bay (Deloitte 2013). Desalination plant(s) would also be required: indeed, even without a renewed mining boom these will be required in the next two decades to accommodate even modest population growth, as current water demand is approaching the available capacity (Deloitte 2013).

Town/country relations

The contrasting urban and rural trends in the hinterland communities have already changed the relative role of their town and dispersed farm elements in each social catchment. Together, these elements operate as a unit to run the voluntary fire service, clubs, church parishes, small hospitals and schools, etc. within the community. However, the severe shrinkage of dispersed relative to town population has reduced the former town/country business relationship and left many shops, banks, stock and station agencies without an adequate local support base. Although the towns have retained their population numbers better than the catchments, with a top-heavy age structure they threaten to become more like retirement and aged care centres than rural service centres, particularly inland. If the current ageing cohorts of long-standing residents pass through the age structure without replacement, with them will go much of the traditional rural culture and connection with the land.

Conclusions

First, following the gradual subsidence of the post-war boom years when regional capital, small towns and the rural workforce were all growing, we conclude that the region did reach something approaching a demographic equilibrium from about 1966 to 1986. During these two decades Port Lincoln and the minor towns grew slowly in constant mutual proportion as the rural population first stabilised and then began to fall, while fertility remained high enough to sustain local populations despite net out-migration.

Second, we find that this somewhat labile balance was changed radically by the exogenous shocks endured during the rural crisis, starting in the middle 1980s and reaching its height in the period from 1986 to about 1994. From around 1986, although fertility remained above the national average throughout the study period, age-selective net out-migration from the hinterland outstripped its ability to replace the 15–24 age groups, and the hinterland workforce entry/exit ratio fell below that of the city. By 2001 it fell below replacement level, and a series of early warning indicators of impending decline began to appear for the region as a whole. The loss of a large part of the hinterland's dispersed farm population and serious ageing of its residue set in motion a lagged chain reaction of ageing in the now disproportionately large township populations, whose top-heavy age structures ensure that these effects must continue for decades to come. Population loss and ageing have affected the hinterland communities differentially, most severely in smaller and inland communities.

Third, will resistance and resilience enable the region to escape from its present downward spiral, with depopulation working inwards from the periphery? Our findings are pessimistic, though inconclusive pending closer investigation at individual community level. Some hope remains. The regional population has continued to grow slowly, Port Lincoln's share having stabilised at around 50 per cent since the 2001 census. Fertility remains high even in the inland communities. At the aggregate level, the hinterland has been bolstered by the rise of knowledge-based aquaculture and fisheries in the coastal communities along with tourism and retirement migration, including an extra-regional component. Inland, some equivalent possibility of knowledge-based economic diversification may yet be found to allow the communities to stabilise at a new, lower level. Other literature previously cited demonstrates the tenacity and resilience of the region's population, as well as the importance of its output to the State economy.

Summarising, on the main question: in Port Lincoln's case, is the increasing concentration of population growth into a single centre likely to promote or impair the long-term sustainability of the regional population as a whole? As yet, potential damage to the functional region from Port Lincoln's growth remains hidden in the depleted age structures, shrunken total populations and loss of employment in the traditional public and private service functions of the hinterland communities. During the (exogenously caused) crisis and recovery years and the millennial drought Port Lincoln's presence as a regional anchor point was probably a net advantage to the region. The migration flows between the city and its region yielded only a small net gain to Port Lincoln, inadequate to account for the city's growth. The increased hinterland-to-city inflows during the worst crisis periods suggest that Port Lincoln's presence allowed at least some of the uprooted rural population to remain in the region. Moreover, much of the impetus for the technical advances in aquaculture was generated there.

Finally, what aspects of Port Lincoln's experience are likely to be shared in other Australian regions? As with other regional capitals, Port Lincoln's threat to its hinterland, and ultimately to its region, is not in direct poaching of hinterland people but indirect in its draining of employment-generating establishments, both public and private. The growing disparity in scale economies and cost-cutting imperatives of firms, though not discussed here, inevitably becomes a circular and cumulative process. The time-bomb of ageing and decline in the peripheral communities suggests that in the absence of strong public policy to promote economic diversification, especially in inland towns, depopulation starting from a functional region's periphery may well outstrip growth in its centre.

Inevitably, the present case study partly reflects national and, increasingly, global forces, but also contains specific geographical and socio-cultural characteristics that themselves shape the broader dimensions and directions of this change, for better or worse. Nevertheless the existing literature amply demonstrates that much of rural Australia is currently affected by processes of structural demographic change similar to those documented above. Our own preliminary findings show that the present study area's age profile in 1981 closely matched that of five comparable Australian functional regions, and by 2011 two of these (Shepparton and Murray Bridge) actually had age structures older than that of Port Lincoln.

Methodologically, the present paper seeks to understand the shifting population fortunes of rural regions by tracing the interaction between demographic trajectories of the various tiers within the regional urban hierarchy—rather than just its apex. Reflecting Graeme Hugo's legacy, our paper highlights the need for truly multi-scalar analyses of the drivers and outcomes of demographic change over space and time.

Notes

1. Staples theory, building on the original work of Innis (1930, 1933), proposes that the economic development of outlying regions is stimulated by remote demand for specific, locally plentiful, primary resources (e.g. fish, furs, minerals, wheat) around which infrastructure for an export industry develops. Watkins (1963, 141) comments that 'Economic development will be a process of diversification around an export base. The central concept of a staple theory, therefore, is the spread effects of the export sector.' However, many factors such as the power of extra-regional vested interests and unevenly contested control of knowledge, wealth and force may impede or prevent such diversification (Comor 2001). Tonts, Martinus, and Plummer (2013) show how, rather than generating spread effects, regions focused on export staples may become locked in to an increasingly specialised and vulnerable economic trajectory.
2. In this region LGAs are spatially identical to Statistical Local Areas (SLAs).
3. For the six intercensal periods, net migration was estimated for each quinquennial age group using the cohort survival method. For each age group results were summed (maintaining sign) over the 30 years. Summing these results for all of the 17 age groups (*ignoring* sign) gives the total net person-movements for the period. Total 1981–2011 net movement by migration in the whole region (−8909, Figure 4A) is less than the combination of Port Lincoln (−4201, Figure 4B) and the seven minor communities (−6037, Figure 4C) due to the intra-regional component of the flows.
4. The migration effectiveness ratio is given by (net migration/gross migration)*100.
5. The index is a summation of the aggregate percentage *shrinkage* of the age groups potentially capable of reproduction (0–44), plus the aggregate *expansion* of the post-reproductive groups

(45+). To allow for non-zero summation, the sign of the percentage differences of the latter group is switched so that all ageing is expressed as negative values, while positive values indicate rejuvenation. The index is cumulative in that the sum of RAI values calculated for several consecutive periods is identical to that calculated for the period in its entirety.

Acknowledgements

The authors gratefully acknowledge the helpful comments of the journal editors and two anonymous referees, and the expert assistance of Mr Jarrod Lange in the preparation of the figures.

Disclosure statement

No potential conflict of interest was reported by the authors.

References

Australian Bureau of Meteorology. 2012. "Rain Deficiencies 1 January 2004 to 31 January 2010." ID Code IGMapAWAPRainDefs. Issued 17.9.2012.

Australian Bureau of Statistics. 1989. "Births, South Australia." Cat. No. 3301.4. 1983-1989 issues.

Australian Bureau of Statistics. 2001. "Regional Statistics South Australia." Cat. No. 1362.4.

Australian Bureau of Statistics. 2007. "Births, Australia." Cat. No. 3301.0, Australian Bureau of Statistics, Belconnen.

Australian Bureau of Statistics. 2008. "Cross-Classified Migration Tables from the 1996, 2001 and 2006 Censuses of Population and Housing." Australian Bureau of Statistics, Belconnen.

Australian Bureau of Statistics. 2011. "Births, Australia, 2008." Cat. No. 3301.0, Australian Bureau of Statistics, Belconnen.

Australian Bureau of Statistics. 2012. "Births, Australia, 2011." Cat. No. 3301.0, Australian Bureau of Statistics, Belconnen.

Australian Bureau of Statistics. 2013. "Deaths, Australia, 2012." Cat. No. 3302.0, Australian Bureau of Statistics, Belconnen.

Australian Bureau of Statistics. 2015. "Cross-Classified Migration Tables from the 2011 Census of Population and Housing via TableBuilder." Australian Bureau of Statistics, Belconnen.

Comor, E. 2001. "Harold Innis and "the Bias of Communication'." *Information, Communication and Society* 4 (2): 274–294.

Deloitte Australia. 2013. "Regional Mining and Infrastructure Planning Project: Interim Report." Accessed February 17, 2016. http://www.dpti.sa.gov.au/_/data/ ... /Eyre_and_Western_maps-Report.pdf.

Essletzbichler, J., and D. Rigby. 2007. "Exploring Evolutionary Economic Geographies." *Journal of Economic Geography* 7: 549–571.

Gray, I., and G. Lawrence. 2001. *A Future for Regional Australia.* Cambridge: Cambridge University Press.

Heathcote, R. L. 1996. "Settlement Advance and Retreat: A Century of Experience on the Eyre Peninsula of South Australia." In *Climate Variability, Climate Change and Social Vulnerability in the Semi-Arid Tropics*, edited by J. Ribot, A. Magalhaes, and S. Panagides, 109–122. Cambridge: Cambridge University Press.

Hefford, R. K. 1985. *Farm Policy in Australia.* St. Lucia: University of Queensland Press.

Innis, H. 1930. *The fur Trade in Canada.* Toronto: Toronto University Press.

Innis, H. 1933. *The Problem of Staple Production in Canada.* Toronto: Ryerson Press.

Jackson, N. 2014. "Sub-national Depopulation in Search of A Theory—Towards A Diagnostic Framework." *New Zealand Population Review* 40: 3–39.

Kroehn, M., A. Maude, and A. Beer. 2010. "Leadership of Place in the Rural Periphery: Lessons From Australia's Agricultural Margins." *Policy Studies* 31 (4): 491–504.

Penfold, L. 2006. "Eyre Peninsula Railway Upgrade Issues." Submission from Liz Penfold MP to SA Public Works Committee. Accessed February 22, 2016. http://www.lizpenfold.com/pdf/EP% 20Grain%20Transport%20SUBMISSION.pdf.

Pritchard, B. and P. McManus, eds. 2000. *Land of Discontent.* Sydney: UNSW Press.

Rail Tram and Bus Union. 2002. "Eyre Peninsula Railway Under Review." Accessed February 17, 2016. http://www.rtbu.nat.asn.au/ISI.html.

Simmie, J., and R. Martin. 2010. "The Economic Resilience of Regions: Towards an Evolutionary Approach." *Cambridge Journal of Regions Economy and Society* 3: 27–43.

Smailes, P. J. 1979. "The Effects of Changes in Agriculture upon the Service Sector in South Australian Country Towns, 1945–74." *Norsk Geografisk Tidsskrift* 33: 125–142.

Smailes, P. J. 1996a. "Accessibility Changes in South Australia and the Country Town Network." In *Social Change in Rural Australia*, edited by G. Lawrence, K. Lyons, and S. Momtaz, 119–138. Rockhampton: Central Queensland University.

Smailes, P. J. 1996b. "Entrenched Farm Indebtedness and the Process of Agrarian Change; A Case Study and its Implications." In *Globalization and Agri-Food Restructuring*, edited by D. Burch, R. Rickson, and G. Lawrence, 301–322. Aldershot: Avebury.

Smailes, P. J. 1998. "Double or Quits: Restructuring and Survival on the Margin of the South Australian Wheatbelt." In *Local Economic Development: A Geographical Comparison of Rural Community Restructuring*, edited by C. Neil, and M. Tykkyläinen, 52–93. Tokyo: United Nations University Press.

Smailes, P. J. 2006. "Redefining the Local: The Social Organisation of Rural Space in South Australia, 1982–2006." PhD thesis, Flinders University, Adelaide. 330–337. Accessed http:// catalogue.flinders.edu.au/local/adt/public/adt-SFU20061005.151832.

Smailes, P. J., N. Argent, T. Griffin, and G. Mason. 2002. "Rural Community Social Area Identification." National Centre for Social Applications of Geographical Information Systems, University of Adelaide, Adelaide.

Smailes, P., T. Griffin, and N. Argent. 2014. "Demographic Change, Differential Ageing and Public Policy in Rural and Regional Australia: A Three-State Case Study." *Geographical Research* 52 (3): 229–249.

South Australia: Jubilee 150 Local Government Executive Committee. (1986) *The Civic Record*, 480. Adelaide: Wakefield Press.

Smailes, P. J., and G. J. Hugo. 2003. "The Gilbert Valley, South Australia." In *Community Sustainability in Rural Australia: A Question of Capital?* edited by C. Cocklin, and M. Alston, 65–106. Wagga Wagga: Centre for Rural Social Research, Charles Sturt University.

Sumerling, P. 1987. "Heritage of Eyre Peninsula: A Short History." SA State Historic Preservation Plan Regional Heritage Survey Series. Department of Environment and Planning, Adelaide.

Tonts, M., N. Argent, and P. Plummer. 2013. "Evolutionary Perspectives on Rural Australia." *Geographical Research* 50 (3): 291–303.

Tonts, M., K. Martinus, and P. Plummer. 2013. "Regional Development, Redistribution and the Extraction of Mineral Resources: The Western Australian Goldfields as A Resource Bank." *Applied Geography* 45: 365–374.

Watkins, M. H. 1963. "A Staple Theory of Economic Growth." *Canadian Journal of Economics and Political Science* 29 (part 2): 141–158.

Labour and Environmental Migration in the Asia-Pacific: in memory of Graeme Hugo

It is with great pleasure, honour and sadness that we present this second special issue of *Australian Geographer* in memory of Graeme Hugo: 'Labour and Environmental Migration in the Asia-Pacific'. This issue follows on from the first special issue, 'Population, Migration and Settlement in Australia' (vol. 47, no. 4). Together, these two issues of *Australian Geographer* mark the broad coverage of many of Hugo's key areas of research focus. Pleasure and honour stem from the privilege of observing the vast extent of Hugo's influence through the sound intellectual contributions made in essays and articles contributed by 36 special issue authors. Sadness stems from missing out on the opportunity to know how Hugo himself may have engaged with these works. Possibly, to find merit in their messages for 'developing appropriate population and development policies' as his 'persuasive concern for social justice' often led him to do (Connell 2015, 273).

As two guest editors born at the time when Hugo was beginning his career in geography, we find it interesting to reflect on Graeme's early areas of research focus. At that time, Hugo was establishing himself as an expert on labour mobility and circular migration dynamics in Asia, alongside his always present focus on Australian demography and population distribution (Connell 2015). In his early works about Asia, Hugo (1985, 75) demonstrated 'the nature and importance of temporary population movements', highlighting their significant—and often undetected and underestimated—social and economic implications for both sending and receiving countries. A focus on the importance of temporary migration and its implications for origin and destination locations, as well as for migrants' identities, rights and well-being, persists to this day. It has emerged as a key theme in papers featured across both of these special issues.

Situating Australia in the context of a rapidly globalising world, Castles' (2016) contribution, in the first special issue, drew attention to the dramatic and very swift rise in temporary skilled and student migration to Australia over the past decade—reflecting on its implications for notions of multiculturalism and transnational citizenship. In the same issue, Breen (2016) reflected on the emotional toll, at a family scale, of misunderstandings relating to temporary skilled migration opportunities. Temporary migration (both skilled and 'low-skilled')—and its dynamic consequences for social, economic and/or environmental changes in both migrant sending and receiving countries—has also emerged as a consideration among three works focused on labour migration in this second special issue (Bedford et al., Walton-Roberts et al., and Dun and Klocker). This common area of focus suggests a persistent need to continue with the work that Hugo commenced decades ago. Explorations of temporary migration—across all forms of skilled, 'low-skilled' and 'unskilled' migration—are needed now as much as ever. A concern with 'what sort of society we want in the twenty-first century—and how immigration can contribute to achieving this' (Castles 2016, 7) was certainly maintained by Hugo until the time of his passing.

Returning to Hugo's Asia-focused work in the mid-1980s, Associate Professor Alan Gamlen (Gamlen et al. 2016) recently explained that Graeme's interest in Indochinese refugees led to his focus on population movements impelled by natural disasters. This, Gamlen said, subsequently amounted to a 'Eureka moment' that culminated in the seminal publication 'Environmental Concerns and International Migration' (Hugo 1996). That publication firmly established Hugo as amongst the first migration scholars (and geographers) to focus on environmental change and human migration. Up to that point, that field had been dominated by policy makers and scholars from environmental and ecology backgrounds. Gamlen—Hugo's successor as the Director of the Hugo Centre for Migration and Population Research (formerly the Australian Population and Migration Research Centre)—noted that Hugo's interests in Indochinese political refugees and environmental migration were connected by his concern with the choices and constraints that influence population movements (Gamlen et al. 2016). Hugo and Chan (1990, 22) proposed the idea of 'natural disaster migrants' and worked through the idea of voluntary *vs* forced migration. In adopting that terminology, their focus was on the immediate factor triggering population movement, rather than 'the deeper underlying long term determinants', although they recognised that many '"natural" disasters have their root causes in long-term political, social, economic or agricultural practices or policies' (Hugo and Chan 1990, 22). This early work clearly identified the complexity of factors involved in migration associated with environmental change, and this is a key ongoing assertion in the field of environmental migration today (see, for example, Black et al. 2011). Of note, Hugo and Chan (1990) also reflected (albeit briefly) on government-initiated resettlement of populations due to volcanic eruptions in Indonesia and dam construction on the Mekong River in Laos and Thailand, further indicating Hugo's early comprehensive scoping out of the modes of movement (structure *vs* agency) linked to environmental change.

Our point here is to acknowledge how Graeme's keen interest in Asia, and his careful analysis of data and empirical studies from that continent, sparked much of his conceptual thinking in the realm of environmental migration. This later expanded into a focus on the Pacific Islands region, with increasing recognition of climate change. The ongoing influence of Hugo's leadership in this field is exemplified by four works published in this special issue focusing on the issues of migration, displacement and resettlement linked to potential or actual environmental change (McAdam, McNamara and Farbotko, Connell and Lutkehaus, and Tan). These papers make conceptual and methodological advances in this field, especially in relation to expressions of agency and forms of resistance (often *against* movement) amongst populations affected by environmental changes.

This second special issue commences with two Thinking Spaces that highlight the complex challenges that environmental destruction and change raise for relations and connections between Pacific Island nations and Australia. 'The High Price of Resettlement: The Proposed Environmental Relocation of Nauru to Australia', by Jane McAdam, highlights the implications of phosphate exploitation on Nauru under former British colonial rule. The environmental destruction wrought by phosphate mining resulted in an offer, by the Australian government, to resettle the entire population of Nauru on Curtis Island, Queensland. McAdam's thorough historical work shows how complex issues relating to sovereignty and identity, as well as cultural and spiritual connections to land, led the Nauruans to reject the offer of resettlement in Australia. Karen McNamara and Carol Farbotko's essay, 'Resisting a "Doomed" Fate: An Analysis of the Pacific Climate Warriors', is in turn focused on a contemporary case of resistance against migration in the context of environmental change. By applying a critical feminist lens to a case of peaceful protest, they show how Pacific Islanders are resisting the inevitability of migration in response to the threats posed to their islands by climate change. As these two complementary essays show, the desire to remain at home—and attachment to place—should not be underestimated in discussions and policy debates relating to climate change and its implications for population movement.

Our own Thinking Space, 'The Migration of Horticultural Knowledge: Pacific Island Seasonal Workers in Rural Australia—A Missed Opportunity?', shifts the focus towards Hugo's work on migration and development. We argue that seasonal and circular migration programmes ought to take further account of horticultural knowledge exchange. Focusing on Australia's Seasonal Worker Programme (SWP), we highlight the agency displayed by individual migrants with regards to knowledge transfer, and point towards potential structural interventions that could support the programme's development outcomes.

In addition to these Thinking Spaces, this special issue contains four journal articles, many of which have been prepared by Hugo's close colleagues and friends. The first two focus on labour migration while the latter two explore resettlement in relation to sudden-onset and slow-onset environmental changes.

The article 'Managed Temporary Labour Migration of Pacific Islanders to Australia and New Zealand in the Early Twenty-first Century' by Richard Bedford, Charlotte Bedford, Janet Wall and Margaret Young provides a rich account of New Zealand's Recognised Seasonal Employer (RSE) scheme and Australia's SWP. The authors provide a detailed overview of these managed, 'low-skilled', temporary labour migration schemes—including some insights from unpublished work by Graeme Hugo. Reflecting on Hugo's (2009) observations regarding 'best practice' in temporary labour migration, Bedford et al. consider the multiple actors and complex systems of relationships enmeshed in such schemes, and the ongoing management required to avoid workers becoming permanently trapped in a cycle of temporary movement.

The following article, 'Care and Global Migration in the Nursing Profession: A North Indian Perspective', aligns with Hugo's research interests at the intersection of care, labour and migration in Asia. Margaret Walton-Roberts, Smita Bhutani and Amandeep Kaur address the changing nature of nursing in India as influenced by the increasing globalisation of the care worker industry. They argue that nursing is now perceived as a higher status profession in India, largely because of the opportunities for upward mobility it affords through international migration. This paper also serves as a reminder of the scale of contemporary migration in Asia, through its provision of statistics about the number of Indian nurses abroad.

The final two journal articles focus on 'planned' resettlement linked to environmental changes. In their paper 'Environmental Refugees? A Tale of Two Resettlement Projects in Coastal Papua New Guinea', John Connell and Nancy Lutkehaus explore two cases of resettlement in that country: from Manam Island, where volcanic eruptions led to the resettlement of the entire island's residents, and from the Carteret Islands, where slower-onset processes of change have led to population displacement and resettlement. By providing rich insights into the islanders' experiences, the authors show that resettlement plans often come to grief due to issues of land tenure in resettlement locations, as well as social tensions, economic factors and a yearning for connection to ancestral lands. By focusing on 'the question of what happens to those from small islands that become uninhabitable but which are part of a larger nation' (p.92), the authors prompt reflection on the challenges of resettlement—even among culturally similar populations.

The final article in this special issue, 'Resettlement and Climate Impact: Addressing Migration Intention of Resettled People in West China', by Yan Tan, offers a new approach to the study of environmental change and migration. Tan's paper focuses on households resettled by the Chinese government—from a region where severe desertification, land salinisation and water shortage overlap—to resettlement locations that are themselves ecologically vulnerable. She explores the factors that influence resettled households' intentions to relocate *again*, or adapt locally. The paper adopts a two-step Probit regression model to assess relocation decisions, contributing a methodological advancement in the environmental migration field. Once more, this paper reminds us of the scale of population movement in Asia, with Tan's study taking place in Hongsibu district, 'the largest single environmental resettlement area in China' (p.102), which has received about 177 000 relocated people over a decade.

In closing this editorial, we would like to reflect on recent significant events that have further recognised Graeme Hugo's legacy, and which have coincided with the timing of our work on this latter special issue. A conference held in Graeme Hugo's honour took place at the University of Liège, Belgium, 3–5 November 2016. The Hugo Conference: Environment, Migration, Politics was 'named after the late Graeme Hugo (1946–2015), formidable scholar in migration studies and a pioneer in the examination of the linkages between environmental changes and migration'.[1] This conference also marked the launch of an international scholarly association for the study of environmental migration and The Hugo Observatory. The Hugo Observatory is based at the University of Liège and is 'a first-of-its-kind research structure […] dedicated specifically to the study of environmental changes and migration'.[2] The Hugo Conference and the Hugo Observatory both underscore the extent of this Australian geographer's highly regarded work on environmental migration and ongoing global influence.

Amongst the last of the PhD theses that Graeme Hugo examined before his passing was Olivia's thesis about the role of agricultural and environmental changes in human migration decisions (Dun 2014). In his examiner's report, reflecting on the approximately 30 country-level studies analysing the links between environmental change and migration contained in the literature review section of Dun's (2014) thesis, Hugo remarked:

> The section on Country-Level Studies … is interesting. I think one of the real messages from this is the diversity and multiplicity of the migration responses and [that] the search for universal 'laws' about how people respond to environmental change is a waste of time.

Of course, Hugo was not arguing against the need for ongoing research. Rather, he was once again drawing attention to the complexity of migration responses that can arise out of environmental change and that context matters.

After over 15 months of work on these special issues, including an intensive 9-month period of editing papers in Hugo's memory, we have an overwhelming sense of awe at what Graeme managed to achieve throughout the course of his working life. We take inspiration from Graeme Hugo's body of work and duly note his indelible influence on our own career trajectories.

Vale Professor Graeme Hugo, AO.

Notes

1. http://events.ulg.ac.be/hugo-conference/conference/
2. http://labos.ulg.ac.be/hugo/ and http://events.ulg.ac.be/hugo-conference/conference/

References

Black, R., W. N. Adger, N. W. Arnell, S. Dercon, A. Geddes, and D. Thomas. 2011. "The Effect of Environmental Change on Human Migration." *Global Environmental Change* 21 (Supplement 1): S3–S11.
Breen, F. 2016. "Australian Immigration Policy in Practice: A Case Study of Skill Recognition and Qualification Transferability Amongst Irish 457 Visa Holders." *Australian Geographer* 47 (4): 491–509.
Castles, S. 2016. "Re-thinking Australian Migration." *Australian Geographer* 47 (4): 391–398.
Connell, J. 2015. "Graeme John Hugo, 1946–2015." *Australian Geographer* 46 (2): 271–279.
Dun, O. 2014. "Shrimp, Salt and Livelihoods: Migration in the Mekong Delta, Vietnam." PhD diss., University of Sydney.
Gamlen, A., D. Bardsley, Y. Tan, and J. Wall. 2016. "Weaving the Strands: A Review of Graeme Hugo's Work on Environmental Migration." Keynote Presentation During the Plenary Session: 'Migration Challenges in a

Transformed Environment' at The Hugo Conference: Environment, Migration, Politics, Université de Liège, Belgium, 3–5 November 2016.

Hugo, G. 1985. "Circulation in West Java, Indonesia." In *Circulation in Third World Countries*, edited by R. M. Prothero and M. Chapman, 75–99. London: Routledge & Kegan Paul.

Hugo, G. 1996. "Environmental Concerns and International Migration." *International Migration Review* 30 (1): 105–131.

Hugo, G. 2009. "Best Practice in Temporary Labour Migration for Development: A Perspective from Asia and the Pacific." *International Migration* 47 (5): 23–74.

Hugo, G., and K. B. Chan. 1990. "Conceptualizing and Defining Refugee and Forced Migrations in Asia." *Southeast Asian Journal of Social Science* 18 (1): 19–42.

Olivia Dun

School of Geography, University of Melbourne, Melbourne, VIC 3010, Australia

Natascha Klocker

Australian Centre for Cultural Environmental Research, School of Geography and Sustainable Communities, University of Wollongong, Wollongong, NSW 2522, Australia

The High Price of Resettlement: the proposed environmental relocation of Nauru to Australia

Jane McAdam

Introduction

Most Australians today know the hot, rocky island of Nauru as a Pacific country to which Australia sends asylum seekers who have come by boat. Far fewer recall proposals 50 years ago to resettle the population of Nauru on an island off the Queensland coast. Extensive and lucrative phosphate mining on Nauru by Australia, the UK and New Zealand throughout the twentieth century devastated much of the 21 km^2 island, and scientists believed it would be rendered uninhabitable by the mid-1990s. With the exorbitant cost of rehabilitating the land, wholesale relocation was considered the only option. But the Nauruans refused to go. They did not want to be assimilated into White Australia and lose their distinctive identity as a people.

This episode adds further complexity to the fraught co-dependency of the Australian–Nauruan relationship, and an incongruous twist to the idea that Nauru might be able to resettle refugees today. In particular, it provides a cautionary tale for perennial discussions about the future relocation of Pacific island communities in the face of climate change.

The Australian–Nauruan relationship[1]

Nauru and Australia have had a long and uneasy relationship. Australia was the administrating power and main beneficiary of phosphate mining in Nauru between 1920 and 1968, during which time some 34 million tons of phosphate were removed, valued at around A\$300 million. Nauru had been a German protectorate from 1886 but was captured by Australian forces in November 1914. Prime Minister Billy Hughes was anxious to annex the territory because of its lucrative phosphate resources, mined by the Pacific Phosphate Company since 1907. These were of great value to the Australian agricultural export industry which needed fertilisers to improve the quality of its soil. Drawing on a message from his cabinet, Hughes explained to the British Colonial Secretary that Nauru's phosphate deposits rendered it 'of considerable value not only as a purely commercial proposition, but because the future productivity of our continent absolutely depends on such a fertiliser'.[2]

Despite Australia pressing its case at the Paris Peace Conference at the end of the war, mandate status over Nauru was informally granted to the British Empire in 1919, and formally in December 1920. The UK, Australia and New Zealand concluded a tripartite agreement (the Nauru Island Agreement in July 1919), pursuant to which Australia was appointed as the Administrator of Nauru, initially for 5 years. Having bought out the Pacific Phosphate Company, the agreement also established a Board of Commissioners represented by each of the partner governments (the British Phosphate Commissioners), in whom all title to the phosphate deposits was vested. This was described 70 years later by the Independent Commission of Inquiry (established to examine the partner governments' dealings with, and responsibilities in respect of, Nauru) as being wholly inconsistent with the very notion of a 'mandate', which was a 'sacred trust' for the benefit of the residents of the mandated territory, *not* for the profit of a company pursuing commercial and other interests of the administering powers (Weeramantry 1992).[3] In 1963, the New Zealand Prime Minister had in fact privately confirmed that 'the main object of the whole exercise was to secure the supply of cheap phosphate to Australia and New Zealand' (Written Statement of Nauru 1991, par. 81).

Nauru was occupied by the Japanese from 26 August 1942 until 14 September 1945, during which time most of the phosphate workings were destroyed. In 1947, Nauru became a trust territory of Australia, New Zealand, and the UK, with Australia again designated as the Administrating Authority. The UN Trusteeship Council consistently raised concerns about Australia's neocolonialist attitude to Nauru, recommending a larger degree of self-governance by the Nauruan population (UN General Assembly 1949). Nauru became self-governing in 1966 and independent in 1968.

Today, the mined-out areas cover almost 90 per cent of Nauru, with limestone pinnacles exposed to a depth of up to 6 m. Rehabilitation has been extremely expensive and slow. The Environmental Vulnerability Index (produced by the South Pacific Applied Geoscience Commission, the UN Environment Programme and partners) classifies Nauru as 'extremely vulnerable'—the highest level of vulnerability.

Early resettlement proposals

The question of resettlement of the Nauruan population was first raised in 1949, when the Australian government stated that the phosphate deposits would be exhausted within 70 years, and all but the coastal strip of Nauru would be 'worthless' (UN General Assembly 1949, 74). By 1956, the Commonwealth Scientific and Industrial Research Organisation (CSIRO) had revised the estimate down to 40 years and ruled out rehabilitation of the land as impracticable.

Prior to 1940, Australian authorities had believed that the rim around the island would provide sufficient land for the Nauruans to reside on indefinitely. But with the increasing phosphate requirements of Australian farmers in the post-war period, and the growth of Nauru's population, it became clear that it was inadequate. Resettlement thus became an increasingly attractive option for Australia, not least because it would facilitate the wholesale mining of the island.

The idea that a community could be relocated to another country was not regarded as far-fetched or fanciful. In fact, it was an already utilised policy tool in this part of the Pacific, and it tapped into a popular sentiment in the early to mid-twentieth century that redistributing the world's population could be a means of reducing resource scarcity and, in turn, conflict (Bashford 2014; McAdam 2015). UN Visiting Missions to Nauru in 1950, 1953 and 1956 pinpointed resettlement as the only viable long-term solution to

Nauru's impending uninhabitability. Possible relocation sites in and around Fiji, Papua New Guinea, the Solomon Islands and Australia's Northern Territory were explored, but were ultimately found to be inappropriate. Nevertheless, the Visiting Missions urged that a plan for gradual resettlement be agreed upon as soon as possible, rather than waiting until the phosphate deposits were exhausted. They noted that attention should be given to equipping younger Nauruans with vocational skills that would assist them to find employment in other parts of the Pacific.

Although Australia supported resettlement from Nauru, it rejected the Visiting Missions' view that phosphate mining made it necessary. While mining had destroyed much of the land, that was not, in the Australian government's view, the chief problem. Rather, it countered, contact with European enterprise as well as the adoption of 'European ways and standards', were the real reasons (UN General Assembly 1953, 114). The UN Trusteeship Council regarded this view as totally illogical. While it recognised that Nauru could not continue to support its people at their current living standards, this was inextricably connected to the environmental destruction caused by phosphate mining, as well as rapid population growth (Trusteeship Council 1961).

By and large, the Nauruans favoured rehabilitation of their land over relocation because it would enable them to remain in their homes and preserve their identity. However, recognising the dismal projections for Nauru's ongoing habitability, Head Chief Hammer DeRoburt mobilised support for resettlement, telling the UN Visiting Mission in 1956 that the Nauruans were now more in favour of total community resettlement in Australia, but were 'opposed to individual, gradual or piecemeal resettlement' (UN General Assembly 1956, 324). This was because Australia at this time favoured 'steadily educating' the Nauruans to a stage where they could 'fit into the economic and social life' of Papua New Guinea (then an Australian territory), or perhaps even Australia, rather than resettling them on an isolated island as a group.[4]

Instead, the Nauruans sought a commitment from the partner governments (UK, Australia and New Zealand) to meet the costs of a new homeland, including the cost of erecting villages, administrative centres, other public institutions and communication systems. In 1955, attempts were made on their behalf by Australia (as Adminstrator) to secure Woodlark Island between Papua New Guinea and the British Solomon Islands Protectorate, but were abandoned when it became clear that the island did not meet their needs—namely, employment opportunities that would enable them to maintain their standard of living, a host community that would accept them, and willingness and readiness on the part of the Nauruans to mix with the host population.

The UN Visiting Missions emphasised that the partner governments had a moral obligation to 'provide the most generous assistance towards the costs of whatever settlement scheme was approved', since they 'had benefited from low-price, high-quality phosphate' (UN General Assembly 1962, 40). Australian Prime Minister Robert Menzies acknowledged the three governments' 'clear obligation … to provide a satisfactory future for the Nauruans',[5] which involved 'either finding an island for the Nauruans or receiving them into one of the three countries, or all of the three countries', while having 'great regard to the views of the Nauruans'.[6]

Australia's position oscillated between group relocation, and individual immigration and assimilation into a metropolitan society. This was perhaps because of the difficulties in finding a suitable single relocation site. By 1959, the Visiting Mission thought that earnest consideration should be given to allowing Nauruans to migrate to one of the partner government countries or to a possession 'where the standard of living was comparable to that enjoyed by the Nauruans' (Memorial of the Republic of Nauru 1990, par. 162, sub-par. 62).

In October 1960, the three partner governments concluded that the most feasible option would be gradual resettlement within their metropolitan territories (Preliminary Objections of Australia 1990, par. 61), predominantly in Australia. At this time, the indigenous population of Nauru was about 2500 people. The idea was for gradual individual or household migration over a period of at least 30 years. This would ease pressure in Nauru by opening up alternative living space, and would provide new opportunities for those who moved. According to Nauruan documents, it was 'never envisaged that all Nauruans would take up the offer. Many would stay, and ... Nauru would always remain a spiritual home for those resettled' (Written Statement of Nauru 1991, par. 19). The terms of settlement were to include citizenship, equal opportunity and freedom of social contact. Young people were to receive education to the fullest extent of their capabilities, plus an allowance of £600 per annum (approximately A$17 000 today[7]) for five years, after which time they would be assisted to find suitable employment. Adults who were able to work, and for whom suitable employment could be found, were to receive their passage, a house, maintenance for 6 weeks, further training or the tools necessary for self-employment, and would be eligible for all social welfare benefits (Trusteeship Council 1961, 685–686).

The Nauruans rejected the offer, arguing that the very nature of the scheme would lead to their assimilation into the metropolitan communities where they settled. In Australia, an editorial in *The Age* expressed the well-meaning, but ultimately paternalistic and disempowering, view that 'the direct route to complete assimilation' was the best way to handle the problem, because '[t]here should be no racial enclaves in Australia and no second-class citizenship for these Pacific people' (cited in Viviani 1970, 142).[8]

In December 1960, the Nauruans requested an island of their own in a temperate zone off the Australian coast. In 1962, Fraser Island was identified as their preferred option. While Australia was willing to entertain the idea, it made clear upfront that sovereignty would not be transferred. When an expert survey concluded that Fraser Island did not offer sufficiently strong economic prospects to support the population, the Nauruans believed that this was simply an excuse on the part of the Queensland government to deny resettlement altogether (see Viviani 1970, 144). Indeed, archival materials suggest that the timber industry was keenly opposed to such a move.[9]

That same year, the Minister for Territories in Australia appointed a Director of Nauruan Resettlement, tasked with 'assiduously [combing] the South Pacific looking for spare islands offering a fair prospect'.[10] As a result, in 1963 the Australian government offered Curtis Island in Queensland (near Gladstone). Land on Curtis Island was privately held, but the government planned to acquire it and grant the Nauruans the freehold title. Pastoral, agricultural, fishing and commercial activities would be established, and all the costs of resettlement, including housing and infrastructure, would be met by the three partner governments—at the estimated cost of £10 million (about A$274 million today)[11] (Preliminary Objections of Australia 1990, par. 63).

While Australia reiterated that 'sovereignty would not be surrendered' (Preliminary Objections of Australia 1990, par. 62), the government agreed to grant the Nauruans freedom of movement and the right to 'manage their own local administration and legislate for their own country', subject to their acceptance of 'the privileges and responsibilities of Australian citizenship' (United Nations Fourth Committee 1963, 565, par. 3). Thus, 'a Nauruan Council would be established with wide powers of local government within the jurisdiction of the Queensland Government' (UN General Assembly 1964, 24), permitting a degree of self-governance. The idea that Nauru could create a 'new Nauru' was dismissed out of hand, an early departmental minute recording that 'our best interests would be served by playing along' with the idea, but never seriously entertaining it.[12]

Nauru again rejected the resettlement offer, deeming the political arrangements to be unsatisfactory. The Nauruan representatives feared that they would not be able to maintain their distinct identity and would be 'assimilated without trace into the Australian landscape' (Memorial of the Republic of Nauru 1990, par. 171).

> Your terms insisted on our becoming Australians with all that citizenship entails, whereas we wish to remain as a Nauruan people in the fullest sense of the term even if we were resettled on Curtis Island. To owe allegiance to ourselves does not mean that we are coming to your shores to do you harm or become the means whereby harm will be done to you through us. We have tried to assure you of this from the beginning. Your reply has been to the effect that we cannot give such an assurance as future Nauruan leaders and people may not think the same as we do.[13]

The Nauruans proposed 'the creation of a sovereign Nauruan nation governed by Nauruans in their own interest but related to Australia by a treaty of friendship' (UN General Assembly 1962, 32). External affairs, defence, civil aviation and quarantine would remain in the hands of Australia. This, the Nauruans argued, would safeguard Australia against anything which might endanger its national security. When Australia refused the offer, Nauru accused the government of not taking its proposal seriously.

Nauru rejects resettlement in Australia

Australia's offer of resettlement was finally rejected by the Nauruans in July 1964. Nauruan and Australian perspectives on the issue reveal quite different views as to why it failed.

The Nauruans claimed that resettlement was offered as a quick-fix solution that would cost the Australians far less than rehabilitating the land. It was, they said, an attempt to break up their identity and their 'strong personal and spiritual relationship with the island', ignoring 'the right of the Nauruan people at international law to permanent sovereignty over their natural wealth and resources' (Written Statement of Nauru 1991, pars. 20, 74). Hammer DeRoburt stated that his people were never 'seeking full sovereign independence' over Curtis Island, but that 'anything which did not preserve and maintain [their] separate identity was quite unacceptable'.[14] In summarising their rejection of Australia's resettlement offer, the Nauruans explained:

> We feel that the Australian people have an image of Nauruans which is quite wrong, but which the Government has made little effort to correct. Australians seem to have a picture of an absurdly small people who want too much from Australia, who want complete sovereign independence, and who are not as grateful as they should be for what Australia is generously offering them.
>
> We feel that most Australians think that the predicament facing the Nauruan people today which has given rise to their need for resettlement elsewhere is due to natural over population

and would-be sophistication of the younger Nauruan generation. We feel that Government propaganda aimed at shifting the blame to natural causes and evolution, is responsible for this unfair emphasis but have met with very little success. Although such factors may be regarded as contributory, it is wrong to attribute the necessity of resettlement wholly or primarily to them. We submit again that the main need for resettlement arises out of the physical destruction of the island and its attendant problems. Four-fifths of our island is phosphate-bearing and therefore in the end that much will be destroyed. (Memorial of the Republic of Nauru 1990, par. 173, citing Nauru Talks 1964, 4–5)

By contrast, the Australian government pinned the failure of the resettlement negotiations precisely on the issue of sovereignty. Seemingly frustrated by what it perceived as a genuine and generous attempt to meet the wishes of the Nauruan people, the Australian government told the UN that 'it would not be able to depart from its decision that it could not transfer sovereignty over territory which was at present part of Australia' (United Nations Fourth Committee 1963, 565, par. 4). As the Independent Commission of Inquiry later found, Australia's resettlement proposal 'violated each and every one' of the objectives of trusteeship under the UN Charter—namely, the promotion of political, economic and social advancement, and the promotion of progressive development towards self-government or independence (Weeramantry 1992, 297, 403).

The issue resurfaced in 2003, when Alexander Downer, then Australia's Foreign Minister, was reported as saying that he was considering the resettlement of Nauru's population in Australia and grants of Australian citizenship. He said he was 'very concerned' about Nauru's prospects because it was 'bankrupt and widely regarded as having no viable future' (Marks 2003). Australian officials regarded the country as unsustainable, noting that it had only been kept running by A\$30 million in funding over the previous 2 years to run Australia's offshore asylum seeker processing centres. The resettlement proposal was dismissed by the President of Nauru, who said it would undermine Nauru's identity and culture.

Planned relocation and resettlement in contemporary debates

In contemporary international discussions, 'planned relocation' has been identified as a possible strategy to assist low-lying Pacific island communities at risk from the impacts of climate change. Yet cultural misunderstandings about the importance of land and identity remain, and highlight the enduring importance of matters such as the right to self-determination, self-governance, the preservation of identity and culture, and the right to control resources. Past experiences in the Pacific show the potentially deep, inter-generational psychological consequences of planned relocation, which may explain why it is considered an option of last resort in that region (McAdam 2014).

As Graeme Hugo so aptly observed, it is essential that climate-change-related movement is understood within a broader historical context, linked to 'existing knowledge of migration theory and practice' (Hugo 2011a, 260). As the resettlement literature shows very clearly, 'time is required to put in place all of the institutions, structures and mechanisms to facilitate equitable and sustainable resettlement', and while 'the desired end point may be decades away, there is an urgency to begin the planning process' (Hugo 2011a, 277).

The absence of bilateral, regional or international migration frameworks means that it is unclear how many Pacific islanders will have the opportunity to move voluntarily in anticipation of longer-term changes to their islands. This, in turn, may affect whether and how any cross-border community relocation might occur. It is very unlikely that any country today would provide a dedicated portion of land capable of housing the whole population of a small island state, which is why any future relocation is more likely to involve smaller settlements in a variety of areas.

When I asked the President of Kiribati, Anote Tong, in 2009 whether he would like to be able to retain some form of self-governance for Kiribati if the whole population ultimately had to leave, he said:

> Quite frankly that's an issue that I've never really focused on. I focus on getting our people to survive. But these issues—I think, at some point in time they will have to be addressed. But if you're scattering your people in different parts of the globe, how do you retain national unity?

The matters that concerned Nauru 60 years ago continue to resonate today.

Meanwhile, of course, Australia has sought to designate Nauru as a country *of* resettlement for a different purpose: for refugees who sought to reach Australia by boat, who have been denied the opportunity to settle in Australia.

Refugees who have settled in Nauru (839, as at 31 January 2016: Department of Immigration and Border Protection 2016) live with a well-founded fear of being assaulted or becoming the victim of a violent crime, including rape. They have highly limited prospects for meaningful employment or social engagement, and face a local culture of resentment. Numerous allegations of abuse, including sexual abuse, have been recounted in reports by the Australian Human Rights Commission (2014), the independent Moss Review (2015), and the Senate Committee on the Recent Allegations relating to Conditions and Circumstances at the Regional Processing Centre in Nauru (2015).

In 2014, over 50 single male refugees resettled in Nauru issued the following statement:

> We are living in a camp in the jungle … We want to tell you that we are here like animals … We came to Australia as asylum seekers for safety and for our future. We stay[ed] in Nauru at IDC [Immigration Detention Centre] for 11–12 months. Now we are accepted as refugees and they have given us safety but what can we do with this if there is no future for us, there is no meaning[?] We all came with our hopes and our wishes to help our family and ourselves. But now we have no future, no hope. In our country the Taliban will come onto the bus and they will slash our throat and finish your life. It will take maybe 10 or 15 minutes for us to die. But the English–Australian men are killing us by pain, taking our soul and our life slowly.[15]

In this case, Australia does not need to find an empty island on which to relocate these people from Nauru. Rather, it has a responsibility to bring them to Australia, resettle them in the community, and provide them with durable and meaningful protection in accordance with its international legal obligations.

Indeed, as Graeme Hugo's seminal study of the long-term contributions made by humanitarian entrants to Australia showed (Hugo 2011b), given the opportunity, refugees can become some of the country's most resourceful and successful people. As the then Immigration Minister, Chris Bowen, observed in his foreword to the study:

Given the often extreme hardship from which humanitarian entrants have come, it is all the more impressive that they are able to achieve so much in such an unfamiliar environment. It is these characteristics—resourcefulness, hard work and determination to improve their lives and the lives of their children—that come through so clearly in this research. And it is these attributes that Australians will recognise as those that will continue to make this country great, long into the future. (Bowen 2011, 4)

Too often, the failure to learn from the past means that destructive policies are repeated. Just as the reinstatement of offshore processing was doomed as a 'solution' to displacement (Gleeson 2016), the relocation of communities away from areas threatened by the damaging impacts of disasters and climate change will be highly fraught unless it is underpinned by a respectful, considered and consultative process in which a full range of views can be voiced and heard (Ferris 2012; McAdam and Ferris 2015). As for Nauru, its own future seems sadly rooted in an unhealthy relationship of co-dependency with Australia, its territory once again exploited at the expense of the vulnerable.

Notes

1. Much of the information in this article comes from UN Trusteeship Council records and the written memorials of Australia and Nauru in the *Case Concerning Phosphate Lands in Nauru (Nauru v. Australia)* before the International Court of Justice. Accessed May 20, 2016. http://www.icj-cij.org/docket/index.php?sum=413&p1=3&p2=3&case=80&p3=5. Only direct quotes are specifically attributed.
2. Prime Minister Hughes to Lord Milner, May 3, 1919, Lloyd George Papers, Beaverbrook Library, London, F/28/3/34, cited in Memorial of the Republic of Nauru (1990, par. 34).
3. In that case, the Nauruans argued that mining had rendered the land 'completely useless for habitation, agriculture, or any other purpose unless and until rehabilitation was carried out' and that in its role as Administrator, the Australian government had 'failed to make adequate and reasonable provision for the long-term needs of the Nauruan people' (Application Instituting Proceedings (Nauru) 1989, pars. 15 and 17 respectively). An out-of-court settlement was ultimately reached, with Australia paying A$107 million compensation and Nauru agreeing not to take any further legal action (Australian Government Department of Foreign Affairs and Trade 2013, 4, fn. 14).
4. Departmental minute dated November 5, 1953 by the Secretary to the Department of Territories to the Minister, Australian Archives ACT CRS A518, Item DR118/6 PT.1; Annexes, vol. 4, Annex 62, cited in Memorial of the Republic of Nauru (1990, par. 569).
5. Letter from Robert Menzies (Prime Minister of Australia) to G.F.R. Nicklin (Premier of Queensland), January 22, 1962. Nauruans—Resettlement in Australia, Series ID 5213, Item ID 842358, January 22, 1962–March 22, 1965, Queensland State Archives, cited in Tabucanon and Opeskin (2011, 342).
6. Attributed to the Melbourne *Herald* in a memorandum submitted by the Nauru Local Government Council to the 1965 UN Visiting Mission: *Trusteeship Council Official Records*, 32nd sess., Suppl. no. 2, Annex 1 (May 2–June 30, 1965), 13, cited in Memorial of the Republic of Nauru (1990, par. 562).
7. Accessed May 20, 2016. http://www.rba.gov.au/calculator/annualPreDecimal.html
8. 'Something to be Proud of', *The Age*, June 26, 1961, 2, cited in Viviani (1970, 142).
9. Letter from the President of the Maryborough & Bundaberg District Timber Merchants' Association to O.O. Madsen (Queensland Minister for Agriculture and Forestry), February 23, 1962, Nauruans—Resettlement in Australia, Series ID 5213, Item ID 842358, January 22, 1962–March 22, 1965, Queensland State Archives, cited in Tabucanon and Opeskin (2011, 346).

10. 'Verbatim Record of Public Sitting', *Certain Phosphate Lands in Nauru (Nauru v. Australia)*, International Court of Justice, General List no. 91, November 18, 1991, 17 (Barry Connell), cited in Tabucanon and Opeskin (2011, 346).
11. Accessed May 20, 2016. http://www.rba.gov.au/calculator/annualPreDecimal.html
12. Departmental minute of November 5, 1953, Australian Archives ACT CRS A518, Item DR 118/6 Pt 1, reproduced in Commission of Inquiry Report Documents 896, cited in Weeramantry (1992, 290).
13. Nauru Talks (1964, 1–2), Annexes, vol. 3, Annex 1: 'Summary of the Views Expressed by the Nauruan Delegation at the Conference in Canberra July–August 1964', cited in Memorial of the Republic of Nauru (1990, par. 171).
14. 'Statement by Hammer Deroburt, OBE, GCMG, MP, Head Chief, Nauru Local Government Council', Appendix 1 to Memorial of the Republic of Nauru (1990, par. 21). However, as Tabucanon and Opeskin (2011, 347) note, *The Age* newspaper at the time stated that Nauru wanted to establish Curtis Island as a sovereign State, tied to Australia by a treaty of friendship, and controlled by Australia only in matters of defence, quarantine, and possibly external affairs and civil aviation: 'Island Offer Rejected by Nauru', *The Age*, August 21, 1964. This was based on the 1962 Treaty of Friendship between New Zealand and Western Samoa: see citation in Viviani (1970, 143). Similarly, Tate argues that the three fundamental conditions of resettlement on Curtis Island were that the Nauruans be granted full independence, enjoy territorial sovereignty over their new homeland, and retain sovereignty over Nauru (Tate 1968, 181, cited in Tabucanon and Opeskin 2011, 347).
15. 'Statement from the More Then [*sic*] Fifty Nauru Refugees' (August 4, 2014), http://www.scribd.com/doc/235771504/Statement-From-the-More-Then-Fifty-Nauru-Refugees (accessed January 6, 2016).

Disclosure statement

No potential conflict of interest was reported by the author.

Funding

This research was funded by an Australian Research Council Future Fellowship [FT110100721].

References

Application Instituting Proceedings (Nauru). 1989, May. *Case concerning Certain Phosphate Lands in Nauru (Nauru v. Australia)*, International Court of Justice.
Australian Government Department of Foreign Affairs and Trade. 2013. *Aid Program Performance Report 2012–13 Nauru*. Canberra: Commonwealth of Australia.
Australian Human Rights Commission. 2014. *The Forgotten Children: National Inquiry into Children in Immigration Detention*. Sydney: Australian Human Rights Commission.
Bashford, A. 2014. *Global Population: History, Geopolitics, and Life on Earth*. New York: Columbia University Press.
Bowen, C. 2011. Immigration Minister, 'Foreword'. In G. Hugo, *A Significant Contribution: The Economic, Social and Civic Contributions of First and Second Generation Humanitarian Entrants: Summary of Findings* (Commonwealth of Australia).
Department of Immigration and Border Protection. 2016, February. 'Operation Sovereign Borders Monthly Update: January 2016.' Accessed March 15, 2016. http://newsroom.border.gov.au/channels/Operation-Sovereign-Borders/releases/operation-sovereign-borders-monthly-update-january-2.
Ferris, E. 2012. 'Protection and Planned Relocations in the Context of Climate Change'. UNHCR Legal and Protection Policy Research Series, PPLA/2012/04.

Gleeson, M. 2016. *Offshore: Behind the Wire on Manus and Nauru*. Sydney: NewSouth Publishing.

Hugo, G. 2011a. "Lessons from Past Forced Resettlement for Climate Change Migration." In *Migration and Climate Change*, edited by E. Piguet, A. Pécoud and P. de Guchteneire, 260–88. Cambridge: Cambridge University Press.

Hugo, G. 2011b. *A Significant Contribution: The Economic, Social and Civic Contributions of First and Second Generation Humanitarian Entrants: Summary of Findings* (Commonwealth of Australia).

Marks, K. 2003. "Australia Moots Radical Future for Bankrupt Nauru." *The Independent*, December 20, 2003. Accessed January 6, 2016. http://www.independent.co.uk/news/world/australasia/australia-moots-radical-future-for-bankrupt-nauru-83312.html.

McAdam, J. 2014. "Historical Cross-Border Relocations in the Pacific: Lessons for Planned Relocations in the Context of Climate Change." *Journal of Pacific History* 49: 301–27.

McAdam, J. 2015. "Relocation and Resettlement from Colonisation to Climate Change: The Perennial Solution to "Danger Zones"." *London Review of International Law* 3: 93–130.

McAdam, J. and Ferris, E. 2015. "Planned Relocations in the Context of Climate Change: Unpacking the Legal and Conceptual Issues." *Cambridge Journal of International and Comparative Law* 4: 137–66.

Memorial of the Republic of Nauru. 1990, April. *Case concerning Certain Phosphate Lands in Nauru (Nauru v Australia)*, International Court of Justice, Pleadings.

Moss, P. 2015. *Review into Recent Allegations Relating to Conditions and Circumstances at the Regional Processing Centre in Nauru: Final Report*. http://www.border.gov.au/Reportsand Publications/Documents/reviews-and-inquiries/review-conditions-circumstances-nauru.pd f#search=moss.

Preliminary Objections of the Government of Australia. 1990, December. *Case concerning Certain Phosphate Lands in Nauru (Nauru v Australia)*, International Court of Justice, Pleadings.

Senate Committee on the Recent Allegations relating to Conditions and Circumstances at the Regional Processing Centre in Nauru. 2015. *Taking Responsibility: Conditions and Circumstances at Australia's Regional Processing Centre in Nauru*. Canberra: Commonwealth of Australia.

Tabucanon, G. M., and B. R. Opeskin. 2011. "The Resettlement of Nauruans in Australia: An Early Case of Failed Environmental Migration." *Journal of Pacific History* 46: 337–56.

Tate, M. 1968. "Nauru, Phosphate, and the Nauruans." *Australian Journal of Politics and History* 14 (2): 177–92.

Trusteeship Council. 1961. "Examination of Conditions in Trust Territories." *International Organization* 15(4): 671–702.

United Nations Fourth Committee. 1963, December 12. UN GAOR, 4th Comm, 18th sess, 1513th mtg, Agenda Item 13, UN Doc A/C.4/SR.1513.

United Nations General Assembly. 1949. *Report of the Trusteeship Council*. UN GAOR, 4th sess, UN Doc A/933 (SUPP).

United Nations General Assembly. 1953. *Report of the Trusteeship Council*. UN GAOR, 8th sess, UN Doc A/2427 (SUPP).

United Nations General Assembly. 1956. *Report of the Trusteeship Council*. UN GAOR, 11th sess, UN Doc A/3170 (SUPP).

United Nations General Assembly. 1962. *Report of the Trusteeship Council*, UN GAOR, 17th sess, UN Doc A/5204 (SUPP).

United Nations General Assembly. 1964. *Report of the Trusteeship Council*, UN GAOR, 19th sess, UN Doc A/5804 (SUPP).

Viviani, N. 1970. *Nauru: Phosphate and Political Progress*. Canberra: Australian National University Press.

Weeramantry, C. 1992. *Nauru: Environmental Damage under International Trusteeship*. Melbourne: Oxford University Press.

Written Statement of Nauru. 1991, July. *Case concerning Certain Phosphate Lands in Nauru (Nauru v Australia)*, International Court of Justice, Pleadings.

Resisting a 'Doomed' Fate: an analysis of the Pacific Climate Warriors

Karen E. McNamara and Carol Farbotko

Introduction

The impacts of climate change have been, and are predicted to be, particularly concentrated in tropical areas such as the Pacific Islands. With rising sea levels, more droughts, and more frequent and intense storm activity now evident across the region, scholars have deemed low-lying countries as likely to be uninhabitable in the future. Hugo (1996, 125) made the case early on that '[I]nternational relocation may provide an enduring solution' for small islands. This pragmatic position—of migration being a vital component of an effective adaptation response—was repeated in much of Hugo's work (see Bardsley and Hugo 2010; Hugo 2010). In this vein, Hugo's work provided a significant contribution to the complex challenge of uninhabitability in the Pacific Islands region, but it is not the only narrative. A growing number of island nation leaders and civil society groups have vocalised their opposition to a scenario whereby resettlement abroad is considered inevitable. This essay provides details of a grassroots network that defies the inevitability narrative and, like the extensive work of Hugo, offers important and critical contributions to the serious challenges facing the Pacific Islands region now and in the future.

Pacific Climate Warriors is a grassroots network of young men and women from various countries in the Pacific Islands region driven by the need to take action to peacefully protect the islands from climate change impacts. They have a clear message: 'We are not drowning, we are fighting!' (350 Pacific n.d., n.p.). One of their most prominent campaigns was a tour of Australia in October 2014 which involved blockading the world's largest coal port (located in Newcastle, Australia) to demand that companies trading in fossil fuels and heavy greenhouse gas (GHG) emitting countries take responsibility for their actions. In this essay we analyse the Pacific Climate Warriors network, in particular their blockade of Newcastle coal port, to show how these activists are resisting narratives of future inevitability of their Pacific homelands disappearing, and re-envisioning islanders as warriors defending rights to homeland and culture. We also reflect on how the Pacific Climate Warriors network is recasting historical patriarchal figures (the male 'warrior') by evoking feminine characteristics, creating a blurring of gender identities. The net effect of such is a robust grassroots movement bound by solidarity and a collective identity. We begin with a brief overview of climate change in the Pacific Islands region and of literature foregrounding a critical feminist perspective on climate change. We then provide a more in-depth discussion of the Pacific Climate Warriors and their role in re-envisioning a Pacific Island future.

Climate change and the Pacific

The Pacific Islands region is made up of 22 countries and territories, all with varying population sizes, politics, physical characteristics and cultures (Barnett and Campbell 2010). Key industries include agriculture, fisheries and tourism, and many of the smaller economies are significantly dependent on remittances and aid (Barnett and Campbell 2010). Many human and economic development challenges exist across the region, which also grapples with the impacts of climate change and globalisation (Hughes 2003; McNamara and Westoby 2014).

Over the last decade, climate change has risen in prominence as a key challenge for governments and communities across the Pacific Islands region. Climate change impacts have included sea level rise, increasing temperatures, shifting rainfall patterns, and intensifying storms and cyclones (Nurse et al. 2014). These have seen a decline in water and food security, human health implications and livelihood hardship due to changing ecosystems (Mimura et al. 2007; Nurse et al. 2014). As highlighted by Hay and Mimura (2013, 1), a vast number of vulnerability assessments have been undertaken in communities across the region over the last decade and all point to the region's high level of 'risk to the adverse consequences of climate change'. The most recent Intergovernmental Panel on Climate Change (IPCC) report states: ' ... the threats of climate change and sea level rise (SLR) to small islands are very real. Indeed, it has been suggested that the very existence of some atoll nations is threatened by rising sea levels associated with global warming' (IPCC in Nurse et al. 2014, 1618). Common depictions of the region as 'most vulnerable', 'sinking', 'drowning', 'doomed' and so on are widespread in the popular press, policy, academia and civil society (some examples include Friends of the Earth 2007; Biermann and Boas 2008; Environmental Justice Foundation 2009; GermanWatch 2014; United Nations University and Bündnis Entwicklung Hilft 2016). However, there is a growing movement across the region resisting not the science underlying these scenarios nor the often well-intentioned publicity of significant climate impacts but a (largely externally imposed) narrative of Pacific Island communities as vulnerable victims.

Dominant characterisations of the Pacific Islands region as 'vulnerable' and 'doomed' can hamper short- and long-term adaptive capacity in the islands, weaken efforts for *in situ* adaptation and draw attention away from the importance of global emissions reductions by heavy GHG emitting nations (Farbotko and Lazrus 2012). Robust internal coping capacities of local Pacific Islander communities have been documented, which help to counter dominant typologies of the region as possessing an already-sealed fate. Bridges and McClatchey (2009, 140) made the case that: 'Over many generations ... atoll cultures have survived major, unpredictable and locally devastating changes that are of the same magnitude as those expected from climate changes.' Moreover, island communities have drawn on Indigenous knowledge and have learnt to conserve their resources and respond to various environmental stressors in new ways. Some of the learnings have included: monitoring changes in plant and animal behaviour which can be key indicators of changes in weather and climate; preparing for extreme weather events by using particular planting and food preservation techniques; and managing natural resources such as water through innovative storage practices (see examples by Lefale 2010; McNamara and Prasad 2014; see also Herman 2016). The Pacific Climate Warriors, through their climate activism, further demonstrate these robust internal coping capacities, as we discuss in this essay.

A critical feminist perspective

Gender is a key climate change issue worldwide, in terms of both framing the phenomenon and experience of it (MacGregor 2009; Tschakert 2012). Graeme Hugo argued that when it comes to questions of climate change, displacement and migration, there is a particular need for gender sensitivity (Hugo 2010). Our critical feminist perspective in this essay assesses, rather than accepts, a dominant paternalism in the science and management of climate change. For example, we highlight that certain responses, such as emotion, are typically marginalised from climate change debates in favour of 'rational' decision making (Farbotko and McGregor 2010). Critical feminist perspectives are vitally important for identifying climate change 'solutions' if we acknowledge humanity as heterogeneous, science as political, and social injustice as being exacerbated under climate change threats (MacGregor 2009). As an issue of cultural politics, the ways in which the feminine is invoked in climate change narratives are important in understanding climate activism, and yet such politics has received little attention to date (Detraz and Windsor 2014).

Within the 25-year history of international diplomacy and more recent peaceful grassroots activism that Pacific people have engaged in to voice their concerns about climate change, use has been made of song, dance, poetry and art, internet activism, action research, faith-based activities, and collaboration with documentary makers, museums and artists (Farbotko and McGregor 2010; Dreher and Voyer 2015; Steiner 2015). Many of these climate change activisms incorporate more traditional Pacific feminine characteristics of nurturing and collaborating, and are typically lacking in adversarialism. It is against this backdrop that we describe the Pacific Climate Warriors network and their recent campaign in Australia. By focusing on the gender politics of the Pacific Climate Warrior we can consider whether a hyper-masculine hegemony is quashed among a marginalised group conducting resistance to the dominant economic and political forces that perpetuate fossil fuel reliance (Chen 2014; Fair 2015; Steiner 2015).

Pacific Climate Warriors

Since the 1990s, Pacific Island leaders have called for reduced global fossil fuel dependency. Moreover, with GHG emissions still rising each year, Pacific communities are acutely aware of the costs of inaction. To advance action, in 2011–12 a group of young Pacific Islanders from a number of different Pacific Island nations formed a network called 350 Pacific (also known as the Pacific Climate Warriors), which is linked to the global non-governmental climate action organisation 350.org. This umbrella organisation, established by environmental author and activist Bill McKibben in 2007, encourages citizens to take action and raise awareness about rising carbon dioxide levels in the earth's atmosphere. The intention is to pressure world leaders and businesses to reduce their GHG emissions to maintain a 'safe' level of 350 parts per million (ppm) atmospheric concentrations of carbon dioxide (we note that atmospheric carbon dioxide concentrations currently surpass 400 ppm). The Pacific Climate Warriors, active in 15 countries[1] across the region, aim to empower young people to understand the implications of a changing climate, how it affects the islands, and to take action to protect the islands.

The motivations of these warriors in joining the network are clearly presented on the 350 Pacific website:

> Every morning, we wake up and the ocean is there, surrounding our island. But now the ocean, driven by climate change is creeping ever closer. Unless something changes, many of our Pacific Islands face losing everything to sea level rise. (Unattributed quote; 350 Pacific n.d., n.p.)

> We have to find ways to keep coal and gas in the ground. People all around the world are recognising this and taking action to challenge the power of the fossil fuel industry. For us Pacific Islanders, there is nothing more urgent or necessary. (Mikaele Maiava, Tokelau Pacific Climate Warrior; 350 Pacific n.d., n.p.)

> It is very important for us to take direct actions against climate change because it is threatening our lives and our islands. Our land is the most valuable treasure in our lives and the impacts of climate change will destroy it. We don't want this to happen and we will not allow it to happen. (Mikaele Maiava, Tokelau Pacific Climate Warrior; 350 Pacific 2014a, n.p.)

> I've seen my people and my islands suffer the impacts of climate change through droughts and floods from high tides … None of us who have felt the impacts of climate change should continue to suffer through them just to fulfil others' interests. We don't deserve to lose our Islands and we will do what we must to ensure we won't. (Milañ Loeak, Republic of the Marshall Islands Pacific Climate Warrior; 350 Pacific 2014a, n.p.)

These accounts demonstrate the warriors' determination and conviction to advocate for rights to their homelands and cultures. They also highlight their calls for justice, which is particularly pertinent in this context given their nations' negligible GHG emissions yet disproportionate burden to deal with the consequences of increasing emissions.

Non-violent direct actions taken by the Pacific Climate Warriors include major campaigns such as the tour in Australia in October 2014 (as discussed below) and other events such as participation at the United Nations Framework Convention on Climate Change international negotiations each year. It is clear from much of the published material on the 350 (Pacific and global) websites that frustrations among the Pacific Climate Warriors are voiced squarely at the fossil fuel industry and industrialised-nation governments, resulting in a number of actions directly targeting these entities. One warrior blames the fossil fuel industry as a threat to the existence of Pacific Islanders:

> The fossil fuel industry is the biggest threat to our very existence as Pacific Islanders. We stand to lose our homes, our communities and our culture. But we are fighting back. (Kathy Jetnil-Kijiner, Republic of the Marshall Islands Pacific Climate Warrior; 350.org 2014a, n.p.)

For these warriors, the motivations are clear: to work together, to take action and to advocate strongly against those who control fossil fuels.

Long before the Pacific Climate Warriors were formed, the (male) warrior/navigator was an embodiment first of patrilineal claims to islands among the original settlers of the Pacific Islands and then of postcolonial reclaiming of sovereignty in the Pacific. While the Pacific Climate Warriors have been acclaimed in climate debates for their clear assertion and exercise of political agency by Pacific youth, it is also the *image* of the Pacific Climate Warrior, at face value a masculine figure, which is of interest. There are a number of high-profile female warriors in the network. Kathy Jetnil-Kijiner, a Marshallese poet, is one of the most well known of the Pacific Climate Warriors who quickly rose to prominence internationally as she was chosen from 544 people (nominated from 115 countries) to speak at the opening ceremony of the United Nations Climate Summit in New York City in September 2014. Another well-known Pacific Climate Warrior is Milañ Loeak, who is the daughter of the current President of the Marshall Islands, Chris Loeak. Milañ has often featured in promotional material for the Pacific Climate Warriors with powerful imagery and messages such as: 'Join me in the fight to keep my home above water' (350.org 2014b, n.p.). Their narratives invoke feminine characteristics, such as maternal nurturing and intensive emotional links to ancestors, which is likely to be influenced by wider Pacific women's work promoting peace (George 2012, 2014):

> I never really allowed myself to feel the full emotion of what losing our islands would mean ... [until] I sat outside in the sun and I wept. My cries were more than my own cries—I felt my ancestors sitting beside me, weeping with me. I heard their echoes, reverberating in my sorrow. I felt their/our anguish over our islands, over the next few generations. I felt the shuffling feet of our future generations—floating adrift, the hopelessness and inability to go on ... I have foreseen the loss and the sorrow that awaits our children and grandchildren, because I have fallen into that abyss ... I will drown the wound in salt. I will do anything to save my islands. (Jetnil-Kijiner 2015, n.p.)

> My passion for climate action is stemmed from the love I have for my family, my fellow Marshallese people, our islands and our culture ... most importantly, share your stories with the children. Teach them. Educate them. People say we need to create a better world for our children, but I believe we should be creating better children for the world that God has already blessed us with. Through our children, we will see the fruition of our work. (Loeak 2015, n.p.)

When ideas about climate victimhood are rejected through mobilisation of a new subjectivity, the Pacific Climate Warrior, the warrior figure is also shedding its combative conceptual lineage and promoting more peaceful pathways to change (George 2012, 2014).

Campaign in Australia

The warriors' most visible campaign to date has been a tour of Australia (8–25 October 2014), conducted with support from 350.org. During the campaign, 30 young Pacific Climate Warriors—both men and women—travelled to Australia from Pacific countries including Fiji, Kiribati, Papua New Guinea, Republic of the Marshall Islands, Tokelau, Tonga, Tuvalu and Vanuatu. The aim of the campaign was to peacefully raise awareness of the urgent need to reduce GHG emissions by providing first-hand accounts of the impacts of climate change on Pacific Island nations, and to directly confront and disrupt the fossil fuel industry.

The visual and symbolic centrepiece of this campaign occurred on 17 October 2014 in Newcastle, Australia—the site of the world's largest coal port. Here, the warriors, with support from hundreds of local Australian residents, paddled out in traditional canoes and kayaks in an effort to prevent coal ships entering and leaving the port for the day.

Based on accounts by 350 Pacific (2014b), only four of the twelve ships that were to pass through the port that day were able to do so (two of these ships were coal ships) while the others were all successfully blockaded by the warriors. The action, which continued for several hours, involved several encounters by canoeists and kayakers with police and police vessels while on the water, with reports of some (protestor) boats capsizing (350 Pacific 2014b). The canoes used during the blockade were built in the Pacific Islands using traditional methods and brought to Australia via cargo ship specifically for the action. Plates 1 and 2 provide images of the warriors and the flotilla on 17 October 2014.

On the day of the blockade, a female warrior led the poignant moment of prayer before the fleet of canoes set sail in Newcastle's coal port. Despite being warriors in their appearance on the day of the blockade, each Pacific participant was also a non-violent direct activist, probably also a dancer, an orator and a youth leader—none of which are particularly gender-specific identities in many parts of the Pacific (Plates 1 and 2). Further emphasising this blurring of gender identities, members of the campaign emphasised that canoe-building is not gender specific but is a community activity—women are responsible for weaving the sails, while men make the hull and rig (Steiner 2015). The canoe has also been invoked by Pacific activists and scholars to realign questions of sustainability and mobility with Indigenous Pacific values (see Hau'ofa 1994; Herman 2016). The Polynesian Voyaging Society is a powerful example of this where we see a revival of ancient canoe-building, navigating and seafaring skills that have emerged as a vehicle of regional cultural revival and reclaiming of oceanic agency and identity. The canoe has become a celebratory sense of peaceful, yet powerful, postcolonial unity, and the Pacific Climate Warrior campaigns should be read in this light.

Plate 1. Pacific Climate Warriors at the 17 October 2014 action in Newcastle.
Source: Jeff Tan Photography and 350 Pacific.

Plate 2. Some of the warriors—both men and women—setting out on a canoe as part of the flotilla at the 17 October 2014 action in Newcastle.

Source: Jeff Tan Photography and 350 Pacific.

The warriors also dispersed across Australia as part of a nation-wide speaking tour to raise public awareness about the plight of their countries. For these smaller supporting events around the country, oral testimony and advocacy were central. Storytelling—which is neither highly masculinised nor feminised in many Pacific cultures—supported and added layers of cultural complexity and context to the dominant figures of male and female warriors in their canoes, tempering the warriors' physicality with accounts of faith, history, tradition, geography and kinship. Overall, the campaign was considered by *The Guardian* (Australia) to be the second most important sustainability campaign of 2014 globally, described as 'the David versus Goliath campaign of the year' (Buckingham 2014, n.p.).

Re-envisioning Pacific futures

As feminist geographers with a long-standing interest in climate change in the Pacific, we argue that the Pacific Climate Warriors are re-envisioning Pacific futures in new ways. As their campaign in Australia shows, the warriors are refusing to accommodate ideas of Pacific Islanders as destined to be passive victims of climate change (see also Farbotko 2005). They are resisting ideas about being '"climate refugees" in waiting' (McNamara and Gibson 2009, 477; see also Farbotko and Lazrus 2012). While the warriors' narratives do not deny that a worst-case scenario for the islands would be that people would have to leave their homes (see Hugo's work on migration as an adaptation response—Hugo 1996; Bardsley and Hugo 2010; Hugo 2010), they are refusing its inevitability. Instead, these warriors are part of a vibrant network of young people who are committed to protecting their islands, cultures and livelihoods, and demanding international action and accountability from the major GHG polluters.

350.org states that a substantial part of its activity involves 'fighting iconic battles against fossil fuel infrastructure' and 'countering industry/government narratives' (350.org n.d., n.p.). The warrior, then, is involved not in a literal battle, nor is there an enemy embodied in a particular individual or group. Rather, the warriors and 350.org are engaged in symbolic, discursive battles. As we have explored in this essay, this Pacific Climate Warriors' campaign in Australia involved much more than warriors in canoes heading out to take on a coal ship, as powerful as such an image may be. Female warriors demonstrated the importance of prayer, kinship and connections to ancestors as tools in climate change activism.

An often external characterisation of the Pacific Islands region is as passive victim of climate change. Within the region there have been varying levels of opposition to such narratives—from island-nation leaders to civil society groups and grassroots networks. The Pacific Climate Warriors is an example of this, whereby the warrior has been used to demonstrate a new vision of Pacific Islanders in climate change narratives. Female warriors are significant in the network generally and in the Newcastle coal port campaign in particular, playing crucial roles as canoeists, orators and faith leaders. Indeed, both men and women took part in oral testimony, advocacy and in all stages of the peaceful action at Newcastle's coal port, blurring boundaries around specific gender identities. This is not to say, however, that vigilance in recognising and responding to the gendered impacts of climate change can be relaxed, and serious concerns around gender inequity across the region, and beyond, remain. But the overwhelming sense is that the Pacific Climate Warriors network is an inspiring group of young Pacific Islanders who are re-envisioning Pacific futures by resisting narratives about the future demise of their homelands and people as well as re-casting historical patriarchal figures (warriors) in new ways. The Pacific Climate Warriors network has used solidarity and symbolism to convey a collective identity and message that is ultimately about fighting for their survival.

Notes

1. Federated States of Micronesia, Fiji, Kiribati, Nauru, New Caledonia, Niue, Palau, Papua New Guinea, Republic of the Marshall Islands, Samoa, Solomon Islands, Tokelau, Tonga, Tuvalu and Vanuatu.

Disclosure statement

No potential conflict of interest was reported by the authors.

References

350.org. 2014a. "The Pacific Climate Warriors challenge the world". *350.org*. Accessed February 13, 2016. http://350.org/the-pacific-climate-warriors-challenge-the-world/.

350.org. 2014b. "A message from the Pacific". *350.org*. Accessed October 24, 2016. https://350.org/a-message-from-the-pacific/.

350.org. N.d. "How we work: Who we are, what we stand for, and where we're going". *350.org*. Accessed February 13, 2016. http://350.org/how/.

350 Pacific. 2014a. "Pacific Climate Warriors from 12 countries blockade world's largest coal port." *350 Pacific*, 17 October 2014. Accessed February 13, 2016. http://350pacific.org/pacific-climate-warriors-from-12-countries-blockade-worlds-largest-coal-port/.

350 Pacific. 2014b. "Update: Pacific Climate Warriors continue to hold Newcastle coal port." *350 Pacific*, 17 October 2014. Accessed February 13, 2016. http://350pacific.org/update-pacific-climate-warriors-continue-to-hold-newcastle-coal-port/.

350 Pacific. N.d. "Pacific Climate Warriors." *350 Pacific*. Accessed February 13, 2016. http://350pacific.org/pacific-climate-warriors/.

Bardsley, D. K., and G. J. Hugo. 2010. "Migration and climate change: Examining thresholds of change to guide effective adaptation decision-making." *Population and Environment* 32: 238–262.

Barnett, J., and J. Campbell. 2010. *Climate Change and Small Island States: Power, Knowledge and the South Pacific*. London, UK: Earthscan.

Biermann, F., and I. Boas. 2008. "Protecting climate refugees: The case for a global protocol." *Environment: Science and Policy for Sustainable Development* 50 (6): 8–16.

Bridges, K. W., and W. C. McClatchey. 2009. "Living on the margin: Ethnoecological insights from Marshall Islanders at Rongelap Atoll." *Global Environmental Change* 19 (2): 140–146.

Buckingham, F. 2014. "Top 10 sustainability campaigns of 2014." *The Guardian*, Accessed February 13, 2016. http://www.theguardian.com/sustainable-business/2014/dec/24/top-10-sustainability-campaigns-2014.

Chen, C. H. 2014. "Prioritizing hyper-masculinity in the Pacific region." *Culture, Society and Masculinities* 6 (1): 69–90.

Detraz, N., and L. Windsor. 2014. "Evaluating climate migration." *International Feminist Journal of Politics* 16 (1): 127–146.

Dreher, T., and M. Voyer. 2015. "Climate refugees or migrants? Contesting media frames on climate justice in the Pacific." *Environmental Communication* 9 (1): 58–76.

Environmental Justice Foundation. 2009. *No Place Like Home—Where Next for Climate Refugees?* London, UK: Environmental Justice Foundation.

Fair, H. 2015. "Not drowning but fighting: Pacific islands activists." *Forced Migration Review* 49: 58–59.

Farbotko, C. 2005. "Tuvalu and climate change: Constructions of environmental displacement in the Sydney Morning Herald." *Geografiska Annaler: Series B, Human Geography* 87 (4): 279–293.

Farbotko, C., and H. Lazrus. 2012. "The first climate refugees? Contesting global narratives of climate change in Tuvalu." *Global Environmental Change* 22 (2): 382–390.

Farbotko, C., and H. McGregor. 2010. "Copenhagen, climate science, and the emotional geographies of climate change." *Australian Geographer* 41 (2): 159–166.

Friends of the Earth. 2007. *A Citizen's Guide to Climate Refugees*. Melbourne, Australia: Friends of the Earth.

George, N. 2012. "'Just like your Mother?' The politics of feminism and maternity in the Pacific Islands." *Australian Feminist Law Journal* 32 (1): 77–96.

George, N. 2014. "Promoting women, peace and security in the Pacific Islands: Hot conflict/slow violence." *Australian Journal of International Affairs,* 68 (3): 314–332.

GermanWatch. 2014. *Loss and Damage: Roadmap to Relevance for the Warsaw International Mechanism*. Germany: GermanWatch.

Hau'ofa, E. 1994. "Our sea of islands." *The Contemporary Pacific* 6 (1): 148–161.

Hay, J. E., and N. Mimura. 2013. "Vulnerability, risk and adaptation assessment methods in the Pacific Islands Region: Past approaches, and considerations for the future." *Sustainability Science* 8 (3): 391–405.

Herman, R. D. K. 2016. "Traditional knowledge in a time of crisis: climate change, culture and communication." *Sustainability Science* 11 (1): 163–176.

Hughes, H. 2003. *Aid has Failed the Pacific (Issue Analysis No. 33)*. Sydney, Australia: The Centre for Independent Studies.

Hugo, G. 1996. "Environmental concerns and international migration." *The International Migration Review* 30 (1): 105–131.

Hugo, G. 2010. "Climate change-induced mobility and the existing migration regime in Asia and the Pacific." In *Climate Change and Displacement: Multidisciplinary Perspectives*, edited by J. McAdam, 9–36. Hart, Oxford, Portland: Hart Publishing.

Jetnil-Kijiner, K. 2015. "A moment of clarity—Why I'm going to Paris COP21." Accessed October 26, 2016. https://kathyjetnilkijiner.com/2015/11/22/a-moment-of-clarity-why-im-going-to-paris-cop21/.

Lefale, P. F. 2010. "Ua 'afa le Aso stormy weather today: Traditional ecological knowledge of weather and climate. The Samoa experience." *Climatic Change* 100 (2): 317–335.

Loeak, M. 2015. "On the road to Paris #COP21 Milañ Loeak Republic of the Marshall Islands." Accessed October 26, 2016. https://tewhareporahou.wordpress.com/2015/10/06/on-the-road-to-paris-cop21-milan-loeak-republic-of-the-marshall-islands/.

MacGregor, S. 2009. "A stranger silence still: The need for feminist social research on climate change." *The Sociological Review* 57 (s2): 124–140.

McNamara, K. E., and C. Gibson. 2009. "'We do not want to leave our land': Pacific ambassadors at the United Nations resist the category of climate refugees." *Geoforum*, 40 (3): 475–483.

McNamara, K. E., and S. S. Prasad. 2014. "Coping with extreme weather: Communities in Fiji and Vanuatu share their experiences and knowledge." *Climatic Change*, 123 (2): 121–132.

McNamara, K. E., and R. Westoby. 2014. "Ironies of globalisation: Observations from Fiji and Kiribati." *The Journal of Pacific Studies*, 34 (2): 53–62.

Mimura, N., L. Nurse, R. F. McLean, J. Agard, L. Briguglio, P. Lefale, R. Payet, and G. Sem. 2007. "Small Islands." In *Contribution of Working Group II to the Fourth Assessment Report of the Intergovernmental Panel on Climate Change, 2007*, edited by M. L. Parry, O. F. Canziani, J. P. Palutikof, P. J. van der Linden, and C. E. Hanson, 687–716. Cambridge, UK: Cambridge University Press.

Nurse, L. A., McLean, R. F., Agard, J., Briguglio, L. P., Duvat-Magnan, V., Pelesikoti, N., Tompkins, E. and Webb, A. 2014. "Small Islands." In *Contribution of Working Group II to the Fifth Assessment Report of the Intergovernmental Panel on Climate Change*, edited by V. R. Barros, C. B. Field, D. J. Dokken, M. D. Mastrandrea, K. J. Mach, T. E. Bilir, M. Chatterjee, et al., 1613–1654. Cambridge, UK: Cambridge University Press.

Steiner, C. E. 2015. "A sea of warriors: Performing an identity of resilience and empowerment in the face of climate change in the Pacific." *The Contemporary Pacific*, 27 (1): 147–180.

Tschakert, P. 2012. "From impacts to embodied experiences: Tracing political ecology in climate change research." *Geografisk Tidsskrift-Danish Journal of Geography* 112 (2): 144–158.

United Nations University and Bündnis Entwicklung Hilft. 2016. *World Risk Report*. Bonn, Germany: United Nations University Institute for Environment and Human Security and Bündnis Entwicklung Hilft.

The Migration of Horticultural Knowledge: Pacific Island seasonal workers in rural Australia—a missed opportunity?

Olivia Dun and Natascha Klocker

Introduction

In 2012, Graeme Hugo wrote the article 'Migration and Development in Low-income Countries: A Role for Destination Country Policy?' for the inaugural issue of the journal *Migration and Development*. That article, which continues to be the journal's most viewed work,[1] used the case of Asian and Pacific migration to Australia to question 'whether policies and practices by destination governments relating to international migration and settlement can play a role in facilitating positive developmental impacts in origin communities' (Hugo 2012, 25). The importance of such structural support for development has been underscored, in relation to seasonal worker programs, by growing evidence that their broader development benefits—beyond the household or family unit—cannot be taken for granted (Basok 2000; Craven 2015; Joint Standing Committee on Migration (JSCM) 2016).

In this essay we take inspiration from the above-mentioned paper (Hugo 2012), as well as an earlier discussion of 'best practice' temporary labour migration for development (Hugo 2009). Reflecting on Australia's Seasonal Worker Programme (SWP), we make a case for the importance of maximising 'development benefits for origin countries via the transfer of remittances, *skills and knowledge*' (Bedford et al. 2017, 39; emphasis added). Remittances have been a regular area of policy and research focus. However, less attention has been directed towards the knowledges and skills that move with seasonal workers as part of this circular and temporary migration process—in which the choice is not reduced to one 'between staying or going' (Methmann and Oels 2015, 53), but both staying and going (often repeatedly).

Here we draw on our own ongoing research with Pacific Island seasonal workers in Australia's horticultural sector, which points towards the potential for the SWP to facilitate the bi-directional transfer of horticultural knowledges and skills.[2] Many seasonal workers have extensive farming experience developed in their countries of origin. Acknowledgement of their farming skills and identities prompts contemplation of how the horticultural knowledge transfers that already happen spontaneously under the SWP could be better supported.

Australia's SWP

The Pacific Seasonal Worker Pilot Scheme (PSWPS) commenced in 2008 and led to the establishment of the Australian SWP in July 2012 (see Bedford et al. 2017 in this issue for a more detailed overview). The SWP's two key objectives are:

(1) 'to contribute to the economic development of participating countries through the provision of employment experience, skills and knowledge transfer, and [through workers] being able to send money back to their home country through remittances' and;

(2) 'to assist Australian producers and employers who are unable to source enough local Australian workers to meet their seasonal labour needs by providing access to a reliable seasonal workforce, able to return in future seasons' (see Durbin in JSCM 2016, 5).

When the program was founded, its key initial purpose was to contribute to Australia's aid program in the Pacific region—addressing labour demands in Australia's horticultural sector was a secondary purpose (Bedford et al. 2017, this issue; Roddam in JSCM 2016). By way of contrast, New Zealand's equivalent Recognised Seasonal Employer (RSE) scheme was implemented in 2007 in direct response to labour shortages in New Zealand's horti-cultural and viticultural sectors, and consequent demands made by key industry bodies in that country (Bedford et al. 2017, this issue).

The SWP is managed by the Australian federal government, with the Department of Employment as lead administrating department. Governments of the nine participating Pacific Island countries,[3] and Timor-Leste, make their own arrangements regarding how they wish to recruit workers to participate in the SWP, and enter into Memoranda of Understanding with the Australian government about labour sending and receiving arrangements. Citizens from participating countries can work for up to 6 months in Aus-tralia in one of the approved SWP employment sectors,[4] with the exception of workers from Kiribati, Nauru and Tuvalu who are permitted to remain in Australia for up to 9 months at a time due to higher travel costs (Hugo 2009; Department of Employment 2016). Between December 2009 and June 2016, 12 787 seasonal worker places (including 1633 workers who participated in the pilot scheme) had been approved under Australia's SWP (Bedford et al. 2017, this issue). In June 2015, the annual cap on the number of SWP participants was lifted—opening the way for growth. The program had a slow start (Hay and Howes 2012; Doyle and Howes 2015), but the most recent annual intake of 4490 workers (for the year 1 July 2015–30 June 2016) has been the highest yet (Bedford et al. 2017, this issue; Howes and Sherrell 2016).

Initial evidence from the multiple submissions and public hearings associated with the recent (2015–16) Australian Parliamentary Inquiry into the SWP, conducted by the Joint Standing Committee on Migration (JSCM), points towards positive changes in the workers' lives. Improvements have occurred primarily through higher income earning opportunities, which have contributed to poverty alleviation, housing improvements and enhanced access to education in countries of origin (Australian Council of Trade Unions (ACTU) 2015; Gibson and McKenzie 2011). However, the JSCM (2016) review raised questions regarding the broader development outcomes associated with the SWP, and whether these can be extended beyond the participating households. It concluded '[a]t the time of preparing this report, no verified empirical data was available showing specific linkage between Seasonal Worker Programme remittances and economic development in Pacific communities' (JSCM 2016, 111). Similar limitations have been recognised in relation to the longer-running New Zealand RSE scheme (see Craven 2015) and Canada's Seasonal Agricultural Worker Programme (see Basok 2000).

The recommendations arising from the JSCM-led inquiry into Australia's SWP (initiated by the federal Minister for Immigration and Border Protection) focused primarily on the program's capacity, impact and barriers in relation to Australia's labour market needs, including the potential for the SWP's expansion beyond the horticultural sector. These labour-focused recommendations suggest a departure from the program's initial focus on Australia's foreign aid commitments and development outcomes in workers' countries of origin. Howe and Reilly (2015, 3) expressed concerns regarding the expansion of the SWP beyond the horticultural sector, which they argued 'places a greater emphasis on its role as a labour market program, and less on the role of assistance to Pacific nations which arises from the special relationship between Australia and Pacific nations'.

Migration and development: the role of seasonal worker programs

The link between migration and development has long been an area of scholarly concern. Discussions have generally been 'polarised around two schools of thought' (Hugo 2012, 26)—one being 'brain drain', which signals a negative development outcome arising from a reduction in human capital in countries of origin; and the second being the sending of remittances by migrant workers. The latter has been viewed positively, especially because remittances reach family members quickly and directly, and can thus have a greater impact on poverty reduction and human well-being than large, bureaucratic development programs and development aid (Skeldon 2008; de Haas 2010).

Complicating the idea of 'brain drain', Hugo (2012) highlighted the positive contributions of diaspora populations to development in countries of origin through remittances, and also through their capacity to act as conduits of information (see also de Haas 2010). Eschewing a purely economic focus, he further noted that migrants 'may return with greater skills and experience than they had before they left and potentially make a greater development contribution' (Hugo 2012, 28). While discussing these ideas in the context of skilled migration, Hugo also espoused these sentiments in relation to so-called unskilled and low-skilled migration (Hugo 2009).

Our contention in this essay is that the SWP, as a temporary and circular migration program which *legislates* return migration, creates important possibilities for knowledge circulation and skills transfer, rather than a one-way 'brain drain'. Similar ideas about the benefits of migration for knowledge transfer have been highlighted by Curtain et al. (2016, 8) who recently argued that '[b]oth low-skilled and skilled Pacific migrants can transmit knowledge and skills to their compatriots both upon return and while abroad'. This potential may be particularly strong when participating seasonal workers are involved in horticulture in their countries of origin and destination—as has been the case for the vast majority of seasonal workers interviewed as part of our research in Robinvale, in rural north-western Victoria, Australia.

In the remainder of this essay we draw attention to the potential for the SWP to contribute to the transfer of horticultural knowledge and skills, using the case of Samson,[5] a seasonal worker from Papua New Guinea (PNG). Our key argument is that this example of knowledge transfer was reliant upon individual initiative—and, indeed, the role of migrant agency in the development process has garnered considerable attention in the academic literature (see, for example, Faist 2008; Skeldon 2008; Castles 2009; de Haas 2010). Mobility is a standard way in which people exercise agency to improve their livelihoods, and 'the poor' are not merely passive victims of the global capitalist system (Castles 2009; de Haas 2010). However, a focus on development being achieved by migrants themselves arguably diverts attention away from the important role of structural aspects of development policy (Skeldon 2008). The SWP is a highly managed migration program involving heavy engagement by governments, institutions and organisations at both ends (sending and receiving). It thus provides important opportunities to support the transfer of knowledges and skills via structural points of intervention (see also Curtain et al. 2016). In making these observations we acknowledge that the SWP is still in its infancy—thus our intent is to point towards future pathways that could strengthen the program's capacity to achieve its stated development objective.

Farmer-led knowledge transfer through Australia's SWP

Between February and June 2015 we conducted five focus group interview sessions in Robinvale with 20 horticultural workers participating in the SWP.[6] These seasonal workers were from PNG (4), Tonga (13) and Kiribati (3) and ranged from first-time program participants to those in their seventh program year (given that they had also participated in the pilot scheme). Many of the seasonal workers involved in our study have been involved in horticulture since their childhoods—and maintain subsistence or small-scale commercial farms in their countries of origin, while travelling to Robinvale on an annual basis to work on almond plantations. We feel that the label 'seasonal worker', and notions of unskilled or low-skilled work, does not correctly describe our research participants (nor the work they do in Australia[7]). The men to whom we spoke are *farmers*. We contend that their involvement in the SWP could be productively reframed as a form of 'farmer exchange'.

Our focus here is on Samson, a farm owner in PNG and seasonal worker in Australia. As shown in Box 1, through repeated participation in the SWP since 2011, Samson has made observations and developed skills that have benefited his own noni[8] plantation in PNG. Both repeat participation and the opportunity to witness practices on Robinvale's almond plantations *firsthand* appear to have been vital to the success of this knowledge transfer process.

The SWP provides important opportunities for farmer-led knowledge transfers—as seen in the case of Samson's noni plantation—particularly as relationships and conversations develop between seasonal workers and their host country counterparts and employers over areas of common interest. The potential for such exchanges to evolve spontaneously has also been noted in relation to New Zealand's RSE scheme. Bedford et al. (2017, this issue) have described joint ventures between seasonal workers and New Zealand farmers, involving 'agricultural production in Vanuatu (coffee growing on Tanna) and vegetable growing in Samoa' (Bedford et al. 2017, 48). Also in relation to the RSE, Gibson and McKenzie (2014) found that workers reported gaining pruning skills during their time in New Zealand. However, these processes have received little research attention to date. As part of our research project we aim to further document the types of horticultural knowledges and skills that move to and fro, from diverse Pacific countries to Australia and back, through the circular migration flows of the SWP. Prompted by Hugo's (2012, 25) call to consider whether the 'policies and practices by destination governments … [are] facilitating positive developmental impacts in origin communities', we now turn to some suggestions for how Australian policies could support horticultural knowledge and skills transfer as an overt component of the SWP.

Box 1.   The noni plantation of a PNG seasonal worker (S = Samson; O = Olivia).

S: You know most of us we come from agricultural backgrounds … My idea of coming here was … not for money … because I've started a little bit of project [in PNG] because we own the land … I am the landowner … So … when working in the almond farm [in Australia] I've seen how the irrigation system is … The distance how they plant the trees, at least I've learnt. So the previous years when I came and went back [to PNG] I started planning because I'm embarking on a noni project … So previously before coming here … I was planting one metre apart and then the result wasn't good, they [the plants] were too close … I came here [to Robinvale] I learned a lot so I went back [to PNG] and say 'No, this is wrong'. So I have to change the idea of planting the noni, so probably three or four metres apart.

O: And now, how is it?

S: Yeah now it's good. It's growing well. I got a distance where it can really bear big fruit compared to the previous ones where I was like planting them 100 [centi]metres apart and the fruit was not really that big. So when I do my spacing to four metres apart the fruit is really big. And that is the difference … I thank the program, the Australia[n] government, and PNG government, for allowing us to come over here, and just learn something and then take it back and implement it back home.

O: So what is the most valuable thing for you to learn here?

S: Oh I think the most valuable thing is … just looking at how the irrigation system and everything is set up in the farm here. So with that knowledge, if I go back and continue to put my mind to that project, at least I think I should come up with a bigger one and … [become] successful.

O: … How important was it that you saw it yourself here in Australia? If somebody came to your farm, your land in PNG and you showed them your noni plants and they told [you] 'Ah you have to make [space] them further apart', if you hadn't have been to Australia and someone came to your farm to tell you, do you think you would have changed?

S: No I don't think. Honestly, I don't think … It was very important for me coming here. What I've seen, I have implemented it. But if I'd not come over to Australia, like as you have said it, someone coming and telling me to do my spacing, I don't think I would believe him because normally I have to experience [things] myself, see things on my own … then I will expand on my experience …

O: [That's] human nature hey.

Structural supports to enhance horticultural knowledge transfer via the SWP

As noted by Hugo (2009, 63, 69), amongst the many parameters for best practice in temporary labour migration is the 'provision of training … [which] provides not only a better workforce for employers but the opportunity for social mobility among the migrant workers' and '[t]he adoption of more "development friendly" migration policies by both sending and receiving countries'. Existing reviews into Australia's SWP, and their resulting reports (discussed below), suggest that seasonal workers' capacity to contribute to positive development outcomes could be better supported through targeted training programs that are relevant to needs in their origin communities. Through their experiences on Australian farms, some of the seasonal workers we interviewed showed interest in fruit, nut and vegetable growing techniques, grafting practices and greenhouses—and were interested in how these practices might be applied in their home communities. Formalised training opportunities would likely be useful in this regard. In making this argument we do not intend to negate the agency of individual migrants. Seasonal workers who are *farmers*, like Samson, are already active participants in the knowledge transfer process. However, we are cognisant of broader critiques in the migration–development literature of approaches to development that arguably let governments 'off the hook' through a focus on individual responsibility. Indeed, a recent report into labour mobility (under the World Bank's Pacific Possible[9] initiative) asserts that both migrant sending and receiving countries need to make improvements with respect to the targeting and delivery of appropriate training (Curtain et al. 2016)—a valid point that we think could be better reflected on with specific reference to the SWP.

The SWP has already incorporated additional training activities, including first aid, English literacy and numeracy, and information technology courses as part of the Add-on Skills Training component of the program (Department of Employment 2015; JSCM 2016). However, the appropriateness of some of these courses has been questioned. As noted by an Approved Employer under the SWP, in a submission to the JSCM-led inquiry, 'some of the courses offered are not very conducive for our employees' learning; e.g. eight hours of numeracy and literacy ... provides ... I would say no lasting benefits for the person attending the course' (Golden Mile No. 1 Pty Ltd in JSCM 2016, 107). Of particular note—given the heavy horticultural focus of the SWP—is that formally integrated horticultural training appears to have been somewhat neglected. This gap was noted by Queensland TAFE, in an undated submission to the Parliamentary Inquiry into the SWP (see JSCM 2016). The submission emphasised the role that such training could play in enhancing the knowledge transfer processes and development objectives of the program:

> Of more interest is how any training that is provided in Australia to the seasonal workers transfers into growing workforce capability in the source countries. The skilling of the workers and the encouragement to share their skills when they return to their source country could be highly beneficial in raising the level of agricultural output and subsequent economic opportunity for the workers and their families. TAFE Queensland is of the view that extension of the SWP should include some base level training that is not dissimilar to the supported training for new migrants to Australia. It would incorporate literacy and numeracy training as well as some of the training *required for the job they are coming for.* (Queensland TAFE n.d., 13; emphasis added)

In its submission, Queensland TAFE (n.d.) also highlighted its affiliation with the Australian Pacific Technical College (APTC). The APTC is an Australian government initiative managed, as part of the Australian aid program, through the Department of Foreign Affairs and Trade (DFAT) (APTC 2016). It delivers technical and vocational training for Australian-standard qualifications to course participants across 14 Pacific Island countries (DFAT n.d.). Such training would be well suited to the circular nature of seasonal worker movement—but there is a key gap in its capacity to add value to the SWP both because it currently does not offer training in horticulture, and because the APTC has not explicitly focused on training SWP migrants. This is a missed opportunity, both in terms of supporting the development outcomes of the SWP and in terms of ensuring that Australian employers have access to workers trained in the skills that they need. Better alignment between the SWP and APTC is the type of structural intervention that could support 'best practice' in training and skills transfer associated with temporary labour migration schemes. Knowing that SWP migrants will generally return to their home countries may also go some way to allaying concerns expressed by some Pacific Island governments that trained residents will migrate to high-income countries (Curtain et al. 2016). And for seasonal migrants who are farmers, like Samson, such connections could provide important support for—and formal recognition of—the knowledge transfer processes in which they are already engaged. While APTC training takes place 'in country', Samson's story underscores the importance of *firsthand* experience on Australian farms. Better links between the APTC and SWP could therefore also benefit the development outcomes associated with the former.

Finally, while not the explicit focus of this essay, efforts to facilitate and support knowledge transfer as part-and-parcel of the SWP should also pay attention to the potential for bi-directional knowledge transfer—that is, the incorporation of seasonal workers' knowledge into the Australian context both in and beyond their workplaces. Our broader research project has documented examples of informal horticultural knowledge exchanges between Pacific Island migrants and Australian residents and farmers in rural northwestern Victoria. As proposed by Queensland TAFE (n.d., 10), it is important to 'recognise the skills that the migrant labour force brings and the knowledge exchange that occurs as a result of informal interaction during the job'. Acknowledging that many seasonal workers involved in the SWP are farmers—and thus re-envisioning aspects of the program as a form of 'farmer exchange'—could be an important first step.

Concluding remarks

Remittances play a vital role in supporting seasonal workers' households in countries of origin. However, we see considerable potential in the SWP for a complementary focus on the exchange of horticultural knowledge and skills. Evidence is beginning to emerge —including through our own ongoing research—of the skilfulness of seasonal migrant workers, who have shown a capacity to transfer practices witnessed on Australian farms onto their own farms in countries of origin. In this essay, we have argued that horticultural knowledge and skills transfer ought to be more formally integrated into the SWP, especially in situations where the seasonal workers are involved in similar activities in countries of origin *and* destination. Taking inspiration from Graeme Hugo, we have made a case that Australia's SWP—which already involves migrants who are highly practised farmers—provides an important opportunity to look beyond labour and remittances. Through this intervention we also seek to counter worrying trends that have emerged from the labour (rather than development) focus of the Australian Parliamentary Inquiry into the SWP. At a time when the program is increasingly being framed in terms of Australian labour shortages and needs (i.e. what migrants workers can do for 'us'), it is worth drawing attention back to the SWP's original emphasis on foreign aid objectives in the Pacific.

Notes

1. See Irudaya Rajan's (2015) editorial tribute to Graeme Hugo in *Migration and Development*.
2. This research is being undertaken as part of an Australian Research Council Discovery Project (DP140101165), on which Professors Lesley Head, Gordon Waitt and Heather Goodall are co-investigators.
3. Fiji, Kiribati, Nauru, Papua New Guinea, Samoa, Solomon Islands, Tonga, Tuvalu and Vanuatu.
4. Initially the PSWPS and SWP applied only to Australia's horticultural sector where seasonal workers are hired to carry out tasks such as picking, packing, thinning and pruning horticultural produce (such as fruits, vegetables and nuts). However, in June 2015, along with the

release of the Commonwealth of Australia's White Paper on Developing Northern Australia 'Our North, Our Future', the SWP was extended to incorporate cane, cotton and aquaculture, the accommodation sector (only in certain Australian locations) and the tourism industry (in Northern Australia only) (JSCM 2016). In February 2016 there was further expansion of the SWP to other areas of the agricultural sector including dairying, livestock, hatcheries as well as broadacre and mixed farming enterprises (JSCM 2016).

5. This is a pseudonym.
6. These participants were working under the labour sub-contractor, Tree Minders, a family-owned labour hire company supplying labour for the horticulture sector. Tree Minders helped to facilitate our recruitment of interview participants.
7. This assertion matches recent calls from Australian fruit and vegetable growers, the National Farmers Federation and the Primary Industries Skills Council to recognise the complexity of intensive horticulture farming and the need for 'skilled horticultural workers' (Martin 2013, n.p.).
8. The noni plant 'Morinda citrifolia L (Noni) has been used in folk remedies by Polynesians for over 2000 years, and is reported to have a broad range of therapeutic effects, including anti-bacterial, antiviral, antifungal, antitumor, antihelmin, analgesic, hypotensive, anti-inflammatory, and immune enhancing effects' (Wang et al. 2002, 1127).
9. 'Pacific Possible is focused on the genuinely transformative opportunities that exist for Pacific Island countries over the next 25 years' in recognition that 'Pacific island countries face unique development challenges'. It seeks to uncover such opportunities by commissioning new research that 'aims to answer the question: What is possible in the Pacific?' (http://www.worldbank.org/en/country/pacificislands/brief/pacific-possible).

Acknowledgements

Our thanks go to the Pacific Island seasonal workers who have given us their time and shared their stories. We are very grateful to Alf Fangaloka of Tree Minders, Robinvale, for facilitating access to Pacific Island seasonal workers. Professor Lesley Head's encouragement to think differently about people's relationships with nature has been influential for our thinking in this essay. We also acknowledge research assistance provided by Ikerne Aguirre Bielschowsky. Finally, our deepest gratitude to Graeme Hugo for his work on migration and development and best practice for temporary labour migration that has provided much guidance, structure and direction regarding points of intervention for improving development outcomes.

Disclosure statement

No potential conflict of interest was reported by the authors.

References

ACTU (Australian Council of Trade Unions). 2015. "ACTU Submission to Parliamentary Inquiry into the Seasonal Worker Program" (Inquiry into the Seasonal Worker Programme: Submission 19). ACTU, Melbourne. http://www.aph.gov.au/Parliamentary_Business/Committees/Joint/Migration/Seasonal_Worker_Programme/Submissions.

APTC (Australia Pacific Training College). 2016. "'About Us' Australia-Pacific Technical College, Suva." https://www.aptc.edu.au.

Basok, T. 2000. "Migration of Mexican Seasonal Farm Workers to Canada and Development: Obstacles to Productive Investment." International Migration Review 34 (1): 79–97.

Bedford, R., C. Bedford, J. Wall, and M. Young. 2017. "Managed Temporary Labour Migration of Pacific Islanders to Australia and New Zealand in the Early Twenty-first Century." *Australian Geographer* 48 (1): 37–57.

Castles, S. 2009. "Development and Migration—Migration and Development: What Comes First? Global Perspective and African Experiences." *Theoria* 56 (121): 1–31.

Craven, L. 2015. "Migration-affected Change and Vulnerability in Rural Vanuatu." *Asia Pacific Viewpoint* 56 (2): 223–236.

Curtain, R., M. Dornan, J. Doyle, and S. Howes. 2016. "Labour Mobility: The Ten Billion Dollar Prize." *Pacific Possible* Development Policy Centre, The Australian National University, Canberra and The World Bank, Washington D.C. http://www.worldbank.org/en/country/pacificislands/brief/pacific-possible.

Department of Employment. 2015. *Add-on Skills Training Fact Sheet (version 19 June 2015)*. Canberra: Department of Employment. https://docs.employment.gov.au/system/files/doc/other/06-15_seasonal_worker_programme_aost_factsheet.pdf.

Department of Employment. 2016. *Seasonal Worker Programme*. Canberra: Department of Employment. https://www.employment.gov.au/seasonal-worker-programme.

DFAT (Department of Foreign Affairs and Trade). n.d. "Joint Standing Committee on Migration Inquiry into the Seasonal Worker Programme: Submission of the Department of Foreign Affairs and Trade" (Inquiry into the Seasonal Worker Programme: Submission 37). Department of Foreign Affairs and Trade, Canberra. http://www.aph.gov.au/Parliamentary_Business/Committees/Joint/Migration/Seasonal_Worker_Programme/Submissions.

Doyle, J., and S. Howes. 2015. *Australia's Seasonal Worker Programme: Demand-side Constraints and Suggested Reforms*. Washington, DC: World Bank Group.

Faist, T. 2008. "Migrants as Transnational Development Agents: An Inquiry into the Newest Round of the Migration–Development Nexus." *Population, Space and Place* 14 (1): 21–42.

Gibson, J., and D. McKenzie. 2011. "Australia's PSWPS: Development Impacts in the First Two Years." *Asia Pacific Viewpoint* 52 (3): 361–370.

Gibson, J., and D. McKenzie. 2014. "The Development Impact of a Best Practice Seasonal Worker Policy." *Review of Economics and Statistics* 96 (2): 229–243.

de Haas, H. 2010. "Migration and Development: A Theoretical Perspective." *International Migration Review* 44 (1): 227–264.

Hay, D., and S. Howes. 2012. "Australia's Pacific Seasonal Worker Pilot Scheme: Why has Take-up been so Low?" [Blog], *Devpolicy*, the Development Policy Centre, Australian National University Canberra. http://devpolicy.org/australias-pacific-seasonal-worker-pilot-scheme-why-has-take-up-been-so-low20120404/.

Howe, J., and A. Reilly. 2015. "Submission to the Joint Standing Committee on Migration inquiry into the Seasonal Worker Programme" (Inquiry into the Seasonal Worker Programme: Submission 36). Public Law and Policy Research Unit, University of Adelaide, Adelaide. http://www.aph.gov.au/Parliamentary_Business/Committees/Joint/Migration/Seasonal_Worker_Programme/Submissions.

Howes, S., and H. Sherrell. 2016. "Seasonal Worker Program Grows by 50 Per cent" [Blog], *Devpolicy*, the Development Policy Centre, Australian National University Canberra. http://devpolicy.org/seasonal-worker-program-grows-50-per-cent-20161116/.

Hugo, G. 2009. "Best Practice in Temporary Labour Migration for Development: A Perspective from Asia and the Pacific." *International Migration* 47 (5): 23–74.

Hugo, G. 2012. "Migration and Development in Low-income Countries: A Role for Destination Country Policy?" *Migration and Development* 1 (1): 24–49.

Irudaya Rajan, S. 2015. "Migration and Development: In Pursuit of Unlimited Debates." *Migration and Development* 4 (1): 1–3.

JSCM (Joint Standing Committee on Migration). 2016. *Seasonal Change: Inquiry into the Seasonal Worker Programme*. Canberra: The Parliament of the Commonwealth of Australia.

Martin, S. 2013. "Lettuce Use Foreign Skills to Grow." *The Australian*, March 1. http://www.theaustralian.com.au/national-affairs/industrial-relations/lettuce-use-foreign-skills-to-grow/story-fn59noo3-1226587987585.

Methmann, C., and A. Oels. 2015. "From 'Fearing' to 'Empowering' Climate Refugees: Governing Climate-induced Migration in the Name of Resilience." *Security Dialogue* 46 (1): 51–68.

Queensland TAFE. n.d. "TAFE Queensland's Response to the Inquiry into the Seasonal Worker Programme" (Inquiry into the Seasonal Worker Programme: Submission 27). http://www.aph. gov.au/Parliamentary_Business/Committees/Joint/Migration/Seasonal_Worker_Programme/ Submissions.

Skeldon, R. 2008. "International Migration as a Tool in Development Policy: A Passing Phase?" *Population and Development Review* 34 (1): 1–18.

Wang, M.-Y., B. J. West, C. J. Jensen, D. Nowicki, C. Su, A. K. Palu, and G. Anderson. 2002. "*Morinda Citrifolia* (Noni): A Literature Review and Recent Advances in Noni Research." *Acta Pharmacologica Sinica* 23 (12): 1127–1141.

Managed Temporary Labour Migration of Pacific Islanders to Australia and New Zealand in the Early Twenty-first Century

Richard Bedford, Charlotte Bedford, Janet Wall and Margaret Young

ABSTRACT

Circular migration was one of several enduring themes in Graeme Hugo's highly productive research career. Although his specialist field was Asian population movement, during the 2000s he became increasingly interested in labour migration in the Pacific Islands. This paper reviews the development of two managed circular migration schemes targeting Pacific labour that emerged following the UN High-level Dialogue on International Migration and Development in 2006. New Zealand's Recognised Seasonal Employer (RSE) scheme and Australia's Seasonal Worker Program (SWP) have attracted international attention as the kind of 'best practice' temporary labour migration schemes that Hugo had in mind when he emphasised the positive contributions that circular forms of mobility could make to development in both source and destination countries. The two schemes have transformed mobility between the participating countries and have played a major role in the negotiations over a free-trade agreement between Pacific Forum countries, including Australia and New Zealand. Although the schemes have been in operation for almost 10 years, this paper argues that they are not becoming 'business as usual'; they embody complex systems of relationships between multiple stakeholders that require ongoing management to ensure that they do not become traps for low-skilled, low-paid 'permanent' temporary workers.

Introduction

In early November 2016, two significant initiatives to foster research on migration were launched at the Hugo Conference (named after the late Professor Graeme Hugo AO, FASSA),[1] University of Liege in Belgium.[2] The first was the creation of the Hugo Observatory at the University of Liege, a multi-disciplinary research institute dedicated specifically to the study of environmental change and migration. The second was the establishment of an international scholarly association for the study of environmental migration. Hugo's contribution to migration studies is enormous and it is gratifying to see the significance of his research being recognised in ways that will keep his legacy alive. The Hugo Conference and the Hugo Observatory acknowledge his seminal work on links between migration and environmental change, and especially the major contributions he has made since the mid-1990s to debates about the relationships between climate change and migration (Hugo 1996; Hugo et al. 2009; Asian Development Bank 2012; Bedford and Hugo 2012).

During the 5 years before his untimely death in January 2015, Hugo was actively engaged with a number of research issues relating to the Pacific Islands. While the Pacific Islands were not sites for his primary research until quite late in his career, his long-standing interests in temporary forms of population movement in the Asia-Pacific region, coupled with his more recent explorations of the relationships between climate change and migration, inevitably led him into closer engagement with research in the Pacific. It is the connections that can be drawn between Hugo's research on temporary labour migration and recent policy initiatives in the Pacific that are the focus of discussion in this paper.

Hugo and the Pacific

Hugo's interest in forms of temporary migration goes back to his doctoral research on population circulation in Indonesia in the early 1970s (Hugo 1975). At the time he was carrying out his fieldwork, Bedford (1971) and Chapman (1970) were completing their doctoral research into a similar form of population movement in the western Pacific, while Skeldon (1974) was working on a similar process in the context of urbanisation in Peru. Over the subsequent three decades, these four geographers interacted in a range of situations and contexts to further understanding of the process of circular migration. For example, their major cross-national study of circulation in population movement was published in contributions to Chapman and Prothero (1985) and Prothero and Chapman (1985), and these were subsequently re-published by Routledge in 2014 in response to renewed interest in circular migration following the United Nations High-level Dialogue on International Migration and Development in 2006.

Hugo's interest in migration processes in the Asia-Pacific region extended to New Zealand, and he developed a strong relationship with New Zealand's small community of demographers, geographers and sociologists specialising in the study of population dynamics and structures (Bedford 2015). He had a special interest in trans-Tasman migration and the New Zealand-born population in Australia, and worked closely with staff and postgraduate students at the University of Waikato's former Population Studies Centre (now National Institute of Demographic and Economic Analysis—NIDEA) on several externally funded research programmes addressing New Zealand's changing population (Hugo 2015).

Hugo was well aware of the different approaches taken by New Zealand and Australia to immigration from a number of Pacific Island countries. He favoured a more proactive approach to the socio-economic and environmental challenges faced by particular groups of Pacific countries, especially the central Pacific atolls and reef islands that are expected to be adversely affected by climate change, and was keen to see some New Zealand initiatives, such as the Pacific Access Category,[3] adopted in Australia (Bedford and Hugo 2012). His interest in the fate of countries such as Kiribati, Tuvalu, and Nauru was stimulated by his concerns about climate change and its impact on populations inhabiting low-lying coral islands that 'will be extremely vulnerable to sea-level rise, high-intensity cyclones and storm surges' (Hugo 2010, 31).

In 2013, Hugo completed a major report for the Australian Agency for International Development (AusAID), namely its Microstate Futures Study on the role of temporary migration to Australia in facilitating development in Kiribati, Tuvalu and Nauru. This confidential report (Hugo 2013) had a significant influence on the decision by Australia's Department of Immigration and Border Protection to approve a 'Pacific Microstates— Northern Australia Worker Pilot Program' linked with the government's *Developing Northern Australia* initiative (Commonwealth of Australia 2015). The 5-year pilot provides up to 250 (50 per year) citizens of Kiribati, Nauru and Tuvalu with access to a multi-year work visa (2 years, with an option of applying for a third year) to work in lower-skilled jobs in Northern Australia (Department of Foreign Affairs and Trade (DFAT) 2015a, 5).

New Zealand's Recognised Seasonal Employer (RSE) work policy, introduced in April 2007, was also of interest to Hugo. At the time of the RSE policy's commencement, there was a renewed focus at the global level on the relationship between migration and development, and the potential for managed circular migration schemes, like the RSE, to deliver positive outcomes for both source and destination countries. Such programmes have a long and chequered history regarding their effectiveness, delivery of decent work conditions and the extent to which they uphold basic principles of human rights for workers (Martin 2002; Ruhs 2002; Castles 2006; Martin, Abella, and Kuptsch 2006), but the concept was the subject of renewed attention at the first United Nations High-level Dialogue on International Migration and Development in 2006 (United Nations 2006).

The background report that was used to set the scene for the Dialogue, issued under the name of then UN Secretary-General Kofi Annan, had been strongly influenced by Hugo's writing and engagement with the UN Population Division, which prepared the report. Hugo (2009) subsequently went on to write about 'best practice' in temporary labour migration, which aims to match migrant workers with appropriate jobs in destination countries, protect workers' rights and welfare, and maximise development benefits for origin countries via the transfer of remittances, skills and knowledge. He argued that there were a number of advantages for lower-skilled workers of retaining a pattern of circular migration in preference to permanent settlement in destination countries (Hugo 2009). Hugo (2009, 9) acknowledged that temporary migration of lower-skilled workers in the Asia-Pacific region has been widely criticised as a 'new form of indentured labour', due in part to the restrictions placed on workers in the destination country and the lack of access to permanent residence. However, he argued that many problems associated with temporary labour migration schemes are due to poor governance; they are not intrinsic features of this form of migration. Temporary labour migration programmes can be successful if they are carefully managed and monitored and if there is effective cooperation between sending and receiving countries.

At the time he died, Hugo was finalising arrangements for the commencement of a multi-year ARC-funded research programme entitled 'Demography and Diaspora: Enhancing Demography's Contribution to Migration and Development'. The proposed research programme included a case study on Tonga that was to be carried out by one of his Tongan PhD students, Alisi Kautoke Holani. Holani's (forthcoming) thesis explores temporary labour migration and sustainable development in the Pacific with special reference to Tonga in the context of the ongoing negotiations over the Pacific Agreement on Closer Economic Relations (PACER) Plus. PACER Plus aims to create jobs, raise standards of living and encourage sustainable economic development in the Pacific region through increased regional trade and economic integration between the participating states.[4] Holani's research into the contribution that Australia's Seasonal Worker Program (SWP) makes to household welfare in Tonga links with Hugo's long-standing interest in circular labour migration as a process that can contribute to development in migrant source communities as well as benefiting employers in the destination country (Hugo 2009).

There are some strong connections between Hugo's interests in access to temporary work opportunities for Pacific Islanders in Australia and climate change adaptation in the Pacific. Access to such opportunities could assist Pacific Islanders to develop livelihoods that may be sustainable under different environmental conditions in the future. The seasonal work schemes in Australia and New Zealand, which have provided temporary work visas for about 70 000 Pacific workers during the past decade, could become one of most significant policies for facilitating adaptation by Pacific families to changing environmental conditions in their own communities. Temporary work overseas is one of a number of strategies that Pacific peoples have adopted to spread risk of economic failure across a range of activities and options (Bedford 1973; Gibson 2015).

The next section of this paper reviews briefly some contemporary dimensions of New Zealand's Recognised Seasonal Employer scheme and Australia's Seasonal Worker Program, with particular attention to the numbers of workers involved. This is followed by some observations on the contributions that these schemes are making, firstly, to increased productivity for participating horticulture and viticulture businesses in destination countries, and secondly, to workers, their families and communities in the islands. This leads to a short section on the need for continued monitoring and management of the complex relationships that underpin such schemes to ensure that they do not become, by default, avenues for permanent seasonal employment for Pacific workers whose families and livelihoods remain in the islands.

In the final section we review briefly the role that the seasonal work schemes have played in the negotiations over new avenues for temporary employment of Pacific Islanders in Australia and New Zealand as part of the PACER Plus regional free-trade agreement that is currently being negotiated. The special provisions for labour mobility that are being negotiated as part of PACER Plus are quite different from the seasonal work schemes, however, in that they are likely to allow workers to transition to other types of visas and possibly, longer-term, to residence. This will inevitably lead to increasing pressure for the seasonal work policies to allow for some carefully managed transitions of highly skilled seasonal workers to permanent residence in Australia and New Zealand, rather than locking them into a pattern of annual, repeated return as 'permanent' temporary migrants.

The seasonal work schemes in 2016

In the nine years since the launch of the seasonal work schemes, about 70 000 temporary work visas for citizens of countries in the Pacific (including Timor Leste for Australia) have been approved under the two schemes (Table 1). In addition, New Zealand approved

Table 1. Total seasonal work approvals, July 2007–June 2016.

| Period/year | Seasonal work visa approvals | | SWP[a] | Total[b] (RSE + SWP) |
| | RSE | | | |
	Pacific	All countries		
2007–08	3477	4426		4426
2008–09	5912	7617		
2009–10	5083	6829		
2010–11	5859	7619		
2011–12	6313	7742	{1633}	{31 440}
2012–13	6814	8175	1473	9648
2013–14	7047	8415	2014	10 429
2014–15	7853	9275	3177	12 452
2015–16	8327	9757	4490	14 247
2007–16	56 685	69 855	12 787	82 642

Notes:
[a]Data for the PSWPS December 2009—June 2012 are included (1633 visa approvals including a small number for Timor Leste workers). These figures are not available for individual years.
The total number of RSE visas for the period July 2008–June 2012 is 29 807 and these, combined with 1633 PSWPS visas, give a total of 31 440 for that period. The SWP figures include workers from Timor Leste in all years.
[b]The RSE "All countries" figures plus the SWP figures. The total for the 56 685 RSE "Pacific" plus the 12 787 SWP workers is 69 472.
Sources: unpublished data provided by the Pacifica Labour and Skills Team, Immigration New Zealand, Ministry of Business, Innovation and Employment, Wellington, New Zealand and the Seasonal Worker Program, Department of Employment, Canberra, Australia.

over 13 000 RSE visas for citizens of Asian countries, especially Malaysia, Indonesia and Thailand.[5] The great majority of the visas for Pacific workers (57 000) have been for work in New Zealand; the Australian seasonal work scheme has evolved more slowly (with about 12 000 visas approved since 2009). This is largely due to the concessions that the Australian government has given to people on Working Holiday Maker (WHM) visas who are prepared to work in rural areas, along with the prevalence of illegal workers in the horticulture industry (Ball 2010; Doyle and Howes 2015; Commonwealth of Australia 2016). Australian horticultural employers have not yet seen the advantages that New Zealand's RSEs have in employing Pacific workers (Hay and Howes 2012; Doyle and Howes 2015). Such advantages include the stability and security provided to employers each season, through the use of an increasingly experienced RSE workforce, and the associated gains in productivity as workers shift from the learning phase in the first year, to having acquired the requisite skills to perform various tasks on the orchard or vineyard (Bedford 2013).

The two seasonal work policies, while addressing a common problem relating to supply of low-skilled labour for the agricultural sector, have had different levels of buy-in from New Zealand and Australian employers from the outset (TNS 2011; Hay and Howes 2012; Doyle and Howes 2015; Curtain 2016). The New Zealand scheme was designed and introduced as part of an industry-led initiative that began in the early 2000s to address labour shortages, particularly in the apple industry, and to reduce the use of illegal labour (Whatman et al. 2005). From 2004, New Zealand horticulture and viticulture enterprises were using existing temporary migration policies to access Pacific labour and were prepared to engage with a managed seasonal migration scheme, crafted to meet their needs, as soon as it was introduced (Bedford 2013). Australian employers, on the other hand, were not accessing Pacific labour before the Pacific Seasonal Worker Pilot Scheme (PSWPS) was introduced in 2009, and employers had to be persuaded that participation in the pilot scheme would be in their best interests. Furthermore, Australia's scheme was a government-led initiative with the primary objective of contributing to its aid development programme in the Pacific. Addressing unmet demands for labour in the horticulture industry was a secondary aim (Bedford 2013; Curtain 2016).

As the RSE approaches its 10th year, and the SWP heads for its 5th season, demand for Pacific seasonal workers in both countries continues to grow. The year ended June 2016 saw record numbers of seasonal workers from the Pacific approved for work in Australia and New Zealand. Australia's Department of Immigration and Border Protection (DIBP) approved 4490 visas (subclass 416) between 1 July 2015 and 30 June 2016, with 2624 going to Tongans and 1198 to ni-Vanuatu workers. These two countries accounted for 85 per cent of the seasonal work visas approved during the year. In the case of New Zealand, statistics for the year ended 30 June 2016 reveal that 9757 visas were approved with just under 4000 visas issued to citizens of Vanuatu (3932), followed by Tonga (1765) and Samoa (1550). These three countries accounted for 87 per cent of the seasonal work visa approvals.

The annual visa approvals, by country, for the year ended June 2016 are shown in Table 2. It can be seen that Tonga sent more seasonal workers to Australia (2624) than to New Zealand (1765) between July 2015 and June 2016. Tongan residents in rural Australia have played a major role in developing links with prospective employers and several Tongans run labour hire companies in Australia that bring in workers from the islands (Gibson and McKenzie 2011; Holani forthcoming; Bedford 2013). By June 2016, Tonga had just under 4400 seasonal workers legally in Australia and New Zealand as well as an estimated 320 who had not returned home at the end of their contracts (300 in Australia and 20 in New Zealand).[6] To date, the only Pacific workers on the limited-purpose RSE and SWP visas to have absconded in any numbers are Tongans. This is an issue that government officials in Tonga, as well as in Australia and New Zealand, are keen to address.

New Zealand's scheme remains restricted to the horticulture and viticulture sector and the RSE is currently capped at 10 500 arrivals per annum.[7] Pacific RSE worker approvals and arrivals, by country, are shown in Table 3. Visa approvals always exceed arrivals because some workers either choose not take up the offer of work in the end or fail to get their visas approved in time to meet the employer's start date for employment. Arrivals are always just below the cap. As the New Zealand Minister of Immigration frequently

Table 2. Seasonal work visa approvals for Pacific countries, July 2015–June 2016.

Country	RSE	SWP	Total
Fiji	104	160	264
Kiribati	173	20	193
Nauru	20	20	40
Papua New Guinea	69	42	111
Samoa	1550	140	1690
Solomon Islands	649	61	710
Tonga	1765	2624	4389
Tuvalu	65	4	69
Vanuatu	3932	1198	5130
Total Pacific[a]	8327	4266	12 593

Note:
[a]Excluding 1430 visas for RSE workers from Asian countries and 224 visas for SWP workers from Timor Leste.
Sources: see Table 1.

Table 3. RSE visa approvals and arrivals, Pacific countries, July 2015–June 2016.

Country	Approvals	Arrivals	% Arrived
Fiji	104	92	88.5
Kiribati	173	162	93.6
Nauru	20	20	100.0
Papua New Guinea	69	68	98.6
Samoa	1550	1454	93.8
Solomon Islands	649	590	90.9
Tonga	1765	1687	95.6
Tuvalu	65	64	98.5
Vanuatu	3932	3726	94.8
Total Pacific	8327	7863	94.4
Total RSE scheme[a]	9757	9276	95.1

Note:
[a]Including workers from Asian countries. The cap for RSE arrivals is currently 9500.
Source: unpublished data obtained from New Zealand's Ministry of Business, Innovation and Employment.

reminds officials as well as employers: the cap is not a target—it is the ceiling for approved arrivals (Woodhouse 2015).

RSE workers are one of several sources of labour that can meet industry needs in New Zealand, including those approved on Working Holiday visas or on study visas as international students. In 2014, the Ministry of Social Development introduced a seasonal work scheme for New Zealanders to encourage more unemployed and underemployed citizens to take advantage of employment opportunities during peak seasons in the horticulture and viticulture industries. This scheme, which included subsidies for travel and pastoral care, placed about 300 unemployed New Zealanders in seasonal jobs with RSE employers during the year ended June 2016 (Tolley 2016).

For most approved employers, RSE workers comprise less than 40 per cent of their temporary workforce during peak seasons. New Zealanders, working holidaymakers ('backpackers') and international students make up the remainder of the seasonal labour force. While the RSE scheme has virtually no capacity for growing the numbers of arrivals, there remains some ability to spread the benefits of employing Pacific seasonal workers across more employers through the use of joint Approvals to Recruit (ATRs). About 2000 Pacific workers in 2015–16 were employed on joint ATRs (Rarere 2016). Officials and employers use joint ATRs to extend the periods of employment for RSE workers within the permitted 7 months for all countries other than Kiribati and Tuvalu (9 months). In both Australia and New Zealand, workers from Kiribati and Tuvalu are permitted to stay longer than others because of the high costs they pay for international transport to reach their work destination (Bedford and Bedford 2010; DFAT 2015a).

The cap of 5000 seasonal workers in Australia's SWP was removed in 2015, and the programme was expanded to the agriculture (aquaculture, cotton and cane) and accommodation industries in specified locations, as well as the tourism sector in Northern Australia.[8] There is considerable scope for expansion of the SWP; the big challenge is to encourage greater employer participation. The higher costs associated with recruiting workers from the Pacific have been identified as a key barrier to entry into the programme. These costs make Pacific workers less competitive against working holidaymakers and other sources of locally available labour (Doyle and Howes 2015). Australian employers need to be persuaded that the costs of recruiting and employing Pacific seasonal workers are outweighed by the increased productivity that comes from having a core supply of reliable and increasingly experienced labour during peak harvesting and pruning periods. Evidence of these productivity gains is discussed further below, but it can be noted here that a small number of Australian employers and contractors have reported positively on productivity gains that they can link to the use of SWP workers (Jenkin 2015; Owen 2015).

The significance of seasonal work opportunities

In his book reviewing health worker migration in the Pacific, Connell (2009, 173) effectively dismissed managed migration schemes, such as the RSE and SWP, as a form of mobility that 'offers barely a Band-Aid' within the context of burgeoning labour forces and demand for employment opportunities in the region. This has proved to be a surprisingly parsimonious perspective on the extent to which managed seasonal labour migration has provided opportunities for low-skilled workers from some Pacific states to gain temporary employment in the labour markets of New Zealand and Australia since 2007.

The significance of the limited-purpose visa as a pathway for Pacific Islanders to access temporary work in New Zealand and Australia varies by country. Table 4 shows the numbers of RSE visas as well as the total temporary work visas (all types, including RSE visas) approved for New Zealand between July 2007 and June 2016 for each Pacific country participating in the RSE scheme, and shows the percentage of total temporary work visas that are accounted for by the RSE scheme. In the case of Fiji, with its large pool of skilled and semi-skilled labour, and its late inclusion in the RSE (2014) and SWP (2015),[9] the seasonal work schemes have not even been a 'band aid'—they have been irrelevant. Over 60 000 Fiji citizens found temporary employment in New Zealand during that period via other work visas, especially the Essential Skills visa for more skilled workers.

For citizens of Papua New Guinea and Nauru, which had pilot projects in the RSE in 2013 and 2014 respectively, seasonal work visas have comprised under 40 per cent of the total temporary work visas their citizens have had in New Zealand since the commencement of the RSE scheme in 2007. Tuvalu, Kiribati, Samoa and Tonga, all island states with access to earlier temporary work schemes in New Zealand, as well as to quotas for residence visas,[10] have between 40 and 53 per cent of their temporary work visa approvals between 2007 and 2016 accounted for by seasonal work visas. Seasonal work opportunities are more than a 'band aid' for these countries, although for Kiribati and Tuvalu the numbers involved in any form of temporary work in New Zealand and Australia are small.

Table 4. RSE work visas and all temporary work visas, New Zealand, July 2007–June 2016.

Country	Total temporary work visas	RSE visas	% RSE visas
Fiji	60 553	135	0.2
Kiribati	2611	1154	44.2
Nauru	108	40	37.0
Papua New Guinea	897	279	31.1
Samoa	22 757	11 088	48.7
Solomon Islands	4301	3751	87.2
Tonga	25 858	13 846	53.5
Tuvalu	1598	673	42.1
Vanuatu	26 268	25 719	97.9
Total RSE Pacific	144 951	56 685	39.1

Source: unpublished data obtained from New Zealand's Ministry of Business, Innovation and Employment.

The most significant contributions that the RSE scheme has made to access to temporary work opportunities in New Zealand have been for Vanuatu and the Solomon Islands. These two countries have had very limited access to any kind of temporary work overseas since the nineteenth-century 'labour trade' with Queensland (Munro 1990) and the movement of people linked with the Anglican Church (the Melanesian Mission) to New Zealand (Mallon 2012). The RSE scheme has been especially significant for Vanuatu in this regard, providing 25 719 seasonal work visas—98 per cent of all temporary work visas that citizens of Vanuatu have obtained since 2007 (Table 4).

Overall, seasonal work visas accounted for 39 per cent of the 144 951 temporary work visas approved for citizens of the Pacific countries participating in the RSE scheme between 2007 and 2016. In the case of Australia, data for the 2012–13 financial year showed that around 47 per cent of all temporary work visas issued to Pacific citizens from countries that participate in the SWP were for seasonal work (Bedford and Bedford 2013). Arguably, the access to temporary work provided by the RSE scheme and SWP could be considered relatively insignificant in the context of overall labour force growth and demand for job opportunities in the Pacific. However, managed migration programmes such as the RSE and the SWP are not going to be replaced easily by other kinds of access to temporary work in Australia and New Zealand, and both schemes provide an important opportunity for offshore employment that is valued by the governments of participating Pacific states.

The seasonal work schemes: 'wins' for employers and workers?

The RSE scheme has gained international recognition as a 'best practice' managed circular migration programme (International Labour Organization 2009; McKenzie and Gibson 2010). Many of the features that Hugo (2009) concluded were 'best practice' in recruitment and selection—pre-departure preparation, pastoral care and monitoring of conditions at the destination, and assistance with re-integration back into home communities—are features of the RSE and its associated Strengthening Pacific Partnerships (SPP) programme as well as the SWP and its Labour Mobility Assistance Program (LMAP).[11]

All employer-led temporary work schemes that operate on the basis of a limited-purpose visa can be criticised because of restrictions on the freedom of workers to choose the length of their employment, the conditions under which they are employed, or to shift to a different employer in search of better conditions or higher wages (Preibisch 2007). Both the Australian and New Zealand schemes have strict constraints around arrival, departure, length of stay, and the eligibility of employers/contractors. There are also mandatory requirements regarding the payment of tax, paying for a share of the international airfare, paying insurance levies, covering accommodation costs and contributing to transport costs to and from the workplace and, if required, within the workplace. Some of these 'fixed' costs, which are paid by the employer upfront and then subsequently deducted automatically from the workers' earnings, are subject to manipulation, especially the local transport and accommodation costs. There have been complaints both in New Zealand and Australia about excessive charges, particularly regarding weekly rates for shared accommodation. These issues have been raised in many of the major studies of both seasonal work schemes (see, for example, theses by Rockell 2015; Bailey 2014;

Bedford 2013 on the RSE; Gibson and McKenzie 2011 and Holani (forthcoming) on the SWP; and Brickenstein 2015 on both schemes).

Notwithstanding these criticisms, however, the schemes are deemed by many stakeholders to be delivering 'wins' for participating employers, workers and their families. There is clear evidence from successive surveys of horticulture and viticulture enterprises in New Zealand that the RSE scheme has delivered major productivity gains for many participating employers. These gains have enabled them to invest in business improvements and expansion at a consistently higher rate than non-participants (Research New Zealand 2015). Participating employers are also employing larger shares of unemployed New Zealanders referred to them by Work and Income New Zealand than non-participating employers, as well as creating more permanent jobs for New Zealanders. Evidence of the crucial role the RSE scheme plays in raising industry standards and productivity, as well as generating additional job opportunities for New Zealanders, contributed to the government's decision to raise the cap on RSE workers from 8000 to 9000 for the year 2014–15 to 9500 for the year 2015–16, and to 10 500 from November 2016.

Data collected in 2011 from nine RSEs on the gross earnings of their seasonal workers (RSE workers, permanent and casual New Zealand workers and working holidaymakers) provided evidence of the higher productivity of RSE workers on the orchard/vineyard, when compared with other groups (Bedford 2013, 2014). Data on gross earnings per worker were collected over a 10-week period and productivity was measured by the quantity of fruit picked/vines pruned when employed on contract or 'piece' rates.[12] The higher earnings of the RSE workers, as shown in Table 5, were due primarily to three factors: their desire to earn as much money as possible while in New Zealand; their higher rate of attendance at work; and their experience of and willingness to undertake agricultural work (most RSE workers were from rural communities in the islands). Respondents to the annual RSE Monitoring Survey also regularly report favourably on RSE worker performance and attitude (Research New Zealand 2015).

Views on the extent to which participation in the RSE scheme has produced 'wins' for the workers are mixed. On the one hand, the gross amounts earned by Pacific RSE workers over several months of seasonal work are significantly higher than the incomes that rural workers can obtain at home. On the other hand, the costs of participation in the RSE for workers and their families are significant. Research conducted by the Ministry of Business, Innovation and Employment (MBIE) (2015) has attempted to assess the effects of standard deductions (income tax, insurance, accommodation, local transport, share of the airfare to New Zealand) on gross incomes, as well as the amount and frequency of

Table 5. Weeks worked and earnings for RSE and non-RSE workers, 2011.

Measure	Pacific RSE workers	Non-RSE workers	All workers
Number in group	418	145	563
Average weeks worked	9.97	7.01	9.21
Maximum earnings (NZ$)	11 378	10 411	11 378
Minimum earnings (NZ$)	3924	168	168
Median earnings (NZ$)	6862	4158	6548
Average earnings (NZ$)	6890	4163	6178
Standard deviation (NZ$)	1265	2509	2056
Coefficient of variation (%)	18.4	32.9	33.2

Source: Bedford (2013, 312).

remittance transfers made by RSE workers over the season. Research has also been done on the impact of remittances (in cash and kind) for participating households and communities, as well as the impact of worker absences on households in Tonga (Holani forthcoming; Rohorua et al. 2009) and Vanuatu (Rohorua et al. 2009; Bailey 2015, 2014; Craven 2015; Rockell 2015). But a question remains: do seasonal workers in New Zealand—especially those from the more distant participating countries such as Kiribati, Tuvalu, Papua New Guinea and the Solomon Islands—make enough money in 6–9 months of seasonal work to cover the full costs of participating (including living costs in New Zealand), as well as having a good return on their labour to compensate for the lengthy absences from home? To our knowledge, a full analysis of the costs and benefits of participation by Pacific workers in the RSE scheme has not been undertaken.

Notwithstanding the lack of clarity around the real costs of participation, large numbers of workers wish to make repeated trips to New Zealand (about 80 per cent of New Zealand's RSE workers have been back for two or more seasons), and the money obtained from seasonal work plays a vital role in improving the livelihoods of participating households and communities. Money earned in New Zealand is used for a variety of purposes: to meet basic needs (such as food, clothing, basic household amenities); to invest in children's education; renovate or build new homes; purchase land or large goods (vehicles, boats, household appliances); support other relatives; and make contributions to the church (see, for example, Rohorua et al. 2009; McKenzie and Gibson 2010; Bailey 2015, 2014; MBIE 2015; Rockell 2015 among others).

RSE workers' remittance transfers back to the islands can be considerable. Recent analysis of data collected in the MBIE-sponsored Remittance Pilot Project 2014/15,[13] for 264 Samoans and 223 Tongans employed by four orchards and two contractors, revealed that, on average, both groups sent home about 30 per cent of their gross earnings via Western Union or another electronic money transfer agency (Table 6). There were two types of transfers: regular remittances during the course of the time they were working, and a large lump-sum transfer, usually including holiday pay and savings, at the end of the period of employment.[14]

On average, Tongans had higher total and regular remittances, as well as larger average transfers each time they sent money home during their employment, than Samoans. However, the average regular remittance made at each transfer by both groups was similar (about NZ$350–370). Tongans remitted on a slightly higher share of the weeks

Table 6. Remittances (NZ$) by Samoans and Tongans over 22–30 weeks, RSE pilot survey, 2014–15.

Measure	Samoans	Tongans
Number of workers	264	223
Average weeks worked	25	26
Percentage of weeks worked remitted	45.2	53.2
Coefficient of variation (%)	43.9	43.4
Average value of each regular remittance (NZ$)	359	371
Coefficient of variation (%)	74.1	55.6
Average total regular remittances while working (NZ$)	3495	4618
Coefficient of variation (%)	60.6	60.2
Average total remittances including final transfer (NZ$)	6089	6176
Coefficient of variation (%)	50.2	52.1

Source: Bedford and Bedford (2016).

they were working (53 per cent) than Samoans (45 per cent). There were large variations in all of the measures of remittances as reflected in coefficients of variation that regularly exceeded 40 per cent. In light of this variability, the averages need to be interpreted with caution.

The remittances reported as cash transfers are only part of the returns from seasonal work going back to families in the islands. As noted above, there are also remittances in kind (goods taken home by the worker or sent home by post or in containers at the end of their time in work), as well as money carried home as cash. These can be quite considerable, as research by Bailey (2015) shows for workers from Vanuatu, and Holani (forthcoming) shows for the Tongan workers she interviewed in the islands and in Australia. The Tongans and Samoans interviewed for the Remittance Pilot Project 2014/15 (MBIE 2015) probably did not have large sums of money to carry home as cash. By the time their remittances and standard deductions had been removed from their gross incomes, workers had only about 33 per cent of the money they earned to cover living expenses and any other purchases while in New Zealand. This worked out, on average, at about NZ$340 a week for Samoans out of average weekly gross earnings of NZ$860, and NZ$297 a week for Tongans out of gross average weekly earnings of NZ$775 (Bedford and Bedford 2016). No comparable data on earnings and remittances for SWP workers were available at the time of writing.

In addition to their earnings, RSE and SWP workers have access to training while in the host countries. In Australia, Registered Training Organisations offer training in English literacy and numeracy, information technology skills and first aid (DFAT 2015b). In New Zealand, training is provided through the Ministry of Foreign Affairs and Trade (MFAT)-funded Vakameasina programme, which is available in most regions where RSE workers are employed, and covers English literacy and numeracy, financial literacy, health and life skills training (Bedford 2013). Non-government organisations, such as Fruit of the Pacific, also offer health-related courses as well as support with community development initiatives.[15] A number of RSE employers have provided direct support to rural communities where they recruit their RSE labour. Such support has included: installing or repairing water pumps to provide clean drinking water; building a floating jetty for landing cargo and people; and building a new kindergarten and medical centre (Bedford 2013; Bailey 2014). There have also been some joint ventures involving agricultural production in Vanuatu (coffee growing on Tanna) and vegetable growing in Samoa. Information about these has been provided by employers at their annual conferences, including one in Samoa in 2015 when participants visited villages where these sorts of ventures were being undertaken.[16]

On balance, we consider that there have been 'wins' for workers through their participation in the RSE and SWP, although there is far less published research available in regards to the latter. But what about the workers' home communities? Again, there are mixed views here. Rockell's (2015) 'critical lens' on the schemes suggests that the benefits for workers and communities in Vanuatu have been overstated by a focus on the more obvious material dimensions of development. Craven (2015) and Bailey (2014) have also drawn attention to mixed benefits to source communities in their fieldwork with ni-Vanuatu seasonal workers. There can be major disruptions to rural social life caused by repeated absences of adult men and women, as well as by the new attitudes that workers bring home. Changing wealth and power relationships in communities are emerging among those who have regular access to seasonal work overseas and those who have not been selected, or who have chosen not to participate in the seasonal work programmes. The Pacific liaison officers who facilitate the RSE scheme have observed these trends for some time, and they are now being given greater priority for research in both New Zealand and Australia. Re-integration of workers back into their communities and breaking the cycle of repeated annual commitments to seasonal work among those who are encouraged by employers to return regularly, are two issues that require greater attention from officials, employers and researchers.

The least-studied aspect of the seasonal work schemes is their impacts on society and economy in the host communities. In some parts of New Zealand and Australia there are significant fluctuations in population linked with seasonal peaks in demand for workers in the accommodation and the agricultural sectors. Little attention has been given to the impacts of influxes of seasonal workers on community services, infrastructure, commerce and social cohesion. In some communities in New Zealand's South Island, for example, the RSE scheme brought significant numbers of Pacific people into their lives for the first time. The small Central Otago town of Alexandra (population 4800 in 2013) hosts over 1000 temporary workers during the fruit picking and pruning seasons, 500 of whom are ni-Vanuatu. Bailey (2014) describes some of the early adjustments that Alexandra's predominantly 'pakeha' or European-descent population made in accommodating several hundred Melanesians, especially during the summer months.

Contemporary, 'best practice' seasonal work schemes are best conceptualised as complex systems of relationships that span individuals (workers, employers, contractors, government officials), organisations (government agencies, industry organisations, unions, insurance companies, accommodation services) and communities (families and wider social groups in the islands and in the destination countries). The first systematic evaluation of the RSE scheme (Evalue Research 2010) included a useful schematic diagram to capture both the mix of relationships and the range of stakeholders (Figure 1). This seasonal work system, like all complex systems of social and economic relations, is dynamic and constantly adjusting to changing circumstances in both the source and destination communities. There is no such thing as 'business as usual' in the contemporary 'best practice' seasonal work schemes.

It is this dynamism that should allow the RSE scheme and the SWP to avoid some of the challenges the Canadian Seasonal Worker Agricultural Program (SAWP) has faced as the use of seasonal workers became 'integral to the enterprises that employ them' (Preibisch 2007, 439). Under Canadian policy, employers are allowed to request the same workers to return each season, and, with repeated trips, workers are becoming highly skilled and experienced. Some workers have spent up to 25 years in the Canadian scheme, employed on seasonal contracts for up to 8 months each year, and without any pathways to permanent residence.

The RSE scheme and the SWP have been designed, from the outset, to provide a short-term source of employment to further the livelihoods of Pacific workers and their families in the islands. However, like the Canadian workers, with repeated trips to New Zealand and Australia each year (some RSE workers have now been to New Zealand for nine consecutive seasons), a number of Pacific workers are becoming highly skilled at specialist tasks on the orchard/vineyard that generate good returns for their commercial employers. Where Pacific workers are performing skilled work, and wish to pursue careers in these

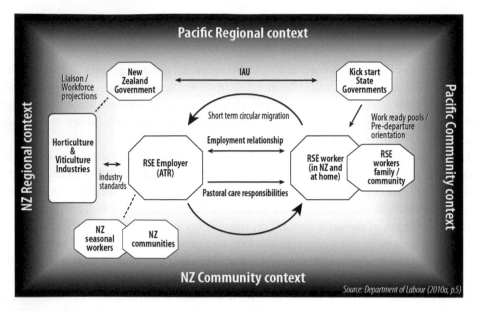

Figure 1. The RSE system. *Source:* Bedford (2013) based on a diagram in Evalue Research (2010, 5).

Note: IAU refers to Interagency Understanding—the form of agreement signed between New Zealand's former Department of Labour and partner Departments/Ministries in the governments of participating Pacific states.

industries longer-term, the seasonal work policies need to be flexible enough to allow for some carefully managed transitions to permanent residence in Australia and New Zealand, rather than locking workers into a pattern of annual, repeated return for seasonal work. In this context, the negotiations surrounding a comprehensive free-trade agreement that would encompass all of the Pacific Islands Forum countries including Australia and New Zealand provide some insights into future labour mobility relationships with Pacific countries.

The seasonal work schemes and PACER Plus

Australia's and New Zealand's seasonal work schemes have become the focus of considerable attention in the negotiations surrounding PACER Plus. Negotiators representing Pacific states have insisted on the inclusion of some special provisions in this agreement for the movement of semi- and low-skilled labour. These would be in addition to the usual General Agreement on Trade in Services Mode 4 (GATS 4) arrangements for movement of people with specialist professional skills that are part of most contemporary free-trade agreements.

A non-binding Labour Mobility Arrangement, relating to the movement of semi- and low-skilled labour that would sit alongside the PACER Plus Agreement, is in the final stages of negotiation. Precise details of this Arrangement have not been made public, but it is known that there will be provision for a Pacific Labour Mobility Annual Meeting (PLMAM) at which specific proposals to increase access to employment opportunities in Australia and New Zealand will be discussed. The initial PLMAM was held in New Zealand in late August 2016, and two pilot projects—the employment of Pacific trades workers in the rebuilding of Christchurch after the devastating earthquakes there in 2010–11, and the employment of offshore fisheries workers in New Zealand's commercial fishing fleet—were a focus of discussion at this meeting.

In addition, Australian negotiators have made it clear that once PACER Plus is signed the SWP will be extended to all Pacific Islands Forum countries.[17] This will provide other Polynesian (Cook Islands and Niue) and Micronesian states (Palau, Federated States of Micronesia (FSM), and the Republic of the Marshall Islands) with seasonal work opportunities, subject to employer demand (DFAT 2015a). However, Cook Islanders and Niueans already have access to the Australian labour market by virtue of their New Zealand citizenship, while Palauans, citizens of FSM and the Marshall Islands have access to the United States through their Compacts of Free Association (Bedford, Burson, and Bedford 2014; Howes 2015). It remains to be seen whether employers will recruit from the northern Micronesia states following the signing of PACER Plus, given the high transport costs involved.

The big gains for the Pacific under the Labour Mobility Arrangement will be in accessing other types of temporary work visas—visas that may allow skilled workers to transition to residence and to be united with their families in Australia or New Zealand. Providing pathways to residence will address a fundamental concern about temporary labour migration schemes, namely the employment restrictions placed on temporary workers and their lack of mobility in the destination country (Preibisch 2007). Hugo (2009) argued that such pathways should be provided, under certain conditions, as an element of 'best practice' in the destination country. If the PLMAM does deliver such opportunities, there will be increasing pressure from horticulture and viticulture employers, as well as highly skilled seasonal workers, for a pathway to longer-term residence that provides them with the same privileges.

Back to Hugo and the Pacific

It is appropriate to conclude by returning to research that Hugo was undertaking on Pacific migration at the time of his untimely death in 2015. Hugo supported greater engagement with Pacific countries through migration and this was reflected clearly in his work for AusAID on the Pacific microstate work pilot and his plans for a major book on migration and climate change in the Asia-Pacific region.[18] He was also a strong supporter of the ideas that two Australian economists had been promoting over two decades in their assessments of Pacific migration for AusAID, namely the merit in adopting policies, like New Zealand's Pacific Access Category, as part of a suite of initiatives to contribute to Pacific development (Appleyard and Stahl 1995; Stahl and Appleyard 2007).

Hugo (2009, 67) acknowledged that 'there is no single best practice in temporary labour migration which is suited to all or even most origin and/or destination countries'. Rather, what is implemented and 'what works' is heavily dependent on context. Moreover, the dynamic nature of schemes such as the RSE and the SWP, which are built on a complex web of stakeholder relationships and that require flexibility among participants to adapt to changing circumstances in source and destination countries, means such programmes require ongoing investment and governmental oversight. He emphasised the need to 'enhance our understanding of what is likely to work under what conditions' and argued that this could be achieved by sharing the lessons learned by individual countries with others (Hugo 2009, 68). Graeme Hugo's enormous experience of migration processes and his pragmatic approach to potential policy options (McAuliffe 2016) would have been greatly appreciated by officials in Australia and New Zealand as they seek to reach an agreed compromise over mobility arrangements linked with PACER Plus, and as they continue to manage a growing number of temporary labour migration schemes that aim to contribute to development in the Pacific.

Notes

1. In 2012 Graeme Hugo, a Fellow of the Academy of the Social Sciences in Australia (FASSA), was awarded the civil honour of Officer in the General Division of the Order of Australia (AO) for his service to population research.
2. http://events.ulg.ac.be/hugo-conference/conference/
3. The Pacific Access Category (PAC) was introduced in 2002 and provides small annual residence quotas for Fiji (250), Tonga (250), Kiribati (75) and Tuvalu (75). Fiji lost its entitlement under the PAC after the 2006 military coup; this was re-established in 2015.
4. A useful introduction to PACER Plus can be found at the website of New Zealand's Ministry of Foreign Affairs and Trade (MFAT) at https://www.mfat.govt.nz/en/trade/free-trade-agreements/agreements-under-negotiation/pacer/ (accessed May 29, 2016). Howes (2014) has prepared a useful outline of the labour mobility dimensions of the PACER Plus negotiations.
5. Data on New Zealand's temporary work visas can be obtained from https://www.immigration.govt.nz/about-us/research-and-statistics/statistics (accessed July 12, 2016).
6. The numbers of Tongans who have absconded or overstayed their seasonal work visas are estimates obtained during conversations with officials in New Zealand's Ministry of Business, Innovation and Employment and Australia's Labour Mobility Assistance Program (LMAP). At the time of writing (October 2016), LMAP was in the process of commissioning research into the Tongan absconder issue in Australia.
7. The RSE scheme has always had a cap on numbers of seasonal workers who are permitted to enter the country each year in order to protect this type of work for New Zealanders. The initial cap was 5000 but this was reached within 2 years. The cap was then raised to 8000, where it remained until July 2015 when it was raised to 9000 and then in November 2015 to 9500 followed by a further increase in November 2016 to 10 500.
8. Expansion of the SWP to include Northern Australia's tourism sector is part of the Australian government's strategy for developing Northern Australia (Commonwealth of Australia 2015). Details of changes to the SWP can be found at https://docs.employment.gov.au/system/files/doc/other/swp_tourism_pilot_fact_sheet_final.pdf and https://docs.employment.gov.au/system/files/doc/other/expansion_of_the_seasonal_worker_programme_-_faqs.pdf (accessed July 2, 2016).
9. Fiji was excluded from participation in both schemes until the end of the 2006 military coup regime and the democratic election of the current government in 2014.
10. From the mid-1970s until 2002, New Zealand had temporary work schemes for Samoan and Tongan citizens. For Kiribati and Tuvalu similar schemes operated from the late 1980s until 2002 and for Fiji from the mid-1970s until the first military *coup d'etat* in 1987. From the late 1960s, there has been a quota of 1100 a year for Samoan citizens seeking work and residence in New Zealand. The Pacific Access Category (PAC), introduced in 2002, provides small annual residence quotas for Fiji (250), Tonga (250), Kiribati (75) and Tuvalu (75). Fiji lost its entitlement under the PAC after the 2006 military coup; this was re-established in 2015. Further information on these arrangements can be found in Bedford (2008) and Mahina-Tuai (2012).
11. Both the RSE scheme and the SWP are supported by capacity-building programmes funded by the Ministry of Foreign Affairs and Trade (MFAT) (New Zealand) and DFAT (Australia),

respectively. MFAT provides financial support for the Strengthening Pacific Partnerships (SPP) project to strengthen the capacity of Pacific countries to participate in labour mobility schemes. Details about the SPP can be found in Nunns et al. (2013). DFAT has contracted Cardno to deliver the LMAP that has been designed to increase the supply and quality of seasonal workers, strengthen linkages with Australian employers and maximise development impacts of the SWP. Further information on the LMAP can be found at http://www.lmaprogram.org (accessed July 15, 2016).

12. Of the 563 workers for whom 10 weeks of gross weekly wages were available, 418 (74 per cent) were Pacific RSE workers while the remaining 145 (26 per cent) included New Zealand regular employees (57), New Zealand casual workers (55) and backpackers (33).

13. The MBIE Remittance Pilot Project was undertaken between November 2014 and June 2015. Data were collected on earnings and remittances for 640 Samoan and Tongan men employed by six RSEs in Hawke's Bay for periods ranging from 8 to 30 weeks. Initial reports on the survey's findings were published late in 2015 (Gounder 2015; MBIE 2015). Additional analysis of the data was undertaken in 2016 to provide a more detailed review of statistics relating to earnings and remittances for 487 (76 per cent) of the 640 workers covered in the pilot project. The 487 workers covered in the analysis were employed continuously for a minimum of 22 weeks (Bedford and Bedford 2016).

14. Information on remittance transfers was collected weekly from the workers by supervisors and pastoral care workers. Evidence of sums transferred came from receipts provided by the money transfer agencies. There were also interviews with 520 workers where information on preferences for remittance transfer agents, plans for use of remittances back in the islands, amongst other topics, was collected. Some of these data are summarised in the reports prepared by Gounder (2015) and MBIE (2015).

15. See, for example, Fruit of the Pacific's recent initiative involving water filters for communities on Tanna (where ni-Vanuatu RSE workers employed by Baygold are recruited from). Available at https://www.facebook.com/FruitOfThePacific/ (accessed July 16, 2016).

16. Information on the RSE Conferences can be obtained from the web under 'RSE Conference xxxx (year)'. Many of the presentations made at the RSE Conference held in Apia in July 2015 are available at http://www.hortnz.co.nz/our-work/people/rse-conference-2015-presentations/ (accessed October 14, 2016).

17. At present there are 10 countries participating in the SWP: Fiji, Kiribati, Nauru, Papua New Guinea, Samoa, Solomon Islands, Tonga, Tuvalu and Vanuatu in the Pacific, as well as Timor Leste in Southeast Asia.

18. During 2014 Hugo was in negotiations with Edward Elgar about a book on climate change and migration in the Asia-Pacific region that drew on a major interdisciplinary study he had led for the Asian Development Bank (ADB) (Asian Development Bank 2012; Hugo et al. 2009). The ADB study included sections on the Pacific Islands that were developed in collaboration with Richard Bedford, extending an existing joint research programme on migration and development in the Pacific region (Bedford and Hugo 2012). There have been discussions amongst Hugo's colleagues about pursuing this book venture, but a decision has been taken recently by Associate Professor Yan Tan (University of Adelaide) to commission papers for a special edition of *Population and Environment* as a way of recognising Hugo's very significant contribution to the study of climate change and migration.

Acknowledgements

We acknowledge the assistance with access to data on Pacific seasonal labour that was provided by officials in Immigration New Zealand within the Ministry of Business, Innovation and Employment, and in Australia's Department of Employment. We are also grateful to a considerable number of officials, employers, industry representatives and Pacific workers who shared their experiences of the SWP and RSE with us. The interpretations we have placed on the information they provided are ours and do not represent the views of the government agencies, businesses or individuals named in this paper. Funding from the University of Adelaide's Department of Geography, Environment and Planning and the University of Waikato's National Institute of Demographic and Economic Analysis supported our field research on the RSE and SWP in the islands, Australia and New Zealand. Graeme Hugo's contribution to our understanding of temporary migration was enormous. We also acknowledge the thought-provoking questions and advice we received from the co-editors of this special issue.

Disclosure statement

No potential conflict of interest was reported by the authors.

References

Appleyard, R. T., and C. W. Stahl. 1995. *South Pacific Migration: New Zealand Experience and Implications for Australia*, AusAID International Development, Issue no. 2, AusAID, Canberra.
Asian Development Bank. 2012. *Addressing Climate Change and Migration in Asia and the Pacific.* Manila: Asian Development Bank.
Bailey, R. 2014. "Working the Vines: Seasonal Migration, Money And Development in New Zealand and Ambrym." Vanuatu (Unpublished doctoral thesis), University of Otago, Dunedin. Accessed July 5, 2016 http://otago.ourarchive.ac.nz/handle/10523/5063.
Bailey, R. 2015. "Using Material Remittances from Labour Schemes for Economic and Social Development." *In Brief 2015/15*, State, Society and Governance in Melanesia Program, ANU College of Asia and the Pacific, Canberra.
Ball, R. 2010. "Australia's Pacific Seasonal Worker Pilot Scheme and its Interface with the Australian Horticultural Labour Market: is it Time for Reform?" *Pacific Economic Bulletin* 25 (1): 114–130.
Bedford, R. D. 1971. "Mobility in transition." (Unpublished doctoral thesis), Australian National University, Canberra.
Bedford, R. D. 1973. "A Transition in Circular Mobility: Population Movement in the New Hebrides 1800-1970." In *The Pacific in Transition: Geographical Perspectives on Adaptation and Change*, edited by Brookfield H.C., 187–227. London: Edward Arnold.
Bedford, R. D. 2008. "Pasifika Mobility: Pathways, Circuits and Challenges in the 21st Century." In *Pacific Interactions. Pasifika in New Zealand—New Zealand in Pasifika*, edited by A. Bisley, 85–134. Wellington: Institute of Policy Studies, Victoria University of Wellington.
Bedford, C. E. 2013. "Picking winners? New Zealand's Recognised Seasonal Employer (RSE) policy and its impacts on employers, Pacific workers and their island-based communities" (Unpublished doctoral thesis), University of Adelaide, Adelaide. Accessed July 5, 2016 http://digital.library.adelaide.edu.au/dspace/handle/2440/82552.
Bedford, C. E. 2014. "New Zealand's Recognised Seasonal Employer Work Policy: Is it Delivering 'Wins' to Employers, Workers and Island Communities?" In *Talanoa: Building A Pacific Research Culture*, edited by P. Fairbairn-Dunlop, and E. Coxon, 78–89. Auckland: Dunmore Press.
Bedford, R. D. 2015. "Graeme John Hugo AO FASSA." *New Zealand Population Review* 40: 145–147.
Bedford, C. E., and R. D. Bedford. 2010. "'Engaging with New Zealand's Recognised Seasonal Employer Work Policy: the Case of Tuvalu." *Asian and Pacific Migration Journal* 19 (3): 421–445.
Bedford, C. E., and R. D. Bedford. 2013. "Managed Seasonal Migration Schemes for Pacific Workers: Just a 'band aid'?" Paper presented at the Inaugural State of the Pacific Conference, State, Society and Governance in Melanesia Program, Australian National University, 25–26 June 2013,Canberra.
Bedford, C. E., and R. D. Bedford. 2016, July. "The RSE Remittance Pilot 2014/15: revisiting the data." Presentation to the RSE Employers' Conference, Napier, Hawke's Bay.

Bedford, R. D., and G. Hugo. 2012. *Population Movement in the Pacific: A Perspective on Future Prospects*, Labour and Immigration Research Centre, Department of Labour, Wellington and Department of Immigration and Citizenship, Canberra. Accessed May 20, 2016 http://www.mbie.govt.nz/publications-research/research/research-index?topic=Migration&type=Report.

Bedford, R. D., B. Burson, and C. E. Bedford. 2014. *Compendium of Legislation and Institutional Arrangements For Labour Migration In Pacific Island Countries*, Report prepared for the International Labour Organization. Accessed June 10, 2016. http://www.ilo.org/wcmsp5/groups/public/---asia/---ro-bangkok/---ilo-suva/documents/publication/wcms_304002.pdf.

Brickenstein, C. 2015. "Impact Assessment of Seasonal Labor Migration in Australia and New Zealand: A win–win Situation?" *Asian and Pacific Migration Journal* 24 (1): 107–129.

Castles, S. 2006. "Guestworkers in Europe: A Resurrection?" *International Migration Review* 40 (4): 741–766.

Chapman, M. 1970. "Population Movement in Tribal Society: The case of Duidui and Pichahila." (Unpublished doctoral dissertation), University of Washington, Seattle.

Chapman, M., and R. M. Prothero1985. *Circulation in Population Movement. Substance and Concepts From the Melanesian Case.* London: Routledge and Kegan Paul. re-published 2014.

Commonwealth of Australia. 2015. *Our North, Our Future: White Paper on Developing Northern Australia.* Accessed July 2, 2016 http://industry.gov.au/ONA/WhitePaper/Documents/northern_australia_white_paper.pdf.

Commonwealth of Australia. 2016. *A National Disgrace: The Exploitation of Temporary Work Visa Holders*, The Senate Education and Employment References Committee. Accessed June 20, 2016. http://www.aph.gov.au/Parliamentary_Business/Committees/Senate/Education_and_Employment/temporary_work_visa/Report.

Connell, J. 2009. *The Global Health Care Chain: From the Pacific to the World.* London: Routledge.

Craven, L. 2015. "Migration-affected Change and Vulnerability in Rural Vanuatu." *Asia Pacific Viewpoint* 56 (2): 223–236.

Curtain, R. 2016. "New Zealand's Recognised Seasonal Employer (RSE) scheme and the Australian Seasonal Worker Program (SWP): why so different?" Unpublished paper presented at a workshop on 'New Research on Pacific Labour Mobility', Crawford School of Public Policy, Australian National University, Canberra, 2 June 2016.

DFAT. 2015a. *Submission of the Department of Foreign Affairs and Trade to the Joint Standing Committee on Migration's Inquiry into the Seasonal Worker Programme, Submission 37.* Accessed July 8, 2016 http://www.aph.gov.au/Parliamentary_Business/Committees/Joint/Migration/Seasonal_Worker_Programme/Submissions.

DFAT. 2015b. "Factsheet—Add-on Skills Training: The Seasonal Worker Programme." Accessed July 12, 2016 https://docs.employment.gov.au/system/files/doc/other/06-15_seasonal_worker_programme_aost_factsheet.pdf.

Doyle, J., and S. Howes. 2015. *Australia's Seasonal Worker Program: Demand-Side Constraints and Suggested Reforms.* Accessed July 14, 2015 https://openknowledge.worldbank.org/bitstream/handle/10986/21491/943680WP0Box380nd0Suggested0Reforms.pdf?sequence = 1&isAllowed=y.

Evalue Research. 2010. *Final Evaluation Report of the Recognised Seasonal Employer policy (2007-2009)*, IMSED Research, Department of Labour, Wellington.

Gibson, J. 2015. "Circular Migration, Remittances and Inequality in Vanuatu." *New Zealand Population Review* 41: 153–167.

Gibson, J., and D. McKenzie. 2011. *Australia's Pacific Seasonal Worker Pilot Scheme (PSWPS): development impacts on the first two years*, Occasional Paper, Department of Economics, University of Waikato, Hamilton.

Gounder, R. 2015. "Development Impacts of Remittances In The Pacific: Economic Benefits of the Recognised Seasonal Employer Work Policy for Samoa and Tonga. Remittances Pilot Project final report." Unpublished report for the Ministry of Business, Innovation and Employment, Massey University, Palmerston North.

Hay, D., and S. Howes. 2012. "Australia's PSWPS: Why Has Take-Up Been So Low?" Powerpoint presentation. Accessed May 19, 2012 https://crawford.anu.edu.au/pdf/events/2012/20120403-making-pacific-migration-work-ppt/2_danielle-hay-and-stephen-howes.pdf.

Holani, A. K. forthcoming. "Temporary Labour Migration and Development for PICs—a policy framework for sustainable economic development" (Unpublished doctoral thesis in preparation), University of Adelaide, Adelaide.

Howes, S. 2014. "PACER Plus and Labour Mobility: How to do a Deal." *DevPolicy Blog from the Development Policy Centre.* Accessed May 16, 2016 http://devpolicy.org/pacer-plus-and-labour-mobility-how-to-do-a-deal-20141124/.

Howes, S. 2015. "A Big Week for Pacific Labour Mobility: SWP Reforms and the Microstate Visa." *DevPolicy Blog from the Development Policy Centre.* Accessed July 10, 2016 http://devpolicy.org/a-big-week-for-pacific-labour-mobility-swp-reforms-and-the-microstate-visa-20150626/.

Hugo, G. 1975. "Population Mobility in West Java." (Unpublished doctoral thesis), Australian National University, Canberra.

Hugo, G. 1996. "Environmental Concerns and International Migration." *International Migration Review* 30 (1): 105–131.

Hugo, G. 2009. "Best Practice in Temporary Labour Migration for Development: A Perspective From Asia and the Pacific." *International Migration* 47 (5): 23–74.

Hugo, G. 2010. *The Future of Migration Policies in the Asia-Pacific region,* IOM Background Paper World Migration Report 2010, International Organization for Migration, Geneva.

Hugo, G. 2013. "Closer Economic Relations and a More Comprehensive Relationship with Australia: A Role For Migration Facilitating Development in Kiribati, Tuvalu and Nauru." Unpublished confidential report for AusAID's Microstates Futures Study, Australian Population and Migration Research Centre, University of Adelaide, Adelaide.

Hugo, G. 2015. "The New Zealand-Australia Migration Corridor." *New Zealand Population Review* 41: 21–44.

Hugo, G., D. K. Bardsley, Y. Tan, V. Sharma, M. Williams, and R. D. Bedford. 2009. "Climate Change and Migration in the Asia-Pacific Region." Unpublished report for the Asian Development Bank, National Centre for the Social Applications of GIS, University of Adelaide, Adelaide.

International Labour Organization. 2009. *Good Practices Database—Labour Migration Policies and Programmes: The Recognised Seasonal Employer (RSE) scheme, New Zealand.* Accessed July 10, 2016. http://www.ilo.org/dyn/migpractice/migmain.showPractice?p_lang=en&p_practice_id=48.

Jenkin, S. 2015. "Benefits of the Seasonal Worker Program: An Employer's Perspective." *DevPolicy Blog from the Development Policy Centre.* Accessed July 10, 2016 http://devpolicy.org/benefits-of-the-seasonal-worker-program-an-employers-perspective-20150305/.

Mahina-Tuai, K. 2012. "A Land of Milk and Honey? Education and Employment Migration Schemes in the Postwar era." In *Tangata O le Moana. New Zealand and the People of the Pacific,* edited by S. Mallon, K. Mahina-Tuai, and D. Salesa, 161–178. Wellington: Te Papa Press.

Mallon, S. 2012. "'Little Known Lives. Pacific Islanders in Nineteenth Century New Zealand'." In *Tangata O le Moana. New Zealand and the People of the Pacific,* edited by S. Mallon, K. Mahina-Tuai, and D. Salesa, 77–96. Wellington: Te Papa Press.

Martin, P. 2002. "Mexican Workers and U.S. Agriculture: The Revolving Door." *International Migration Review* 36 (4): 1124–1142.

Martin, P., M. Abella, and C. Kuptsch. 2006. *Managing Labor Migration in the Twenty-First Century.* New Haven: Yale University Press.

McAuliffe, M. 2016. "Thinking Space: Migration Moderate, 'Master Weaver' and Inspirational Team Leader: Reflecting on the Lasting Legacy of Graeme Hugo in Three Spheres of Migration Policy." *Australian Geographer* 47(4): 383–389 (in press).

McKenzie, D., and J. Gibson. 2010. *The Development Impact of a Best Practice Seasonal Worker Policy,* Policy Research Working Paper 5488, Impact Evaluation Series No. 48, The World Bank, Washington DC.

Ministry of Business, Innovation and Employment. 2015. *The Remittance Pilot Project. The economic benefits of the Recognised Seasonal Employer work policy and its role in assisting development in Samoa and Tonga.* Accessed February 15, 2016 http://www.employment.govt.nz/er/rse/rse-remittance-pilot-project.pdf.

Munro, D. 1990. "The Origins of Labourers in the South Pacific: Commentary and Statistics." In *Labour in the South Pacific*, edited by C. Moore, J. Leckie, and D. Munro, xxxix–li. Townsville: James Cook University of Northern Queensland.

Nunns, H., H. Roorda, C. E. Bedford, and R. D. Bedford. 2013. "Mid-term Evaluation of the Strengthening Pacific Partnerships Project." Report for the Ministry of Foreign Affairs and Trade and the Ministry of Business, Innovation and Employment, Wellington.

Owen, G. 2015. "Benefits of the Seasonal Worker Program: A Recruiter's Perspective." *DevPolicy Blog from the Development Policy Centre*. Accessed July 10, 2016 http://devpolicy.org/benefits-of-the-seasonal-worker-program-a-recruiters-perspective-20150226/.

Preibisch, K. 2007. "Local Produce, Foreign Labour: Labour Mobility Programs and Global Trade Competitiveness in Canada." *Rural Sociology* 72 (3): 418–449.

Prothero, R. M., and M. Chapman. 1985. *Circulation in Third World Countries*. London: Routledge and Kegan Paul. republished 2014.

Rarere, G. 2016, July. "RSE Update Report." Presentation by the Manager, Pacifica Labour and Skills Team, Settlement Protection Attachment Branch, Immigration New Zealand at the Recognised Seasonal Employers' Conference, Napier, Hawkes Bay.

Research New Zealand. 2015. *RSE Monitoring Survey 2015: Executive Summary Report*, Report prepared the Ministry of Business, Innovation and Employment. Accessed July 12, 2016 http://employment.govt.nz/er/rse/monitoring/employers-survey-2015.pdf.

Rockell, D. G. 2015. "Pacific Island Labour Programmes in New Zealand: An Aid to Pacific Island Development? A critical lens on the Recognised Seasonal Employer policy" (Unpublished doctoral thesis), Massey University, Palmerston North.

Rohorua, H., J. Gibson, D. McKenzie, and P. G. Martinez. 2009. "How do Pacific Island Households and Communities Cope with Seasonally Absent Members?" *Pacific Economic Bulletin* 24 (3): 19–38.

Ruhs, M. 2002. "The Potential of Temporary Migration Programmes in Future International Migration Policy." *International Labour Review* 145 (1&2): 7–36.

Skeldon, R. 1974. "Migration in a Peasant Society the Example of Cuzco, Peru" (Unpublished doctoral dissertation), University of Toronto, Toronto.

Stahl, C., and R. Appleyard. 2007. *Migration and Development in the Pacific Islands: Lessons From the New Zealand Experience*. Canberra: Australian Agency for International Development (AusAID).

TNS Social Research. 2011. *Final Evaluation of the Pacific Seasonal Worker Pilot Scheme*, Report prepared for the Department of Education, Employment and Workplace Relations. Accessed July 13, 2016 https://docs.employment.gov.au/system/files/doc/other/pswps_-_final_evaluation_report.pdf.

Tolley, A. 2016, July. "Keynote Address." Presentation by the Minister for Social Development at the Recognised Seasonal Employers' Conference, Napier, Hawkes Bay.

United Nations. 2006. *International Migration and Development. Report of the Secretary-General*. Agenda Item 54(c), Sixtieth Session of the General Assembly of the United Nations, 18 May 2006, A/60/871.

Whatman, R., K. Wong, R. Hill, P. Capper, and K. Wilson. 2005. "From Policy to Practice: The Pure Business Project." Proceedings of the 11th ANZSYS/Managing the Complex V Conference-Systems Thinking and Complexity Science: Insights for Action, Christchurch, 5-7 December 2005.

Woodhouse, M. 2015, July. "Address to the 2015 Recognised Seasonal Employer Conference." Presentation to the RSE Employers' Conference, Apia, Samoa.

Care and Global Migration in the Nursing Profession: a north Indian perspective

Margaret Walton-Roberts, Smita Bhutani and Amandeep Kaur

ABSTRACT

Globalisation, supply–demand dynamics, uneven development, enhanced connectivity including the better flow of information, communication and the reduced cost of travel have encouraged the global integration of nursing labour markets. Developed regions of the world have attracted internationally educated nurses (IENs) because of growing healthcare needs. India, along with the Philippines, has become a key supplier of nurses in the global economy. Traditionally the supply of nurses was heavily regionalised in south India, especially Kerala, but of late Punjab, in north India, has played an increasing role in nurse training and migration as the profession has become more respected and more international. This paper uses survey and interview data to detail the recent interest in nursing as a channel for independent female international migration from Punjab, and to examine how migratory ambitions have developed over the last decade in parallel with the changing status of nursing as an internationally respected profession. We identify growing interest in international migration for nursing students and their increased intention to pursue employment opportunities in Australia and New Zealand. This research highlights how nursing and care migration are increasingly structured by international circuits of training and employment, and how such circuits alter migrant and occupational geographies on the ground in sending regions.

Introduction

Women account for an increasing proportion of all migrants, reaching almost half of today's 191 million international migrants (International Organization for Migration (IOM) 2005). In 2013, women comprised 48 per cent of all international migrants worldwide (International Migration Report 2013). Many women are migrating independently of partners or families (Timur 2000), and are becoming agents of economic change as they enter the international labour market and participate in new channels of global wealth distribution (IOM 2003). However, female migrants are over-represented globally in health and personal care sectors of the global economy that are often undervalued (Folbre 2012).

While the specific policy context of migrant care workers' integration into global labour markets displays variability, there is arguably a structural trajectory of convergence in terms of migrant care workers being incorporated into national systems facing restructured or diminished social welfare spending (Williams 2012; Yeates 2009). In the face of these constraints, international opportunities are seen to provide enhanced benefits. We explore this care and migration intersection by focusing on Indian-trained nurses (specifically from Punjab in north India) and changing global migration circuits.

Nurses represent the largest and most internationalised and feminised section of the health professions globally, with nursing in most nations usually dominated by women (Kodoth and Jacob 2013). Developed regions of the world have demanded well-trained nurses because of an ageing population, the increased prevalence of chronic diseases, shortage of primary care physicians and the use of nursing in managing complex clinical cases (Gostin 2008). In Australia, for example, Graeme Hugo (2009) projected that ageing would see demand for residential and non-residential care workers increase by over 200 000 between 2001 and 2031. This demand, he argued, could not be fulfilled by the domestic labour market and would likely result in increased immigration of care workers. Demographic changes are thus creating global markets for nursing and other care workers (Kingma 2006). Such processes of change are connecting regions of the world through increasingly globally oriented models of training, skills and professional mobility. This is certainly relevant to the case of India, which will soon have the largest and youngest workforce the world has ever seen (World Bank 2015).

The issue of health worker shortages and the use of migration to fill that need has long been an area of study. Recent work has indicated that despite the global financial crisis and the use of the World Health Organization's (WHO) voluntary ethical recruitment codes to restrict higher income countries recruiting health workers from countries with severe health worker shortages, the international mobility of health workers remains a key solution to worker shortfalls in many developed nations (WHO 2014). The USA, UK, Canada, Australia, Saudi Arabia and Ireland are major recipient countries of internationally educated nurses (IENs)[1] (Kline 2003). The Philippines, India and Nigeria are top source countries for nurses migrating to the largest market for IENs, the USA (Chen et al. 2013). Immigrant nurses in the USA made up 4 per cent of the 2.7 million nursing workforce in 2000 (Xu and Kwak 2005). Of the total 167 000 foreign-born registered nurses in the USA, approximately 5 per cent (8350) are from India (Jose et al. 2008). In Australia, India is the second largest provider of foreign-born health workers after the UK (WHO 2014).

India has been a key provider of internationally trained medical graduates globally for several decades (Hazarika et al. 2009; Mullan 2005). In the case of Indian nurses, the combination of push, pull and network factors shape nurses' emigration (Walton-Roberts 2015a). Indian nurses look for overseas opportunities because of poor working and training conditions and the lack of opportunities for professional growth and skill development at home. They aspire to work and live overseas where they can benefit from improved quality of life and professional development (Kodoth and Jacob 2013). Estimates suggest 25 000 domestically trained nurses across India leave every year for better paying jobs overseas (*Times Now* 2016). Aspirants are seeking more secure modes of entry into foreign countries that provide a pathway for permanent residence. Enrolment in postgraduate nursing education in Canada, the UK, Australia and New Zealand is seen as a promising route to permanent migration to these desirable destinations (Walton-Roberts 2015c). Migrants' social networks have also promoted certain destinations over others (Kodoth and Jacob 2013).

The emigration opportunities for nurses have created new business opportunities for commercial and personal networks focused on global nurse recruitment and placement. The booming private medical industry in India is playing a significant role in this process. For example, the Apollo Institute of Health Sciences represents a network of Indian hospitals that have developed a for-profit Global Nurse Program to prepare nurses specifically for export to the USA, UK, and Australia (Brush 2008). In response to the growing demand in the health sector more broadly, the current Indian government has invested substantially in training Indian youth for both the domestic and international markets, and nursing is one of the sectors in which the government of India has invested (Walton-Roberts 2015a). A further important dimension of the ongoing globalisation of health care is the influence that increasing international nurse migration and associated emerging business opportunities exact on professional standards and qualifications for health workers (Segouin, Hodges, and Brechat 2005).

What influence will this global demand for nursing have on global care migration and gender relations in a country such as India? As Hugo (1995) identified in the case of Indonesia, the shifting gender dimensions of Asian international migration have profound social policy implications. Indeed, he argued that the whole field of population geography needed to pay greater attention to the changing gender dynamics of Asian migration and especially its consequences for women (Hugo 1996). The focus on a feminised skilled profession, such as nursing, also contributes to debates regarding the gendered nature of skilled migration policy and structural forms of inequality (Boucher 2016).

Based on surveys and interviews gathered in Punjab, India, we examine the influence of increasingly feminised international migration from India in terms of the culture and geography of nursing, paying particular attention to how nursing is integrated into the changing landscapes of international opportunity and care migration. We begin with an overview of the occupational history of the nursing profession in India to contextualise how gender, mobility and social change are read through the nursing profession in that nation. We then examine the culture of migration in Punjab, our study site, before offering survey results and conclusions.

Indian nursing education and status: the culture of care and the spectre of Nightingale

During the Victorian period (1850s to early 1900s) the 'Lady with the Lamp' and philosopher of modern nursing Florence Nightingale exerted great influence over nursing and the promotion of public health across India (Vallée 2007). Nightingale's position has been recontextualised within a longer history of the role of religious orders in nursing, leading to the argument that the religious identity associated with nursing care has informed the subsequent development of the occupation (Nelson 2001). In the case of India, the role of religious orders and colonial hierarchy in nursing has had a significant influence over the occupation, arguably creating a heavily feminised and subservient frame of reference that continues into the present (Nair and Healey 2006). The legacy of nursing's subservient position within the Indian nation has been central to Indian nurses' struggle for improved professional status and compensation (Timmons, Evans, and Nair 2016).

The development of nursing as a profession in India has been informed by the international interests of developed states for some time. In a similar manner to the case of the Philippines, where the production and professionalisation of nursing was partly a product of American imperialism (Choy 2003), the Indian case is also marked by international interests. We have already mentioned the work of Nightingale, which represents the colonial influence over the development of the nursing profession (Abraham 2004). Reddy (2015) has also recounted how Indian nursing was shaped through investments of empire via the work of Protestant Medical Missions (there were 7 in India in 1858 growing to 280 in 1905) (Rafferty and Robinson 2005); the Dufferin Fund (established in 1885) and then the Rockefeller Foundation (whose International Health Division established programmes in India in the 1930s). Reddy (2015, 79) argued that 'Rockefeller wealth proved critical to the establishment of professional nursing within scientific medicine and its attendant hospital system, an outcome that took the trained nursing led by Nightingale nurses one step further up the ladder of respectability.' Indeed, Reddy (2015) noted that as recently as 1994, half of India's 40 nursing colleges were directed by graduates of the Vellore School of Nursing, a key Rockefeller-funded nursing college established in 1900 in south India.

Today, Nightingale still exerts a profound influence over Indian nursing schools and colleges, acting as one of the major referents essential for nursing practice, education and research in India. Nightingale's nursing ideals see the profession as one that has a degree of autonomy within the medical profession in terms of regulation and training. Nightingale argued that the goal of nursing is to foster health within the patient, that nursing education should be controlled by the school, not the hospital, and that nurses should be educated by nursing staff who specialise in education (Selander and Crane 2012). Despite such visions, recent research has suggested that Indian nursing colleges are limited in their ability to develop and control curriculum improvement to this end (Evans, Razia, and Cook 2013). Nursing education in India is undermined by weak curriculum and faculty training, and a lack of innovative teaching methods, especially in regards to clinical training (Tiwari, Sharma, and Zodpey 2013). The tradition of nursing in India, underpinned by Nightingale's philosophy that the 'Goal of nursing is to foster health within the patient' (Selander and Crane 2012, n.p.), has seen the idea of care and service as the vocational core of the profession. Recent changes, however, have seen the nursing vocation become the core means by which individuals enter international labour markets (Khadria 2007), and thereby advance Indian middle-class desires for transnational social mobility (Radhakrishnan 2011). The international migration of nurses from India has traditionally been studied from the perspective of nurses trained in south India (George 2005). In this paper we expand this analysis to reveal similar patterns of interest in international opportunities in nursing, but from the perspective of north India.

Recasting nursing in an age of global care mobility

Nursing was traditionally seen as 'dirty' work often associated with the menial jobs that lower castes typically undertake in Indian society (Nair and Healey 2006). However, in contemporary Indian society the public's view of the nursing profession has changed due to its wider scope, enhanced earning potential and increased professionalisation (Percot 2006). Alongside this change, India is playing an increasingly important role in supplying nurses for the international market (Hawkes et al. 2009). While the Philippines is renowned the world over for the provision of often highly qualified English-speaking nurses (Choy 2003), India has also become a key supplier. According to the US-based Commission on Graduates of Foreign Nursing Schools (CGFNS 2005), in the 1990s India was ranked sixth in terms of the number of registered nurse applicants aspiring for US visas, but from 2003 onwards India has been ranked second after the Philippines. According to the National Council of State Boards of Nursing (NCSBN) in the USA, 628 IENs from India took the US national nursing registration exam in 2016, which was the second highest number of IENs after the Philippines (NCSBN 2016). In Ireland and the UK the number of Indian nurse migrants is similar to those from the Philippines (Matsuno 2009). The increased profile of Indian-trained nurses within the international nursing system has resulted in the occupation being recognised, within Indian society, as offering greater potential for overseas opportunities compared to other professions such as engineering or academia (Gill 2011).

The perception of the nursing profession in India is changing. Increasingly it is being viewed as a professional occupation that provides an opportunity for personal growth; and offers bright prospects abroad as well as economic security for the family (Patidar et al. 2011). As the motivation for entering the profession in India is increasingly informed by overseas options, the nurse-training sector is likely to transform in profound ways. In response to and reflective of the growing interest in nursing, the need to improve the nursing education system has been acknowledged in India (Garner et al. 2014). The Indian government recently approved the establishment of 260 government nursing schools at the district level to meet national shortages of nursing staff and also increased the central government's budgetary allocation for nursing education (Senior 2010). Currently, across India, there are 7377 nursing institutes; these include those established by the government and the private sector (Walton-Roberts 2015a).

The density of registered nurses in India was 1.65 per 1000 persons in 2012, and in Punjab, the site of our research, it was 2.39 per 1000 persons, suggesting Punjab produces and employs a greater number of nurses compared to most states in India. While Punjab has surpassed the national average in this regard, the density of registered nurses is less than many southern states, including Tamil Nadu, Kerala and Karnataka (Table 1). Though Kerala has a long tradition of training nurses for migration (Nair 1996), recent trends show that Punjab is an upcoming state in terms of its capacity to train nurses given the increasing number of nurse education institutes appearing in the state within the past decade (Figures 1–3). For instance, the growth in the number of institutions in Punjab offering Post-basic Bachelor of Science (Nursing) courses increased by over 8000 per cent from 2004 to 2012, compared to a national increase of 1547 per cent (Figure 3).

Further details exploring the rise of international nursing migration from Punjab can be found in our earlier study (Bhutani, Gupta, and Walton-Roberts 2013). In this paper we further examine how the culture of migration in Punjab is adapting to international nursing opportunities, and how these migratory geographies have altered over the last decade or so.

Table 1. Number of Registered Nurses in India (as on 31 December 2012) and population of India (2011) broken down by state

State	Auxiliary nurses and midwives (ANM)	General nurses and midwives (GNM)	Lady health visitors (LHV)	Total Registered Nurses[a] (ANM + GNM + LHV)	Total population (2011)	Ratio of Registered Nurses and midwives per 1000 population
Andhra Pradesh	121 159	168 947	2480	292 586	84 580 777	3.459 (5th)
Assam[b]	22 495	16 371	170	39 036	31 205 576	1.250 (15th)
Bihar	7501	8883	511	16 895	104 099 452	0.162 (23rd)
Chhattisgarh	3190	4608	1352	9150	25 545 198	0.358 (20th)
Delhi	3122	39 791	NA[g]	42 913	16 787 941	2.55 (8th)
Gujarat	36 874	89 460	NA	126 334	60 439 692	2.090 (10th)
Haryana	15 837	20 015	694	36 546	25 351 462	1.441 (13th)
Himachal Pradesh	10 798	9939	499	21 236	6 864 602	3.093 (7th)
Jharkhand	3405	1998	137	5540	32 988 134	0.167 (22nd)
Karnataka	51 109	187 053	6840	245 002	61 095 297	4.010 (2nd)
Kerala	28 979	136 341	8144	173 464	33 406 061	5.192 (1st)
Madhya Pradesh	31 528	100 361	1605	133 494	72 626 809	1.838 (11th)
Maharashtra[c]	33 158	93 032	566	126 756	112 374 333	1.127 (17th)
Manipur	1034	2452	NA	3486	2 570 390	1.640 (12th)
Meghalaya	867	2365	112	3344	2 966 889	1.127 (16th)
Mizoram	1774	2350	NA	4124	1 097 206	3.758 (3rd)
Odisha	59 225	72 461	238	131 924	41 974 218	3.142 (6th)
Punjab[d]	18 152	45 801	2584	66 537	27 743 338	2.398 (9th)
Rajasthan	24 175	45 762	850	70 787	68 548 437	1.032 (18th)
Tamil Nadu[e]	54 635	202 949	11 112	268 696	72 147 030	3.724 (4th)
Tripura	1036	1266	148	2450	3 673 917	0.666 (19th)
Uttar Pradesh	30 767	25 748	2763	59 278	199 812 341	0.296 (21st)
Uttarakhand	1111	387	11	1509	10 086 292	0.149 (24th)
West Bengal[f]	56 782	50 409	12 363	119 554	91 276 115	1.309 (14th)
India	618 713	1 328 749	53 179	2 000 641	1 210 193 422	1.653

Notes:
[a]Registered Nurse: after acquiring the required qualifications, candidates are able to register themselves with their respective state nursing council as a registered nurse/midwife.
[b]Figure for Assam includes figures for Arunachal Pradesh and Nagaland.
[c]Figure for Maharashtra includes figure for Goa.
[d]Figure for Punjab includes figure for Jammu Kashmir.
[e]Figure for Tamil Nadu includes figures for Andaman, Nicobar Island and Pondicherry.
[f]Figure for West Bengal includes figure for Sikkim.
[g]NA: not available.
Source: Indiastat (2012) and Census of India (2011).

Context: the culture of migration in Punjab

Punjab figures among those states of India with a large diaspora population. People in Punjab have a very strong desire to settle abroad and to have a well-established Punjabi culture overseas (Mooney 2006). Punjab certainly qualifies as a transnational space, one that, over at least a century, has been subjected to intense international migration, creating a territory that continues to be at the centre of multiple transnational networks linking overseas migrants and their relatives back to Punjab (Walton-Roberts 2004). International migration from India has typically been led by men, but during the 2000s Indian women have emerged as lead migrants to OECD countries (Sharma 2011). In Punjab, female international migrants were traditionally sponsored by male kin, but now there is evidence that they are choosing the nursing profession as a means to move abroad independently with the ambition not only to settle there permanently but also later to sponsor and

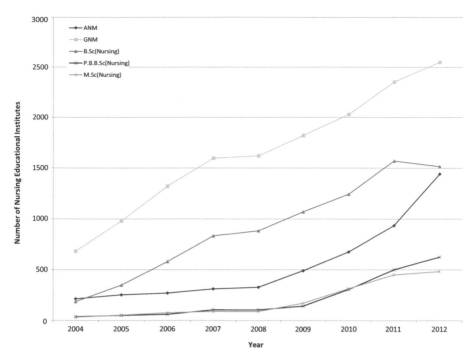

Figure 1. India: total number of nursing educational institutes recognised by the Indian Nursing Council (2004–12), broken down by course of study (ANM: Auxiliary Nursing and Midwifery; GNM: General Nursing and Midwifery; B.Sc. (Nursing): Bachelor of Science in Nursing; P.B.B.Sc. (Nursing): Post-basic Bachelor of Science in Nursing; and M.Sc. (Nursing): Master of Science in Nursing). *Source*: Indian Nursing Council (2004–12).

support other family members to join them (Walton-Roberts 2015b). At the same time there has been a significant increase in the number of nursing colleges in Punjab, and north Indian-based immigration agencies and travel agents are also guiding their clients to adopt educational migration and the nursing profession as one route towards permanent settlement abroad (Walton-Roberts 2015c). The case of Punjab reflects wider processes of educational and economic change that have shaped international migration opportunities for women across Asia:

> [B]ecause of such developments as increases in levels of formal education, the virtual achievement of universal education in Asia and the penetration of mass media into the region, there is an increasing awareness and knowledge of these [economic] disparities and of the opportunities of earning higher incomes in other countries. (Hugo 2006, 167)

There is also a culture of migration embedded in the health sector more broadly. As Connell (2014, 77) has discussed in the case of the Pacific Islands:

> Through education and the social context of education, a more evidently medical culture of migration (but extended into nursing, dentistry and allied activities) overlays and has become intertwined with an existing, more pervasive culture of migration.

The intersection of two traditions of spatial and occupational migration culture, in Punjab and nursing, provides the context for our analysis. We contend that nursing in the Indian context is integrated into international opportunities at its very core; in the curriculum

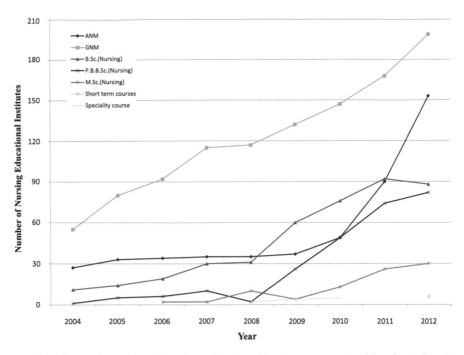

Figure 2. Punjab: total number of nursing educational institutes recognised by the Indian Nursing Council (2004–12) broken down by course of study. *Source*: Indian Nursing Council (2004–12).

Note: while figures for Punjab also include figures for Jammu and Kashmir, more than 90 per cent of nursing institutes are located within Punjab.

(internationalised), the history of nursing (colonial), the education system (expanding, increasingly privatised and outward oriented), and in terms of its growing popularity (embedded in notions of mobility, both spatial and social). We see these tendencies intensified in the location of our study, Punjab, and we argue that this region displays similar traits to the Pacific Islands that Connell highlighted, exemplifying the power of education and global communications that Hugo (2006) noted more generally across Asia. We will highlight how we understand this to be occurring in Punjab through our research data.

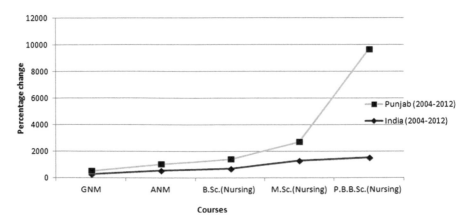

Figure 3. Percentage change in institutes offering nursing courses in India and Punjab, 2004–12. Source: Indian Nursing Council (2004–12).

Methodology

The present study is based on both primary and secondary data. In the context of India, as specified by the Indian Nursing Council (INC), institutes offering Auxiliary Nursing and Midwifery (ANM) and General Nursing and Midwifery (GNM) diplomas are known as nursing schools, while institutes offering Bachelor of Science in Nursing (B.Sc. (Nursing)), Master of Science in Nursing (M.Sc. (Nursing)), and Post-basic B.Sc. (Nursing) are termed nursing colleges. Surveys and interviews were conducted in face-to-face meetings with nursing students at 13 government and private nursing

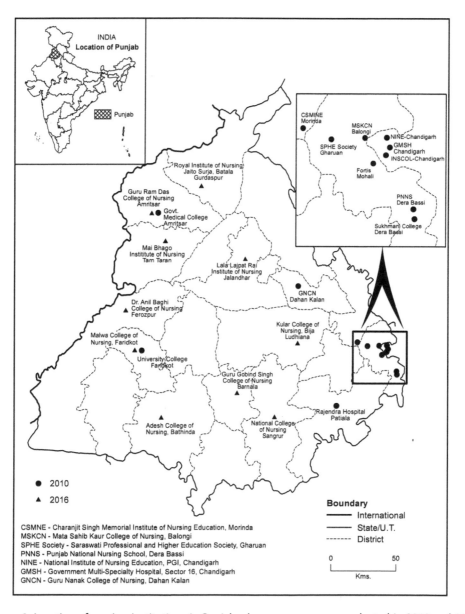

Figure 4. Location of nursing institutions in Punjab where surveys were conducted in 2010 and 2016.

schools and colleges in 2010. This included 1000 nursing students pursuing diplomas or degrees in different districts of Punjab including Amritsar, Patiala, Faridkot, Shaheed Bhagat Singh Nagar (popularly known as Nawanshahr), Sahibzada Ajit Singh Nagar (commonly known as Mohali), Rupnagar and Chandigarh (the capital of Punjab). In April 2016, an additional 717 nursing students were surveyed across 10 different government and private nursing colleges and schools in 10 districts of Punjab (Figure 4). The latter survey was conducted to examine emerging trends and it posed the same set of questions as the initial survey regarding nursing students' demographic characteristics, educational background, and occupational and postgraduate migration plans. In addition, semi-structured questionnaires were used to gather data from students and staff in the 23 nursing schools and institutions regarding why they entered nursing, their changing perceptions of nursing, and international opportunities offered by the profession.

All data were gathered at the respective educational institutions and, before conducting the surveys, permission for data collection was obtained from the principal of each educational institution. After permission was granted we invited students to volunteer to answer the surveys and/or join a small focus group. The 2010 research team comprised postgraduate students and research scholars from the Department of Geography at Panjab University and Wilfrid Laurier University. The research team was engaged in pre-fieldwork ethics and research methods training, and the initial field visits were led by the senior academics/research scholars from both institutions. Subsequently, teams of researchers organised meetings and conducted the surveys and fieldwork during 2010 while the third author conducted the survey in 2016. Key informant interviews with selected faculty members and officials of key nursing educational institutions in Punjab and the Punjab Department of Medical Education and Research were also carried out during 2010–14. Focus group discussions and semi-structured interviews with nursing students were conducted in English and Punjabi. Data on the demographic profile of nursing students were analysed using SPSS statistical analysis software.

Discussion and results

Demographic and occupational profile: young, female and internationally-oriented

Of the 1000 students surveyed in the first survey (2010), those in the 20–29 age group formed about 75 per cent of the total respondents. This youthful demographic profile illustrates one of the main resources that India offers in terms of the production of qualified nurses. The majority of those surveyed were enrolled in the Bachelor of Science in Nursing (45 per cent), followed closely by General Nursing and Midwifery diploma students (40 per cent). The remainder were in Post-basic B.Sc. (Nursing)—a 1-year programme of study to bring diploma-level nurses up to the B.Sc. (Nursing) level—(9 per cent), and undertaking Master of Science in Nursing studies (6 per cent). While India has developed plans to phase out the General Nursing and Midwifery diploma education stream, the programme is still very active and is the course offered at the largest number of training institutes in Punjab (see Figures 1 and 2).

The vast majority of those surveyed were women, with only 2 per cent of the sample being men. Nursing continues to be seen as women's work, a profession which supports feminine characteristics of nurturing, caring and gentleness. Until recently the training landscape in India was also set up with an expectation that women would be the main candidates for nursing programmes, since residential colleges have fewer facilities for co-educational training and residence. In terms of religion, the majority of the students surveyed (64.4 per cent) were Sikhs, followed by Hindu students (31.2 per cent). According to the 2011 census, Punjab is home to 64 per cent Sikhs, 34 per cent Hindus and 2 per cent Muslims.

In the 2010 survey, when asked why they had chosen nursing, just over half of the sample indicated that it was their personal interest and/or because they were advised by their friends, parents and relatives to enter the profession. About 15 per cent indicated that nursing was the next option after failing the Medical Board exams (which are the state exams taken by high school leavers to determine whether they qualify for state medical (physician) training spots). The remaining students gave other reasons for choosing nursing as a career such as, to secure a spouse of Indian origin already living overseas (Non-Resident Indian (NRI)). Traditional motivations for nursing, such as the conventional attitude of care or to serve humanity, were identified by only 5 per cent of the sample, despite the strong discourse of care that is traditionally embedded in nursing in India.

When students were asked whether they intended to go abroad after the completion of their programme, more than two-thirds replied positively. This percentage is similar to results found in Kerala (Walton-Roberts 2010). While this is an indication of desire, not actual migration, the fact that almost 66 per cent expressed an interest in overseas migration is valuable for the purposes of assessing how influential international migration opportunities are for nursing students, and provides some indication of both the motivations for entering nursing and the potential outflows that could occur once students graduate. In addition to stated desire, though, overseas migration from India is a major reality in nursing, with one teacher who was interviewed revealing that in 2005, 30 of her former classmates from a total 50 students had migrated overseas within 3–4 years of their graduation. The nursing students who wished to emigrate gave various reasons for their choice, including: upgrading skills (30 per cent), to earn money (27 per cent), to seek better professional status (24 per cent), to seek adventure (3 per cent), and to improve marriage options (2 per cent).

In terms of the destinations that students mentioned, amongst the 66 per cent of those who intended to go overseas, a large percentage (30.4 per cent) opted for Canada as their desired destination followed by the USA (12.7 per cent), UK (4.2 per cent), Australia (3.4 per cent) and New Zealand (1.2 per cent). However, 14.4 per cent stated their interest in multiple countries (Figure 5). In 2010, Toronto in Canada was chosen by a majority of the survey respondents, indicating a clear and specific migratory network present in their imagined career trajectories. We suspect that the specificity of this mapping is most likely due to the pre-established networks of relatives and kin overseas, highlighted by the fact that over 44 per cent of the sample indicated they already have a relative living overseas. Punjabi-trained nurses demonstrated less interest in emigration to Gulf countries, mentioning social and religious restrictions there (especially in Saudi Arabia), the lack of permanent residence and/or citizenship opportunities, and fewer educational opportunities

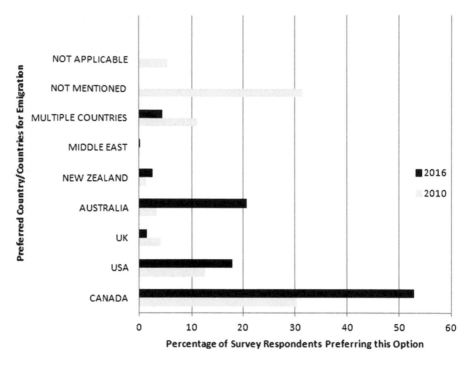

Figure 5. Preferred international emigration locations of Punjabi students studying nursing and mid-wifery in 2010 and 2016. *Source*: authors' own field surveys with nursing students in 2010 (*n* = 1000) and 2016 (*n* = 717).

available for their children (family visas are generally not available to married women migrating to Gulf countries). Thus, the problems faced by Indian nurses in the Gulf countries provoke Punjabi-trained nurses to look for employment in developed countries that offer forms of permanent residence (Percot 2006).

The data collected in 2016 indicated an increase in the share of students intending to migrate and an enhanced interest in Australia and New Zealand as potential markets. For example, of the 717 students surveyed in 2016, 82.2 per cent expressed their desire to move abroad after completing their studies. Amongst these, 20.4 per cent wanted to move to Australia and 2.5 per cent wished to go to New Zealand after completion of their studies. So we observe that in contrast with 2010 data there has been an increase in both the number of students willing to move overseas generally and to Australia and New Zealand in particular. Recent cohorts of nurse migrants have taken jobs in Ireland, Australia and New Zealand where they are offered annual pay packages in the range of Rupees 30–50 lakh (US$45 000–75 000) (*Times Now* May 6, 2016). This increased interest in Australia is also reflected in the proportion of foreign nurses from India already in Australia, which increased from 2 per cent in 2001 to 8 per cent in 2011, while the proportion of overseas-born nurses from the UK decreased from about one-third (36 per cent) to about one-quarter (26 per cent) between 2001 and 2011 (Australian Social Trends, April 2013). It is interesting to note that during the period 2006–11 there was a significant rise in the number of foreign nurses in Australia. In 2011 there were 38 903 more foreign-trained nurses working in Australia than in 2006, representing a 19.4 per cent increase over this time period. This includes an increase in the number of Indian-born nurses from 1503 to 6200 (representing a 313 per cent increase from 2006 to 2011) (Negin et al. 2013). Further, from 2007 to 2011, of the 4683 nurses and midwives who arrived in Australia from South Asia, 81 per cent (3798) were from India (Negin et al. 2013). Similarly, in 2010, out of a total of 1364 internationally qualified nurses who applied for registration in New Zealand, 405 (30 per cent) were from India (Nursing Council of New Zealand 2013).

The training and employment context: increasing international orientation, collaboration, opportunity and competition

The capital of Punjab, Chandigarh, has become a key service market for potential international migrants (Gill and Walton-Roberts forthcoming). There is an increasing number of agencies based in the city that specialise in the international educational mobility of nurses. During our interviews with key informants between 2011 and 2016 we met with one of these agencies. The international scope of their networks and the level of interest in overseas opportunities were evident in their organisational set-up:

> We have a partnership with 20 colleges in Canada, USA, UK, Australia and New Zealand which facilitates student migration to these countries after the completion of their pro-grammes. [We provide] one year study visa, with a further extension of three years, to nursing students after the completion of their [Indian nursing programmes]. Every year around 1200 nursing students move abroad through [our program] and out of these 1200 nursing students, around 80 per cent of the applicants are from Punjab. (Councillor, INSCOL Academy, Chandigarh, May 2016)

International opportunities that Indian-trained nurses can benefit from are presented in terms of the training and enhanced salary candidates will receive:

> Last year 12 students moved to Australia through our college integrated programme with a University in Australia. After their migration to Australia, through this integrated pro-gramme, these students would be earning $35 per hour as a trainee and after the completion of their studies [Australian registration qualifications]; they would start earning $55 per hour. (Faculty member, Indian Medical Institute, Jalandhar, Punjab, April 2016)

The promise of an enhanced salary is also matched with the potential to settle permanently overseas, earn an income and establish a family:

> Four years back one of my unemployed friends in India moved to Australia after her mar-riage. Now she has cleared RN (Registered Nurse) examination and is earning $40 per hour. (Faculty member, Lala Lajpat Rai Institute of Nursing, Jalandhar, April 2016)

The international opportunities that nursing provides have been registered in terms of the data we collected, indicating that the majority of nursing students we surveyed in 2010 were intent on moving overseas when they completed their studies, and 5 years later that share had increased. Over the same time period a greater diversity in the range of possible destinations of interest also emerged. This included a key shift towards Australia and New Zealand as more popular locations than the traditional key markets of the USA, UK, and Canada. One important similarity between all of these locations (especially when compared to Gulf nations) is the opportunity to secure permanent, rather than temporary, settlement. The preference for permanent settlement and the opportunity to study as well as earn an income were identified as a key factor of importance by the majority of the students and instructors involved in our study.

The status of nursing as a profession that opens international doors has certainly realigned the meaning of nursing as a global vocation in the minds of students over our survey period. This has had a related influence on Indian employers. In an interview with a director of nursing education in a large corporate hospital in Chandigarh, Punjab, the alignment of training and international opportunities was outlined:

> A: So they pay, students are much more interested in going there [overseas] and completing means earning as well as education.

> Q: And have you seen increased interest in taking up nursing education in Punjab?

> A: Yeah, very much, very much … this is for the last four, five years … it's much- *ajho, ajho, ajho* [come, come, come]. It's much more … now you see hundreds of students have come up in Punjab. (Director Nursing Education, Corporate Hospital, Chandigarh, June 2010)

The increased interest in nursing in Punjab has not necessarily translated into an excess of qualified candidates for local employers, however, who find that the quality and quantity of available nursing staff does not reflect the increased student interest in the nursing profession:

> Q: Does the Hospital offer employment to all of its graduates?

> A: Yes … [but] the number of [graduates] are very less than [the number of students] joining. I told you either they are going abroad for studies or they are going abroad for marriage or they are going abroad because they are getting better salary …

> Q: So what percentage of your graduates are staying?

> A: Maybe 40 to 30 of the best 50 [leave], maybe three or four stay … they prefer to go [overseas] for higher studies rather than serving [here]. (Corporate Hospital Education Director, Chandigarh, June 2010)

Students similarly reveal the same pattern of interest in departing from India, and the following are responses typical of those shared by the students we interviewed:

> Q: Why had you chosen nursing as a career?

> A: Good scope, demand in foreign [countries] and most of the relatives staying abroad suggested to go for nursing if want to settle abroad. (GNM student, Sukmani College, Dera Bassi, June 2010)

> Q: Do you think the nursing profession may help you in moving abroad? If yes, then how?

> A: Yes … Due to vast scope, value and demand of nursing abroad. (B.Sc. (Nursing) student, Malwa College of Nursing, Faridkot, May 2016)

Nursing has become internationally oriented in terms of its attraction to young women and men in India. The nursing and midwifery education system as a whole has had to orient itself in terms of providing opportunities for international collaborative training and managing the flow of workers into and out of the training and employment system. Our survey results indicate the responsiveness of students training in India to international opportunities. There has been an increased interest in international opportunities, and responsiveness to changing destinations for employment and migration opportunities over the past 15 years in North India.

Conclusion

Our paper has explored the international orientation of the nursing profession in Punjab, a state in north India. Our exploratory examination of the transnational shifts underway in terms of ideas of occupational mobility in care professions in north India suggest we should, as Graeme Hugo (1996) compelled us, continue to focus on 'Asia on the move', especially in terms of its gendered dimensions. As Hugo (2006) remarked, it is education and the proliferation of global media throughout Asia that have created knowledge of the opportunities that lie overseas. As Connell (2014, 80) has also made plain: 'Acquiring education and training in the health sector is tantamount to acquiring cultural migratory capital. In many places it may be the most effective means of acquiring a "passport".' What is so interesting and important about this process in India, and especially Punjab, is understanding how far the adaptation and awareness of these migratory opportunities are causing or paralleling changes to the discourse of the nursing profession in India. We have indicated that the density of teaching resources across India reflects important imbalances, and national nursing needs are serviced by specific regions with greater training resources. Kerala has long provided that supply, but the growth evident in the number of nurses trained in Punjab appears to have emerged in the context of international, not national, demand.

Our data reveal that the majority of nursing students intend to move overseas once they have completed their training in India. This tendency has intensified over the last 5 years between our first set of data collection in 2010 and the second in 2016. In addition to the increased level of intention to move overseas, the destinations that are most popular have also shifted, with increasingly more students noting Australia and New Zealand as markets of interest. The geographical variability and flux that our research reveals indicates how responsive nursing students are to perceptions of international career opportunities, and how the training and education system in India intersects with these dimensions. More research on the transnational connections between sites of training in India and potential sites of employment overseas is needed to understand how effective this transnational transfer of skills actually is. Gender is a key factor in how this process plays out, since other occupations offering international opportunities may not be as appropriate for young women to enrol in. The changing status of nursing in north India should also be examined in light of the new opportunities it offers to women (and men) and the potential outcomes of this occupational shift for gender relations in the home and in the workplace. Moreover, there are more regional dimensions to explore in terms of how feminised skilled migration might alter gender power relations in the sending context, especially in relation to family migration decisions. The significant demographic and life-cycle changes such gendered reorientations infer, and the transnational global connections embedded in this care migration, are all areas of study that Graeme Hugo would have encouraged us to pursue.

Note

1. We use the term internationally educated nurses (IENs) to refer to nurses whose initial training was received in a country other than the one they currently work in.

Acknowledgements

We would like to thank Olivia Dun and Natascha Klocker, the guest editors of this issue of *Australian Geographer* in honour of Graeme Hugo, and two anonymous reviewers for their insightful comments on earlier versions of this paper.

Disclosure statement

No potential conflict of interest was reported by the authors.

Funding

This research was funded through a Social Science and Humanities Research Council of Canada grant file number 410-2010-1013.

References

Abraham, B. 2004. "Women Nurses and the Notion of Their Empowerment." Discussion Paper No. 88, Kerala Research Programme on Local Level Development Centre for Development Studies, Thiruvananthapuram.

Australian Bureau of Statistics. 2013. "Australian Social Trends." *Doctors and Nurses*, Accessed August 26, 2016. http://www.abs.gov.au/AUSSTATS/abs@.nsf/Lookup/4102.0Main±Features20April±2013.

Bhutani, S., P. Gupta, and M. Walton-Roberts. 2013. "Nursing Education in Punjab and Its Role in Overseas Migration." In *Readings in Population, Environment and Spatial Planning*, edited by K. D. Sharma, H. S. Mangat, and K. S. Singh, pp. 203–214. Panchkula: Institute for Spatial Planning and Environment Research.

Boucher, A. 2016. *Gender, Migration and the Global Race for Talent*. Manchester: Manchester University Press.

Brush, B. L. 2008. "Global Nurse Migration Today." *Journal of Nursing Scholarship* 40 (1): 20–25.

Census of India. 2011. Accessed November 13, 2016. http://www.censusindia.gov.in/2011census/population_enumeration.html.

CGFNS (Commission on Graduates of Foreign Nursing Schools). 2005. "The World of Experience (Annual report)." Accessed August 24, 2016. http://www.cgfns.org/wp-content/uploads/2005_CGFNS_International_AnnualReport.pdf.

Chen, P. G., D. I. Auerbach, U. Muench, L. A. Curry, and E. Bradley. 2013. "Policy Solutions to Address The Foreign Educated and Foreign-Born Health Care Workforce in The United States." *Health Affairs* 32 (11): 1906–1913.

Choy, C. 2003. *Empire of Care: Nursing and Migration in Filipino American History*. Durham, NC: Duke University Press.

Connell, J. 2014. "The Two Cultures of Health Worker Migration: A Pacific Perspective." *Social Science and Medicine* 116: 73–81.

Evans, C., R. Razia, and E. Cook. 2013. "Building Nurse Education Capacity in India: Insights From a Faculty Development Programme in Andhra Pradesh." *BMC Nursing* 12: 1–8.

Folbre, N. 2012. "Should Women Care Less? Intrinsic Motivation and Gender Inequality." *British Journal of Industrial Relations* 50 (4): 597–619.

Garner, S. L., L. Raj, L. S. Prater, and M. Putturaj. 2014. "Student Nurses' Perceived Challenges of Nursing in India." *International Nursing Review* 61 (3): 389–397.

George, S. 2005 *When Women Come First: Gender and Class in Transnational Migration*. Berkeley: University of California Press.

Gill, R. 2011. "Nursing Shortage in India with Special Reference to International Migration of Nurses." *Social Medicine* 6 (1): 52–59.

Gill, H., and M. Walton-Roberts. forthcoming. "Placing the Transnational Migrant: Social Inclusion and Exclusion." In *Urbanization in a Global Context*, edited by Alison Bain and Linda Peake. Oxford University Press.

Gostin, L. O. 2008. "The International Migration and Recruitment of Nurses: Human Rights and Global Justice." *JAMA* 299 (15): 1827–1829.

Hawkes, M., M. Kolenko, M. Shockness, and K. Diwaker. 2009. "Nursing Brain Drain From India." *Human Resources for Health* 7 (5): 1–2.

Hazarika, I., H. Nair, S. Bhattacharyya, and D. Gupta. 2009. *Country Report: India*. New Delhi: Public Health Foundation of India.

Hugo, G. 1995. "International Labor Migration and the Family: Some Observations From Indonesia." *Asian and Pacific Migration Journal* 4 (2-3): 273–301.

Hugo, G. 1996. "Asia on the Move: Research Challenges for Population Geography." *International Journal of Population Geography* 2 (2): 95–118.

Hugo, G. 2006. "Immigration Responses to Global Change in Asia: A Review." *Geographical Research* 44 (2): 155–172.

Hugo, G. 2009. "Care Worker Migration, Australia and Development." *Population, Space and Place* 15 (2): 189–203.

Indian Nursing Council. 2004–12. Accessed November 13, 2016. http://www.indiannursingcouncil.org/Statistics.asp.

Indiastat. 2012. Socio-economic Statistical Information about India. State-wise Number of Registered Nurses in India (As on 31.12.2012). Accessed November 22, 2016. http://www.indiastat.com/table/health/16/nurses/207043/636446/data.aspx.

International Migration Report. 2013. Accessed November 11, 2016. http://www.un.org/en/development/desa/population/migration/publications/migrationreport/docs/MigrationReport2013.pdf.

IOM (International Organization for Migration). 2003. *World Migration 2003—Managing Migration– Challenges and Responses for People on the Move*. Geneva. Accessed November 29, 2015. https://publications.iom.int/books/world-migration-report-2003-managing-migration.

IOM (International Organization for Migration). 2005. *World Migration 2005—Costs and Benefits of International Migration'*. Geneva. Accessed November 29, 2015. https://publications.iom.int/books/world-migration-report-2005-costs-and-benefits-international-migration.

Jose, J., M. Q. Griffin, E. R. Click, and J. J. Fitzpatrick. 2008. "Demands of Immigration among Indian Nurses who Immigrated to the United States." *Asian Nursing Research* 2 (1): 46–54.

Khadria, B. 2007. "International Nurse Recruitment in India." *Health Research and Educational Trust* 42 (3 Part II): 1429–1436.

Kingma, M. 2006. *Nurses on the Move: Migration and the Global Health Care Economy*. Ithaca, NY: Cornell University Press.

Kline, D. S. 2003. "Push and Pull Factors in International Nurse Migration." *Journal of Nursing Scholarship* 35 (2): 107–111.

Kodoth, P., and T. K. Jacob. 2013. "International Mobility of Nurses from Kerala (India) to the EU: Prospects and Challenges with Special Reference to the Netherlands and Denmark." (CARIM-India Research no. 2013/19). Accessed December 29, 2015. http://www.iimb.ernet.in/research/sites/default/files/WP%20No.%20405.pdf.

Matsuno, A. 2009. "Nurse Migration: The Asian Perspective." *ILO/EU Asian Programme on the Governance of Labour Migration Technical Note*. Accessed December 2, 2015. http://www.ilo.org/wcmsp5/groups/public/---asia/---ro.

Mooney, N. 2006. "Aspiration, Reunification and Gender Transformation in Jat Sikh Marriages From India to Canada." *Global Networks* 6 (4): 389–403.

Mullan, F. 2005. "The Metrics of the Physician Brain Drain." *New England Journal of Medicine* 353 (17): October 27 1810–1818.

Nair, P. R. G. 1999. "Return of Overseas Contract Workers and Their Rehabilitation and Development in Kerala (India)—A Critical Account of Policies, Performance and Prospects." *International Migration* 37 (1): 209–242.

Nair, S., and M. Healey. 2006. "A Profession on the Margins: Status Issues in Indian Nursing." *Indian Journal of Gender Studies*, New Delhi: SAGE Publications. Accessed December 2, 2015. http://www.cwds.ac.in/ocpaper/profession_on_the_margins.pdf.

National Council of State Boards of Nursing (NCSBN) Exam Statistics and Publication. 2016. Accessed August 26, 2016. https://www.ncsbn.org/exam-statistics-and-publications.htm.

Negin, N., A. Rozeal, B. Cloyd, and Alexandra L. C. Martiniuk. 2013. "Foreign-born Health Workers in Australia: An Analysis of Census Data." *Human Resources for Health* 11: 1984. Accessed October 20, 2016. https://www.ncbi.nlm.nih.gov/pmc/articles/PMC3882294/pdf/1478-4491-11-69.pdf.

Nelson, S. 2001. *Say Little, Do Much: Nursing, Nuns, and Hospitals in the Nineteenth Century Philadelphia*. Philadelphia: University of Pennsylvania Press.

Nursing Council of New Zealand. 2013. *The Future Nursing Supply Projections 2010–2035*. Accessed August 26, 2016. http://www.nursingcouncil.org.nz/News/The-Future-Nursing-Workforce.

Patidar, A. B., J. Kaur, K. S. Sharma, and N. Sharma. 2011. "Future Nurses' Perception Towards Profession and Carrier Plans: A Cross Sectional Survey in State Punjab." *Nursing and Midwifery Research Journal* 7 (4): 175–185.

Percot, M. 2006. "Indian Nurses in the Gulf: Two Generations of Female Migration." *South Asia Research* 26 (1): 41–62.

Radhakrishnan, S. 2011. *Appropriately Indian: Gender and Culture in a new Transnational Class*. Durham, NC: Duke University Press.

Rafferty, A. M., and J. Robinson, eds. 2005. *Nursing History and the Politics of Welfare*. London: Routledge.

Reddy, S. 2015. *Nursing and Empire: Gendered Labor and Migration from India to the United States*. Chapel Hill: University of North Carolina Press.

Segouin, C., B. Hodges, and P. H. Brechat. 2005. "Globalization in Health Care: is International Standardization of Quality a Step Toward Outsourcing?" *International Journal for Quality in Health Care* 17 (4): 277–279.

Selander, L. C., and P. C. Crane. 2012. "The Voice of Florence Nightingale on Advocacy." *The Online Journal of Issue in Nursing* 17 (1). Accessed November 10, 2015. http://nursingworld.org/MainMenuCategories/ANAMarketplace/ANAPeriodicals/OJIN/TableofContents/Vol-17-2012/No1-Jan-2012/Florence-Nightingale-on-Advocacy.html.

Senior, K. 2010. "Wanted: 2.4 Million Nurses, and That's Just in India." *Bulletin of the World Health Organization* 88 (5): 327–328.

Sharma, R. 2011. "Gender and International Migration: The Profile of Female Migrants From India." *Social Scientist* 39 (3/4): 37–63.

Times Now. 2016. "Nursing Brain Drain in Maharashtra as Emigration Hits 5-year High." 6 May 2016. Accessed August 26, 2016. http://www.timesnow.tv/india/article/nursing-brain-drain-in-maharashtra-as-emigration-hits-5-year-high/40703.

Timmons, S., C. Evans, and S. Nair. 2016. "The Development of the Nursing Profession in a Globalised Context: a Qualitative Case Study in Kerala, India." *Social Science & Medicine* 166: 41–48.

Timur, S. 2000. "Changing Trends and Major Issues in International Migration: An Overview of the UNESCO Programmes." *International Social Science Journal* 52 (165): 255–268.

Tiwari, R. R., K. Sharma, and S. P. Zodpey. 2013. "Situational Analysis of Nursing Education and Work Force in India." *Nursing Outlook* 61 (3): 129–136.

Vallée, G., ed. 2007. *Florence Nightingale on Social Change in India. Collected Works of Florence Nightingale*. Volume 10. Waterloo: Wilfrid Laurier Press.

Walton-Roberts, M. 2004. "Returning, Remitting, Reshaping: Non-Resident Indians and the Transformation of Society and Space in Punjab, India." In *Transnational Spaces*, edited by P. Crang, C. Dwyer, and P. Jackson, 78–103. London: Routledge.

Walton-Roberts, M. 2010. "Student Nurses and Their Post Graduation Migration Plans: A Kerala Case Study." In *India Migration Report 2010*, edited by S. Irudaya Rajan, 196–216. London: Routledge.

Walton-Roberts, M. 2015a. "International Migration of Health Professionals and the Marketization and Privatization of Health Education in India: From Push–Pull to Global Political Economy." *Social Science & Medicine* 124: 374–382.

Walton-Roberts, M. 2015b. "Transnational Health Institutions, Global Nursing Care Chains, and the Internationalization of Nurse Education in Punjab." In *Migrations, Mobility and Multiple Affiliations: Punjabis in a Transnational World*, edited by S. I. Rajan, V. J. Varghese, and A. K. Nanda, 296–316. Oxford: Oxford University Press.

Walton-Roberts, M. 2015c. "Femininity, Mobility and Family Fears: Indian International Student Migration and Transnational Parental Control." *Journal of Cultural Geography* 32 (1): 68–82.

Williams, F. 2012. "Converging Variations in Migrant Care Work in Europe." *Journal of European Social Policy* 22 (4): 363–376.

World Bank. 2015. Accessed November 13, 2016. http://www.worldbank.org/en/country/india/overview.

World Health Organization. 2014. *Migration of Health Workers—WHO Code of Practice and the Global Economic Crisis*. edited by A. Siyam and M. Dal Poz. Geneva: World Health Organization.

Xu, Y., and C. Kwak. 2005. "Characteristics of Internationally Educated Nurses in the United States. Findings from the 2000 National Sample Survey of Registered Nurses." *Nursing Economics* 23 (5): 233–238.

Yeates, N. 2009. *Globalising Care Economies and Migrant Workers: Explorations in Global Care Chains*. Houndmills: Palgrave Macmillan.

Environmental Refugees? A tale of two resettlement projects in coastal Papua New Guinea

John Connell and Nancy Lutkehaus

ABSTRACT

Environmental change in small islands may be associated with migration as a means of adaptation. Both Manam and the Carteret Islands in Papua New Guinea (PNG) have experienced rapid- and slow-onset changes, respectively. These have been accompanied by the forced migration and 'temporary' resettlement of the Manam population and attempts at resettlement by Carteret Islanders. Neither has proved successful, thwarted by 'host' landowners, the impossibility of gaining adequate access to land and land rights, and government inactivity. Settlers have been perceived as outsiders and rival claimants to valuable coastal resources. Inability to resettle in national contexts raises issues of ambiguity, identity and citizenship. The problems experienced by quite small population groups moving short distances in similar cultural contexts are indicative of the potential future problems facing environmental migrants in other contexts.

Introduction

Small islands, common in the Pacific region, are particularly vulnerable to threats of various kinds, whether economic or environmental, since they have limited topographical and ecological variation and weak infrastructure provision due to their relative political and economic insignificance (Lewis 2009; Connell 2013a). The economic marginalisation and environmental fragility of small island economies in contemporary Pacific Island states such as Papua New Guinea (PNG), and the emergence of disadvantaged 'outer islands' has resulted in significant migration from such places (Connell 2015).

At the same time, there is a growing concern internationally that environmental change in small Pacific islands will increase demands for migration and that in particularly difficult circumstances a need for migration may require formal resettlement programmes. Such circumstances may enable migration to be perceived as adaptation (Black et al. 2011). While environmental change in this region is no new phenomenon, it has been widely assumed that current climate change is intensifying pressures on island environments, economies, and livelihoods. That migration will be increasingly necessary, especially from atoll states, has been long recognised by some (Connell 1993). The atoll state of Kiribati has purchased an island in Fiji in anticipation of future needs and both Kiribati and Tuvalu have developed new migration policies that may offer future 'migration with dignity' (Connell and Kagan 2015). Similar concerns apply to small islands within larger states, such as PNG. This paper examines the environmental and socio-economic changes that have affected the populations of Manam Island (Madang) and the Carteret Islands (Bougainville) and so led to pressures for resettlement, and the extent to which these experiences offer models or cautionary tales.

This paper builds on the work of Graeme Hugo, who early recognised the growing relationship between environmental changes and migration, the extent to which migration, either temporary or permanent, might be a survival strategy, and the relationship between migration and local coping strategies (Hugo 1996). He subsequently examined the outcomes and consequences of past resettlement projects and programmes in order to establish more effective programmes for a future in which these seem more likely to be needed (Hugo 2011, 2013). It is apparent, firstly, that climate change has continued without substantial global intervention, and, secondly, that widespread nationalistic and populist opposition to international migration has made the task of devising effective resettlement projects more difficult.

Here we develop several themes that Hugo raised about the relationship between climate change (and more general environmental change) and migration, such as how this relationship is perceived by different stakeholders, the range of constraints that can hinder successful adaptations, and the differences and similarities between the effects of sudden *vs* gradual environmental change and its impact on migration. Lessons drawn from these constraints were that multiple factors are necessary for successful resettlement: adequate funding, careful planning and governance, empowerment of settlers and engagement with destination communities, use of existing social networks, reconstruction of settler livelihoods and the re-establishment of social and cultural capital (Cernea 1997; Hugo 2011). Elsewhere, Hugo emphasised that 'one of the clear lessons from decades of resettlement experience is that successful resettlement is not cheap', hence international support is crucial (Hugo 2013, xxxviii; McDowell 2013). In particular, we focus here on access to land, a constraint that is inevitably a central issue in relocation and resettlement.

This paper compares two case studies of small island populations in PNG affected by real or perceived environmental change: Manam, where islanders have been displaced after volcanic eruptions, and the Carteret Islands, where islanders sought to move because of perceived climate change impacts and the destruction of island livelihoods. The former case required an immediate response, as the island became uninhabitable; the latter enabled scope for planning in the face of slow-onset changes. In effect, Manam Islanders have experienced forced migration since they have not been officially allowed to return to Manam even though the immediate threat of volcanic eruption has ceased, while Carteret Islanders have struggled to achieve migration. Socio-economic and environmental structures and changes on Manam and the Carteret Islands are quite different. However, both case studies raise issues about the social construction of environmental hazards, migration as adaptation, the role of place in cultural identity, the challenges of environmental change to development in fragile nations, and ultimately the rights and entitlements of citizenship in such nations.

Information about Manam and the Carteret Islands in this paper stems largely from the authors' engagement with these locations over the past four decades. In particular, both authors engaged in fieldwork in Manam and the nearby coast in 2015 which involved interviews with Manam Islanders, both on Manam and the mainland, as well as government officials and coastal villagers living near the Manam settlements. Both have visited Manam in previous years beginning in 1978. Lutkehaus lived on the island in 1978–79 and has made numerous visits since then. Connell has worked in Bougainville since 1974 and had extended discussions with Carteret Islanders and mainlanders in 2015 and 2016. Further details of the methodology are available in Connell (2016a) and Connell and Lutkehaus (2016).

Manam

Manam is a high volcanic island of Madang Province, some 12 km from the mainland of New Guinea (Figure 1), with a lengthy history of volcanic activity. One of many active volcanoes along the Pacific 'Ring of Fire', it has erupted on numerous occasions and is regarded as one of the two most active volcanoes in PNG (Johnson 2013). Partly as a consequence, Manam Islanders have long maintained social relationships with villagers on the mainland, along the north coast of Madang Province and also southwards near the mouth of the Sepik River (Lutkehaus 1985; Lipset 1997). In the past, when Manam erupted, islanders made the 12 km trip across to the mainland by canoe. Sometimes only two or three villages out of the island's fifteen villages were forced to evacuate, but even when the entire island population was evacuated, it was only temporarily. Islanders would stay for several months with their hereditary exchange partners (*taoa*) on the mainland, who offered food and a place to stay until volcanic activity died down and they could return to Manam (Lutkehaus 1995). Conditions were different, however, when the volcano erupted in December 2004 and January 2005. All of the islanders were displaced and this time volcanologists warned the government that the volcano posed an ongoing threat, and recommended nobody be allowed to return permanently to Manam.

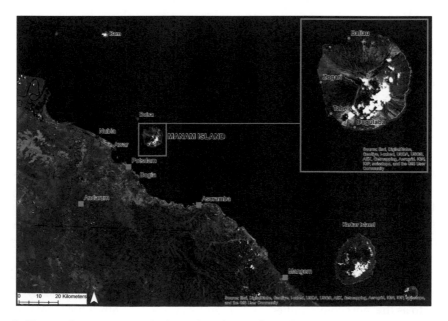

Figure 1. Manam. Source: map made for authors by Stephanie Duce.

Only once before and briefly (1957–58) had the entire Manam population been evacuated. Never before had some 10 000 Manam Islanders arrived simultaneously on the mainland. In the past, migration had always been temporary and reciprocity for the mainlanders' hospitality was made through exchanges of food, pigs, and canarium nuts with one's *taoa*. Without the resources to make the appropriate reciprocal gestures, since their gardens and nut-bearing trees had been destroyed by volcanic ash and their pigs abandoned during the evacuation, the islanders now lacked the necessary resources to make appropriate reciprocal gestures and they quickly overstayed their welcome with their *taoa*. International aid agencies assisted the PNG government in settling the displaced islanders in temporary 'care centres' at Potsdam, Asuramba and Mangem. These 'care centres' were established hastily on three small former coconut plantations along the north coast of Madang which had been alienated from its traditional owners during the German colonial period. Now 'owned' by the PNG government (Figure 1), this land was the traditional land of the islanders' mainland *taoa*. For the past 11 years the Manam Islanders have been officially classified as Internally Displaced Persons (IDPs) by the Office of the UN High Commissioner for Refugees (UNHCR). Most have continued to stay in the care centre grounds, living in makeshift 'temporary' settlements in houses constructed of traditional materials. They became environmental migrants, forced to leave their homes, gardens, cash crops, traditional fishing grounds and ancestral burial sites because of the volcanic eruption, but also were forced to wait years to be permanently resettled on the mainland because of the potential environmental danger the volcano poses. Meanwhile, their population has grown rapidly. Not surprisingly, with 10 000 men, women, and children, along with their dogs, canoes, and other material possessions, suddenly forced to settle along a coast already relatively densely populated, social tension and then violence erupted between the uninvited Manam Islanders and their coastal neighbours, who resented the islanders' extended presence on what they considered their traditional land (Connell and Lutkehaus 2016).

Underpinning most resettlement problems is the role of land. Although resettlement of the Manam Islanders took place on land that had been alienated a century earlier for colonial plantations, the descendants of the early landowners never lost their sense of land ownership, as did landowners in similar circumstances elsewhere in PNG (where such descendants often sought to rescind 'agreements' made in earlier colonial times). The mainland coastal Madang descendants retained traditional authority over that land and had somewhat reluctantly consented to its use for care centres on the assumption that this would be temporary. With the demise of the plantation system post-independence in 1975, most assumed that the land had not been permanently alienated from them and would soon revert to them for the use of their own growing populations, especially since the government was not using that land. Long-term use of the land for care centres was therefore both unanticipated and unwelcome. As one local landowner stated: 'We initially gave this land to Europeans to be used as plantation land, and we derived benefits from employment and services. We did not give this land to others, to be occupied without benefit to us.' Continued settlement by Manam Islanders and the subsequent lack of any sense of closure thus concerned local villagers, ironically officially referred to as the 'host communities', and became exacerbated with subsequent social conflict and violence.

Islanders were resettled on the basis of their Manam village affiliation and replicated their island settlement patterns, as much as possible, on the narrow strips of coastal land available to them in the care centres. However, houses became overcrowded and the 'symbolics of space' were altered, reducing the status of the traditional hereditary chiefs, who had lived in larger structures and built imposing men's houses (*haus tambaran*) on Manam. Still waiting for a permanent resettlement plan, the frustrated Manam Islanders have cleared what areas of the resumed land they could, built more houses, planted food gardens and trees, but were given little government assistance. Because Manam chiefs were seldom consulted about prospective resettlement plans, their traditional authority was further undercut, 'reducing their morale and leaving a feeling of worthlessness' (Mercer and Kelman 2010, 417). Agricultural practices and plant combinations have changed in a different environment from Manam and little space was available for livestock; chickens have become less important, and it has taken longer to acquire pigs for socially significant events, hence social cohesion and social relations have been altered. Diets have changed, prompting nostalgia for island food.

Access to services on the mainland such as fresh water, sanitation and education was worse than it had been on Manam, and there have been no compensatory employment possibilities. Moreover, the care centre land was inadequate for the agricultural needs of the growing IDP population. It allowed no scope for cash cropping, while islanders had little access to coconut palms and limited rights to marine resources. As the Manam population in the care centres has grown, increasing food shortages intensified conflict over resources and land use as islanders sought to obtain timber for fuel and building materials, pandanus and coconut palms for thatch, and hunt on nearby land.

Tensions increased and violence eventually erupted between the Manam Islanders and the local population, as islanders struggled to adapt to the mainland, where it quickly became apparent that 'host villages' were not particularly welcoming. The uninvited islanders were blamed for all manner of problems. Violence and petty crime increased alongside alienation and social anomie, partly because of weakened leadership and growing frustration at the impossibility of return, and later over the lack of a decision on their future. Both settlers and hosts had growing populations, greater needs and more frustrations. More concentrated settlement and inadequate access to land created problems and conflicts.

Within 6 months of their displacement in 2005, it was 'already evident that the stay of the Manam Islanders in the care centres was causing tension with local communities over land use issues' (United Nations Human Rights Commission 2010, 13). As early as 2006, 'host' villages near the Asuramba care centre asked their local MP to evict the Manam Islanders. While conflicts often had a particular catalyst—encroachment on land, chopping down valuable trees, taking firewood from the wrong place, sexual relationships between youths, and so on—in many respects, as one local landowner phrased it: 'the longer you stay together a small issue becomes a big issue'. By mid-2005 Manam Islanders had overstayed any limited welcome, disputes intensified and festered, eventually resulting in the deaths of about 20 people. After one particularly violent encounter between mainlanders and Baliau (Manam) villagers in 2009, close to 1000 people were sent back to Manam, where they resettled at their traditional Baliau village site even though the government provided no facilities, such as education, health care or transportation to and from the mainland. That accentuated a slow ongoing 'return to Manam', and over time more islanders, though conscious of risk, returned to Manam.

By 2015 several thousand Manam Islanders had returned to Manam by default or by choice as rehabilitation of gardens proved effective, and fruit trees and coconut palms came back into production. Beyond the mainland violence, social factors have played a part in the decision to return: 'care centre life was hard; we felt we were inside a fence and freedom was restricted; here we have freedom'; 'the landowners' customs are not the same as ours; we share, we work together, we build our houses collectively'. Beyond that, the role that Manam and the surrounding sea play in spiritual and cultural life is more important than any volcanic threat. As in many Pacific Island cultures (Kahn 1996), 'place'—the land where one's ancestors dwelled—is of utmost spiritual and social value to the collective identity and continued vitality of the Manam people. Many islanders believe that it is important for deceased relatives, and for their living descendants, to be buried on Manam, so that their souls reside there, providing continuity between the living and the dead (Lutkehaus 2016). That belief has continued and is a powerful incentive to return. The returnees stressed the better access to land on Manam and their view that it was healthier and cheaper on the island: 'everything is free here', 'it's hard work but it's home'. Indicative of the determination to re-establish livelihoods on Manam was the returnees' use of their own resources to reopen a primary school, so that village children did not have to move to the mainland for education, and to support a dispensary. There was a particularly strong sentiment that 'the government has not helped us thus far; there is always a delay, always they tell us there is no money' and 'donors promise things but they do not appear either'. Despite the lack of services, most were satisfied with their revived lives on Manam and had no wish to return to the tensions and uncertainties of the mainland, preferring to cope with the impact of ongoing volcanic activity than with food shortages and conflicts in the care centres.

Since 2005, islanders have also regularly returned to Manam to make copra for cash, restore and cultivate gardens, gather timber and bury the dead. En route they fish between the mainland and Manam because of their lack of use-rights to the coastal waters fringing the mainland. Within a few months of the eruptions in 2004–05, Manam chiefs delegated certain villagers to return to live permanently in their former villages, rebuild houses, regenerate food-bearing trees, such as *galip* almonds and betel nuts, and raise pigs—activities that were difficult in the care centres—and to be custodians of the land. A constant continuing connection exists between the care centres and the island.

The critical factor in the changed circumstances of the Manam Islanders was the volcanologists' decision that the Manam volcano now poses too great a potential danger for the islanders to return permanently. The volcanologists claim, through their scientific expertise, that the volcano is a permanent environmental hazard (Mulina, Sukua, and Tibong 2011). Consequently, national and provincial authorities have withdrawn provision of social services, including transportation to and from the island, schools and school teachers, mail services and health care. Nonetheless, the national government has been slow to respond to the need to find a durable solution to the islanders' displacement. Although it established a Manam Resettlement Authority (MRA), little has happened and early workings of the MRA were plagued with corruption and inefficiency (Connell and Lutkehaus 2016).

Finally, in 2011 the Madang provincial government located land some 30 km inland from the coast at Andarum that it has proposed as a resettlement site (Figure 1). The site was chosen, firstly, because of the ongoing land disputes at the care centres. Other coastal land is either occupied or regarded as prime real estate for tourist development. Secondly, the Andarum landowners were willing to lease some land to the government in order to gain an access road to the coast. The islanders in turn will lease the land from the government. The move to Andarum will entail an enormous transformation in Manam culture. Islanders will be far from Manam and removed from their marine livelihoods. This change will result in transformations in such basic issues as work, diet, and gender relations, especially men's roles in society. In April 2016 the Manam Resettlement Authority Bill was passed by the PNG parliament to enable the move to Andarum, but substantial income will be needed to put in place infrastructure in a still largely inaccessible place, suggesting that many more years of uncertainty, tension and frustration are to come.

A decade after the islanders were evacuated from Manam their lives remain in limbo, with the majority still in what were meant to be temporary care centres, much to their own anger and frustration and that of local landowners. Islanders who had returned to Manam were denied any of the amenities that other citizens of Madang Province and PNG were entitled to. Indeed, islanders are pointedly referring to their own plight as 'refugees in our own land' relative to what they perceived as the superior treatment afforded to alien refugees on Manus.

After more than a decade away from Manam, the islanders' identities have fragmented and changed. While some Manam villagers always left the island for jobs at mine sites and in urban centres such as Madang and Port Moresby, the eruptions were catalysts for a type of mass migration and social change different from in the past. Although at the height of the 2004–05 eruptions a handful of villagers refused to leave the island, and some people have now returned to Manam, at present, the majority remain at the care centre sites. A new generation of children and youths, most of whom have visited the island at some point, have nonetheless begun to develop new 'coastal' or 'displaced' identities. The untethering of society from place has entailed fundamental changes in cultural and ethnic identity. Moreover, differing attitudes towards resettlement emphasise that there is no consensus on a return to Manam or permanent resettlement on the mainland. Over time, most Manam Islanders have decided that having 'two Manams'—the island (the 'real one') and also Andarum—is the best possible outcome for their future. Some islanders prefer a third option, and wish to stay at the care centres, where they have been for a decade and their children have grown up. Ongoing tensions, and the absence of rights to land, make this unlikely.

Exactly who should make moral judgements over what risks Manam Islanders might be expected to bear, and at what scale, and what services might be provided and where, is ultimately unclear. Such questions—involving ethics, human rights, responsibility and justice—have played a considerable part in contributing to inertia, frustration and a deep sense of instability and ambiguity. Not only have the Manam Islanders become impoverished materially as a result of their displacement, they have also suffered socially and culturally, and continue to face a radical transformation of their cultural identity from an island people to an inland people, from proud exchange partners with their mainland *taoa* to diminished, landless migrants. The loss of cultural status and identity is all the more dramatic and irreparable because of the symbolic and spiritual significance that the Manam volcano plays in their ritual, religious, and even their economic life (Lutkehaus 1995).

The Carteret Islands

The Carteret Islands make up a coral atoll of Bougainville Province, about 85 km off Bougainville Island (Figure 2). The Carteret Islands is a reasonably typical coral atoll structure consisting of six main small islands. None are higher than 2 m above sea level and the surface area of the atoll is about 0.7 km^2 (70 ha). The islands were settled from the high island of Buka (Bougainville) some three or four centuries ago, and islanders have linguistic and cultural similarities with the people of north Buka especially.

The Carteret Islands is the most densely populated island in PNG, and the most densely populated (non-urbanised) atoll in the Pacific. The atoll currently has a population of about 1100, with another 300 islanders living elsewhere, mainly in urban Buka. It has achieved

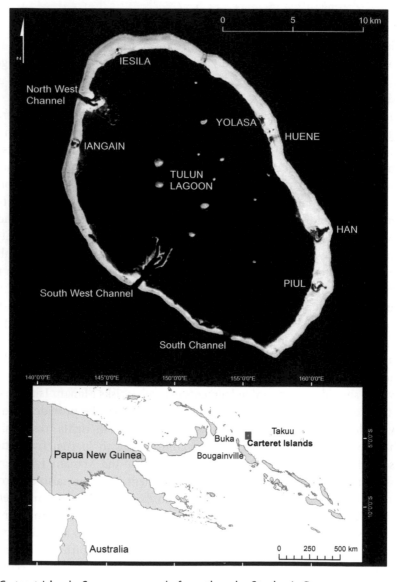

Figure 2. Carteret Islands. Source: map made for authors by Stephanie Duce.

iconic status because of the media focus on the island and the widely reported needs of the islanders to be resettled as the 'first environmental refugees' (Connell 2016a, 2016b).

In the post-Second World War years, and especially since the 1960s, as the island population grew, it periodically experienced food shortages of varying severity, associated with excessive rainfall, growing human and pig populations and the increased conversion of coconuts into copra to generate income, so reducing local food availability. Between the 1960s and 1980s, administration patrol reports regularly commented on food shortages and made occasional reference to malnutrition (Kukang et al. 1987). Some men worked off the island, often on ships, providing the main source of income.

As early as the 1950s the Catholic mission had offered land on two small islands off Buka for resettlement but nothing came of this. At the end of the decade, the colonial administration was so concerned about the growing population on the Carterets and adjoining atolls that they recommended that land be acquired on Bougainville for resettlement, but islanders rejected this, fearing the loss of their cultural identity (O'Collins 1990, 250). In 1965 the islanders themselves made their first request for land and resettlement. While the provincial administration sought to find land, nothing happened until the late 1970s (see below), after the independence of PNG.

By the 1990s little land in the Carterets was cultivated and the main foods were fish and coconuts, alongside imported flour and rice. Taro pits had been flooded and abandoned. A 2002 agricultural survey observed a 'serious and chronic food supply problem' so severe that malnutrition was significant (Bourke and Betitis 2003, 15). Food shortages had further contributed to land disputes and theft (Rakova 2014). Marine livelihoods were also problematic. While diets included fish and other marine species, the Carterets never managed to establish successful commercial fishing projects that would have generated cash incomes, despite a brief period of successful beche-de-mer sales. With few marine products and little copra, no commercial livelihoods or incomes were generated in the islands.

The problem of adequate incomes and food consumption was emphasised further during the 1990s when the closure of the substantial copper mine on Bougainville and the ensuing Bougainville crisis and civil war brought an end to employment on Bougainville and the return of many Carteret Islanders, which in turn meant a massive downturn in remittances, intensifying local problems. In 2008, wind waves swept over the atoll, further damaging the remaining agricultural system, prompting sustained global media interest (that largely misrepresented the crisis as the outcome of sea level rise) and renewed demands for resettlement.

Formal resettlement from the Carteret Islands has had two phases. After early colonial proposals came to nothing, a Bougainville provincial government project began in 1984, north of Arawa in Bougainville, as part of a larger settlement for migrants from the other Bougainville atolls (Takuu, Nukumanu, Nuguria and Nissan). The land (next to the provincial corrective institution) had already been alienated and was no longer used by the Department of Primary Industry, and so became the site for resettlement after unsuccessful attempts had been made to obtain additional alienated land elsewhere.

Just nine households, from the smaller islands in the Carterets, were established on this land, over a 5-year period. After confronting a new environment and new crops (notably sweet potatoes), garden land proved productive and was enough for households to send surplus produce back to the Carteret Islands. However, the settlers had no coconut palms, had secured only limited fishing rights and inadequate land for planting cash crops, despite attempts to secure additional land. They thus had no means of gaining cash incomes. Local village leaders objected to granting fishing rights to the settlers, and settlers were blamed for stealing coconuts, garden food and marine resources (Rakova 2014, 274). Conflicts with other atoll groups were already anticipated (Kukang et al. 1987; O'Collins 1990). Within a couple of years two of the original settler families had returned to the Carterets, frustrated with long delays in finalising land for cocoa and copra production: 'people were afraid that they would lose their rights to the little land that was available on the islands and ultimately would be left without land rights at all' (O'Collins 1990, 267). This entire phase of settlement stopped abruptly in 1989 with the outbreak of conflict in Bougainville. All the resettled households returned to the Carteret Islands during the crisis, adding to local pressures, the land was resumed by its traditional owners and the settlement was never revived. In 1997 a second group of eight households was resettled at Hanahan on Buka (close to the ancestral origins of the Carterets people) but they experienced land disputes with nearby landowners and soon returned to the Carteret Islands.

In the twenty-first century, after the resolution of the Bougainville crisis, a second resettlement phase began in 2007. That phase grandly proposed to relocate half the island's population by 2020. Neither the government of PNG nor that of the now Autonomous Region of Bougainville gave priority to obtaining land for resettlement, and it was left to a local Carteret Islands organisation, Tulele Peisa (Sailing the waves on our own), to seek out new land with some support from the Catholic church: a 'home-grown adaptation strategy' (Rakova 2014, 270). Tulele Peisa regularly reported relocation being needed because of 'rising sea levels'. The lack of state-provided land meant that '[a]n ideal opportunity for securing land for some of the world's first climate change displaced persons [sic] was lost' (Leckie 2012, 18). Obtaining any land, beyond the 81 ha donated by the Catholic church on Tinputz, on the east coast of Bougainville, let alone appropriate land, was extremely difficult, and Tulele Peisa was unable to gain access to any of more than 20 cocoa and coconut plantations ruined in the crisis. At least five failed attempts at land negotiations took place with various landowner groups in Bougainville (Rakova 2014, 274). Moreover, Tulele Peisa had no financial resources to secure land or develop a settlement. House-building on the former mission land began in 2008 but by 2011 no more than 10 households had relocated to Bougainville, and there were frequent disputes with nearby landowners.

Resettlement was quite unlike the kind of migration that had occurred in other atoll contexts. There were no financial resources for clearing land and constructing houses, and local landowners were sceptical over the needs of the islanders. Many islanders themselves were not enamoured of resettlement since living in Bougainville required abandoning some aspects of culture (notably material culture) and adapting to new livelihoods in a different physical, cultural and linguistic environment, where fishing would be problematic and swamp taro impossible to grow. As Carteret Islanders have said in recent years: 'We don't want to lose our ground. Losing our island is losing our lives, losing our identity, losing our custom and whatever we have' so that 'most of our culture will have to live in memory'. An anticipatory nostalgia thus builds on an emotional geography: 'I'll miss the sea, the fish, the palm trees, coconuts. I do belong to the island. I feel sorry for the island' (quoted in Connell 2016b). Ambiguity marks attitudes to resettlement.

Despite cultural similarities, resettlement has been unusually difficult because of land tenure, but accentuated by tensions inherent in a post-conflict society, characterised by scarce employment, growing local pressures on land as population has grown, the absence of kin who might have provided social support and housing, the lack of an effective institutional sponsor and the resources to adequately plan resettlement. After more than half a century of consideration of resettlement, from the colonial to the post-colonial era, albeit on either side of a political crisis, almost nothing had been accomplished.

Between Manam and the Carteret Islands: some parallels

Whereas the resettlement of the Manam Islanders was imposed on the local mainland population, after the extenuating circumstances of the volcanic eruptions, the Carteret Islanders have failed to impose themselves on the Bougainville population. Despite their different temporal trajectories, the land outcome for both groups has been almost exactly the same. Both moved, or sought to move, over quite short distances but were effectively rejected by culturally similar people on the grounds that land was scarce and needed by them. Land was given, but never enough for the needs of settlers and never by the contemporary landowners, but only from land already alienated by plantations, the administration and the Catholic Church. Even unused plantation land could not be resumed by anyone other than the customary owners. Both resettlements are thus in abeyance: a seemingly infinite pause.

Moreover, settlers have experienced gaps in knowledge and experience in the new sites—ecologically different from their home islands. Social tensions have followed as land has become the most crucial asset and the PNG government less authoritative, less effective and more contested. Land shortages have resulted in tensions over land, marine resources, employment and access to services (already delivered inadequately in PNG), while settlers have been blamed for theft, violence and general abuse of the hospitality of unwelcoming 'hosts'. Settlers, meanwhile, have had few effective means of diversifying their livelihoods and thus reducing their dependence on their hosts, or at least on their hosts' land.

Land issues are further complicated—in both cases—by uncertainties about the nature of group ownership of particular tracts of land, by government indecision, and by uncertainty and division between landowners concerning the future of the land. In the PNG context, where traditional modes of land tenure are fragmented and fluid, the outcome has been greater pressure on settlers, especially where growing coastal population concentrations have led to a 'coastal squeeze' (Connell 2013a). Preferred sites for resettled islanders are invariably coastal, as they enable continued access to marine resources and the formal costs of infrastructure provision are less. However, such sites are more valuable and under greater pressure for multiple uses.

Already difficult, both situations have been further complicated by neither group being enthusiastic about the move. At least a third of Carteret Islanders do not want to move; and many Manams would prefer their home island. Seeking to hold on to home in itself can reduce the potential success of resettlement. Settlers and potential settlers are caught between the certainties of land tenure in island homes—albeit unstable or infertile—and the insecurities of access to land elsewhere. Ambiguity is absolute, and there is defiance as much as compliance to the notion of resettlement (Lipset 2013). There is an ambivalence in leaving home places, but not becoming emplaced elsewhere. Indeed, settlers have experienced difficulties in retaining 'community' with the practice of 'culture' and leadership (Lutkehaus 2016) becoming more difficult, hence the emergence of new and problematic notions of identity.

Such circumstances are far from unusual, and numerous Pacific parallels exist across more than half a century (Connell 2014). As early as 1946, when 1300 Niuafo'ouans required resettlement to the large Tongan island of Tongatapu: 'they were viewed with no small concern and hostility by local residents … [stemming] from the fear of increased pressure on limited resources, especially land and jobs, and from social prejudice' (Rogers 1981, 158). Fights erupted between settlers and established residents; only after 18 years had the settlers 'muted *much* of the original criticism and animosity of their neighbours' (Rogers 1981, 160; emphasis added). Half a century later, in Solomon Islands, exactly the same issues resurfaced when people were resettled away from the Gold Ridge mine to coastal sites. Tensions emerged with the local people, despite their sharing the same language, broad notions of *kastom* and even clan membership, and the settlers were effectively forced to return inland (Nanau 2014). Elsewhere in Solomon Islands, in the colonial 1960s, despite no overt hostility between Gilbertese settlers and Solomon Islanders, 'there were feelings of resentment on the part of many Melanesians, who viewed the Gilbertese as having taken land and jobs that ought to belong to Melanesians' (Knudson 1977, 223). These issues are still unresolved half a century later, with land tenure remaining the critical issue (Donner 2016). Land was potentially available in one of two other Solomon Island contexts for nearby kin, but the theory had yet to be tested (Monson and Fitzpatrick 2016).

Similar problems occurred on Lihir Island (New Ireland province, PNG), with forced migrants from a mine site being unable to gain access to adequate garden land, and perceiving themselves as unwelcome 'outsiders', even amongst their kin and near neighbours (Hemer 2016). These problems also recurred with resettlement of Ambrymese in Efate, Vanuatu, displaced by volcanic eruption, despite some land having been previously alienated (Tonkinson 1979). Similarly, the resettlement of Makian Islanders away from their dangerous volcanic island in Maluku Utara province of Indonesia, to areas where they had historic social ties, was largely unsuccessful, even with adequate land being available, because of emotional ties to home. Hence many Makianese returned illegally (Lucardie 1985), much as did Manam Islanders. More recently, land acquisition plagued resettlement from Rabaul (PNG) after the 1994 eruption, despite a long history of volcanic activity (Martin 2013). Even seemingly vacant land is never without owners.

Some traditional owners, such as the Motu-Koitabu in Port Moresby, in a post-colonial reassertion of traditional legal regimes, have fought strenuously to prevent further settlement and to gain more adequate compensation for land alienation (Connell and Lea 2002, 131–132). Elsewhere in coastal PNG, landowners have renegotiated leases with urban settlers, sometimes resulting in landowners attaching more stringent regulations on land; for example, banning some economic strategies such as fishing and the establishment of trade stores and restricting gardening (Numbasa and Koczberski 2012). Such changing practices attest to increased competition for land for housing and agriculture, as well as access to marine and other resources.

By contrast, the resettlement of Banabans from the Gilbert Island chain (now part of the Republic of Kiribati, but then under British colonial rule) on Rabi Island (Fiji) in the late 1940s, to make way for mining, was a rare exception in avoiding many settlement problems. This was successful because capital was available, the colonial government had a moral obligation, Rabi Island was without other claimants, Banabans were consulted well before relocation and acquired their own government on Rabi with autonomy and legal authority, thus replicating familiar social systems and governance (Connell and Tabucanon 2016). A rather different resettlement brought Tuvalu Islanders to the nearby island of Kioa in the 1940s and 1950s, and with a similar degree of success (McAdam 2013). Even so, land tenure in Kioa and Rabi is still ambiguous and over generations nostalgia for and emotion over Banaba has never disappeared.

Settlers involved in these diverse examples from the Pacific have invariably remained 'outsiders' even after generations. Consequently, they are less likely to actually settle, either returning from whence they have come (however difficult that may be), or moving onwards. In both Manam and the Carterets that has resulted in some settlers returning 'home'. Failures of settlement have also been a catalyst for onward migration, emphasising issues of an evolving or disappearing island identity. Migration as successful adaptation and a standard means of diversifying livelihoods has largely been thwarted.

Conclusion: the 'struggle for land'

Migration as an adaptive response to environmental change has become unusually difficult, as environmental change exacerbates conflict and uncertainty. Throughout the Pacific, where most societies are still broadly agricultural, migration in response to environmental hazards is complicated and exacerbated by land issues based on traditional group affinities and is thus often bound up with local opposition and resentment towards resettlement (Brookfield and Brown 1963). In the two cases discussed here, these typical problems were further aggravated, particularly in Manam, when short-term solutions were immediately imperative, and then fossilised into inappropriate long-term solutions; they remained latent for the Carteret Islands over five decades of different phases of failed resettlement.

Landowners have been increasingly reluctant to cede land to others, however moral and worthy their claims, even when they share kinship ties or exchange relationships, especially as their own populations and needs have grown. Neighbours may often be enemies, hence even cultural similarities are not enough to make the worthy goal of moving nearby successful (Lipset 2013). Throughout Melanesia, and the wider Pacific, resettlement has been contested, sometimes violently, by local landowners and tensions have not disappeared as land comes under greater pressure. In post-colonial times land boundaries have become fixed, and land is seen as too valuable for displaced people to lease or purchase, even within the same cultural area (Connell 2013b, 2014; Martin 2013). In the midst of uncertainty little scope exists for settlers to engage in some form of 'insurgent citizenship' as 'principle collides with prejudice over the terms of national membership and the distribution of rights' (Holston 2009, 246). Global concerns over how difference and identity can be accommodated, and where entitlement and environmental justice might fit in, are writ all too small, amidst a Melanesian version of populism.

As Hugo pointed out, environmental stress adds another driver to the complex rationale for migration, and it is the relatively poor who have been most disadvantaged by deteriorating environmental and economic circumstances (Hugo 2013). Dramatic environmental displacement, such as that experienced by the Manam Islanders, forces an already marginalised small island population into even more difficult circumstances as IDPs. They and the Carteret Islanders may not be environmental refugees but rather ecological migrants, where a range of factors are influential. Slow-onset changes are no one's priority (especially where 'host' landowners may perceive an economic rationale) where few actually want—or even necessarily need—to go. Ironically, Manam demonstrates how resettlement is particularly unsuccessful at speed while the Carteret Islands demonstrate how slow-onset problems are largely ignored. In both contexts environmental, economic and socio-political issues are intertwined and entangled.

Balancing the rights and needs of customary owners and migrant citizens represents a critical challenge to a fragile state such as PNG, and there has been a marked reluctance of government, both national and provincial, to intervene in customary land matters. Politicians deliver promises rather than plans. Land is a ubiquitous and problematic feature of development throughout the Pacific, and land management is unusually challenging in Melanesia, yet it is at the core of sustainable national development. Land issues thus contribute to creating an exclusionary version of civil society, and opposition effectively denies the right of mobility to a small but significant proportion of the population (Stead 2015). If notions of human rights, such as the right to adequate food, water, self-determination and cultural identity, exist in Pacific Island states, how can they be accommodated within the state? It is unsurprising, therefore, that Hugo (2013) pointed to the need for international support for settlers to claim rights.

Multiple studies exist that define the rights and political status of those whose national territory may become submerged, and what should happen to them if they become stateless (e.g. McAdam 2012; McAdam et al. 2016; Suliman 2016). Less attention has been paid to the question of what happens to those from small islands that become uninhabitable but which are part of a larger nation. The assumption has been that national resettlement is feasible—but the Manam and Carteret examples show that this is not necessarily true. This is a complex question concerning placelessness not statelessness, and the particular disadvantages of an often inefficient and economically fragile post-colonial state that cannot develop adequate policies and practices. It is particularly significant in countries such as PNG where overseas migration is extremely limited.

Social relationships are written in the ground and editing or removing the writing is almost impossible, establishing a geometry of power that absolutely marginalises potential settlers. In Manam and the Carteret Islands migration as adaptation has largely failed: they offer cautionary tales not models. If these resettlements have failed, within the same country and alongside people of similar cultures, as they have done elsewhere, then resettlement almost certainly poses even greater problems when over long distances, amongst quite different peoples, in very different ecological zones, and involving much larger numbers. These examples raise real issues of concern for other Pacific contexts, especially where slow-onset environmental changes challenge any concept of forward planning.

While two decades ago Hugo pointed out that eradicating poverty, reducing population growth and using the environment more sustainably would reduce the need for environmental migration (Hugo 1996, 126), that more optimistic perspective has faded. Three years ago Hugo concluded that it was urgent to make decisions about the management of enforced mobility since environmental change was becoming more significant and because some of the changes required urgent 'substantial institutional, structural and cultural change which will take time to achieve' (Hugo 2013, xl). With the passage of time, Hugo's conclusion has not only become more valid but making migration as adaptation successful for migrants from small islands, dramatically or steadily forced from their homes, and ensuring a durable solution have become more challenging. No empty lands exist, Pacific populations are generally growing quickly and coastal land especially is increasingly contested.

Disclosure statement

No potential conflict of interest was reported by the authors.

References

Black, R., S. Bennett, S. Thomas, and J. Beddington. 2011. "Climate Change: Migration as Adaptation." *Nature* 478: 447–449.

Bourke, M., and T. Betitis. 2003. *Sustainability of Agriculture in Bougainville Province, Papua New Guinea*. Canberra: ANU.

Brookfield, H., and P. Brown. 1963. *Struggle for Land*. Melbourne: Oxford University Press.

Cernea, M. 1997. "The Risks and Reconstruction Model for Resettling Displaced Populations." *World Development* 25 (10): 1569–1587.

Connell, J. 1993. "Climatic Change: A New Security Challenge for the Atoll States of the South Pacific." *The Journal of Commonwealth & Comparative Politics* 31 (2): 173–192.

Connell, J. 2013a. *Islands at Risk. Environments, Economies and Contemporary Change*. Cheltenham: Edward Elgar.

Connell, J. 2013b. "Soothing Breezes? Island Perspectives on Climate Change and Migration." *Australian Geographer* 44 (1): 465–480.

Connell, J. 2014. "Population Resettlement in the Pacific: Lessons from a Hazardous History." In *Migration, Land and Livelihoods: Creating Alternative Modernities in the Pacific*, edited by G. Curry, G Koczberski, and J. Connell, 13–28. Abingdon: Routledge.

Connell, J. 2015. "Vulnerable Islands: Climate Change, Tectonic Change, and Changing Livelihoods in the Western Pacific." *The Contemporary Pacific* 27 (1): 1–36.

Connell, J. 2016a. "Last Days in the Carteret Islands? Climate Change, Livelihoods and Migration on Coral Atolls." *Asia Pacific Viewpoint* 57 (1): 3–15.

Connell, J. 2016b. "Nothing there Atoll? Farewell to the Carteret Islands." In *Appropriating Climate Change*, edited by T. Crook and P. Rudiak-Gould. Warsaw: de Gruyter. (in press).

Connell, J., and S. Kagan. 2015. *Creating Pathways for Pacific Island People at Risk of Climate Change Displacement*. Suva: ILO.

Connell, J., and J. Lea. 2002. *Urbanisation in the Island Pacific. Towards Sustainable Development*. London: Routledge.

Connell, J., and N. Lutkehaus. 2016. "Escaping Zaria's Fire? The Volcano Resettlement Problem of Manam Island, Papua New Guinea." *Asia Pacific Viewpoint* (in press).

Connell, J., and G. Tabucanon. 2016. "From Banaba to Rabi: A Pacific Model for Resettlement?" In *Global Implications of Development, Disasters and Climate Change*, edited by S. Price and J. Singer, 91–107. London: Routledge.

Donner, S. 2015. "The Legacy of Migration in Response to Climate Stress: Learning from the Gilbertese Resettlement in the Solomon Islands." *Natural Resources Forum* 39: 191–201.

Hemer, S. 2016. "Emplacement and Resistance: Social and Political Complexities in Development-Induced Displacement in Papua New Guinea." *Australian Journal of Anthropology* (in press).

Holston, J. 2009. "Insurgent Citizenship in an Era of Global Urban Peripheries." *City and Society* 21 (2): 245–267.

Hugo, G. 1996. "Environmental Concerns and International Migration." *International Migration Review* 30 (1): 105–131.

Hugo, G. 2011. "Lessons from Past Forced Resettlement for Climate Change Migration." In *Migration and Climate Change*, edited by E. Piguet, A. Pécoud, and P. de Guchtenière, 260–288. Cambridge: Cambridge University Press and UNESCO.

Hugo, G. 2013. "Introduction." In *Migration and Climate Change*, edited by G. Hugo, xv–xlii. Cheltenham: Edward Elgar.

Johnson, R. 2013. *Fire Mountains of the Islands: A History of Eruptions and Disaster Management in Papua New Guinea and the Solomon Islands*. Canberra: Australian National University E-Press.

Kahn, M. 1996. "Your Place and Mine: Sharing Emotional Landscapes in Wamira, Papua New Guinea." In *Senses of Place*, edited by S. Feld and K. Basso, 167–196. Santa Fe: School of American Research Press.

Knudson, K. 1977. "Sydney Island, Titiana, and Kamaleai: Southern Gilbertese in the Phoenix and Solomon Islands." In *Exiles and Migrants in Oceania*, edited by M. Lieber, 195–241. Honolulu: University Press of Hawaii.

Kukang, T., J. Selwyn, A. Siau, E. Tade, and M. Wairiu. 1987. "Atolls Resettlement Scheme, North Solomons Province." In *Rapid Rural Appraisal Case Studies of Small Farming Systems*, edited by M. O'Collins, 65–89. Port Moresby: Department of Anthropology and Sociology, University of Papua New Guinea.

Leckie, S. 2012. "The Management of Climate Displacement." *Forced Migration Review* 41: 17–18.

Lewis, J. 2009. "An Island Characteristic. Derivative Vulnerabilities to Indigenous and Exogenous Hazards." *Shima* 3 (1): 3–15.

Lipset, D. 1997. *Mangrove Man: Dialogics of Culture in the Sepik Estuary*. Cambridge: Cambridge University Press.

Lipset, D. 2013. "The New State of Nature: Rising Sea-levels, Climate Justice, and Community-based Adaptation in Papua New Guinea (2003-2011)." *Conservation and Society* 11 (2): 144–157.

Lucardie, R. 1985. "Spontaneous and Planned Movements among the Makianese of Eastern Indonesia." *Pacific Viewpoint* 26 (1): 63–78.

Lutkehaus, N. 1985. "Pigs, Politics and Pleasure: Manam Perspectives on Trade and Regional Integration." *Research in Economic Anthropology* 7: 123–144.

Lutkehaus, N. 1995. *Zaria's Fire: Engendered Moments in Manam Ethnography*. Durham, NC: Carolina Academic Press.

Lutkehaus, N. 2016. "Finishing 'Apui's Name: Birth, Death and the Reproduction of Manam Society." In *Mortuary Ritual and the Historical Moment in the Pacific*, edited by E. Silverman and D. Lipset, 135–158. New York: Berghahn.

Martin, K. 2013. *The Death of the Big Men and the Rise of the Big Shots*. New York: Berghahn.

McAdam, J. 2012. *Climate Change, Forced Migration and International Law*. Oxford: Oxford University Press.

McAdam, J. 2013. "Caught Between Homelands." *Inside Story*, March 15.

McAdam, J., B. Burson, W. Kalin, and S. Weerasinghe. 2016. "International Law and Sea-Level Rise: Forced Migration and Human Rights." Fridtjof Nansen Report 1/2016, Sydney.

McDowell, C. 2013. "Climate-Change Adaptation and Mitigation: Implications for Land Acquisition and Population Relocation." *Development Policy Review* 31: 677–695.

Mercer, J., and I. Kelman. 2010. "Living Alongside a Volcano in Baliau, Papua New Guinea." *Disaster Prevention and Management: An International Journal* 19 (4): 412–422.

Monson, R., and D. Fitzpatrick. 2016. "Negotiating Relocation in a Weak State. Land Tenure and Adaptation to Sea-level Rise in Solomon Islands." In *Global Implications of Development, Disasters and Climate Change*, edited by S. Price and J. Singer, 91–107. London: Routledge.

Mulina, K., J. Sukua, and H. Tibong. 2011. Report on Madang Volcanic Hazard Awareness Program, 6-29 September 2011, Rabaul Vulcanological Laboratory.

Nanau, G. 2014. "Local Experiences with Mining Royalties, Company and the State in the Solomon Islands." *Journal de la société des océanistes* 138-9: 77–92.

Numbasa, G., and G. Koczberski. 2012. "Migration, Informal Urban Settlements and Non-market Land Transactions: A Case Study of Wewak, East Sepik Province, Papua New Guinea." *Australian Geographer* 43 (2): 143–161.

O'Collins, M. 1990. "Carteret Islanders at the Atolls Resettlement Scheme: A Response to Land Loss and Population Growth." In *Implications of Expected Climate Changes in the South Pacific Region*, edited by J. Pernetta and P. Hughes, 253–269. Port Moresby: UNEP.

Rakova, U. 2014. "The Sinking Carteret Islands. Leading Change in Climate Change Adaptation and Resilience in Bougainville, Papua New Guinea." In *Land Solutions for Climate Displacement*, edited by S. Leckie, 268–290. London: Routledge.

Rogers, G. 1981. "The Evacuation of Niuafo' ou, an Outlier in the Kingdom of Tonga*." *The Journal of Pacific History* 16: 149–163.

Stead, V. 2015. "Homeland, Territory, Property: Contesting Land, State, and Nation in Urban Timor-Leste." *Political Geography* 45: 79–89.

Suliman, S. 2016. "Rethinking about Civilizations: The Politics of Migration in a New Climate." *Globalizations* 13: 638–652.

Tonkinson, R. 1979. "The Paradox of Permanency in a Resettled New Hebridean Community." *Mass Emergencies* 4: 105–116.

United Nations Human Rights Commission. 2010. *Protecting the Human Rights of Internally Displaced Persons in Natural Disasters—Challenges in the Pacific*. Suva: Office of the High Commissioner for Human Rights.

Resettlement and Climate Impact: addressing migration intention of resettled people in west China

Yan Tan

ABSTRACT

The relationship between climate change and human displacement is an important topic of global concern. China is a special case due to a high level of government control enforcing the ecological migration of millions of people since the mid-1980s. Little research has addressed how resettled people adapt to climate impacts in ecologically vulnerable resettlement areas and what factors influence their intentions to relocate again or adapt locally. Employing a social-ecological system approach, this study builds a conceptual econometric framework which differentiates two steps that drive migration intention at the household level. The study uses this approach to examine the role of both contextual and household factors in motivating the migration intentions of resettled people in the largest environmental resettlement area of Ningxia Hui Autonomous Region, China, where household survey data were collected in 2012. This framework enabled an analysis, first, of how local contextual factors and household factors shape the severity of climate impacts on households and, second, how these factors interact with the experience of climate impacts to further influence a household's migration intention as a response to climate impacts. The results show that some contextual factors (such as limited use of water-saving techniques, little practice of cultivating aridity-resistant crops, and lack of government support), strong local social networks and being in receipt of low rates of financial remittances have significant associations with adverse climate impacts experienced by resettled households, and also with their anticipated further relocation to respond to these impacts.

Introduction

Climate change is rarely the only, or main, cause of migration but, as explicitly pointed out by the late Professor Graeme Hugo, it is 'a new and increasingly significant driver among a constellation of several dynamic forces impinging on mobility with which it interacts' (Hugo 2013, xxv). In many parts of China, climate change has presented major challenges to rural people's livelihoods and their living environment (Millennium Ecosystem Assessment 2005; Li et al. 2013; National Development and Reform Commission 2013). This is especially the case in ecologically vulnerable zones (Ministry of Environmental Protection

2008) where local environments have an intrinsically low environmental carrying capacity to support human settlement (Meng, Zhang, and Zhou 2010). Many parts of western China, in particular, have experienced severe environmental deterioration that is closely related to climate change: desertification, water scarcity, and soil erosion (Wang, Chen, and Dong 2006; Li et al. 2012). Environmental degradation leads to poverty and, in turn, poverty deepens environmental degradation across extensive rural areas of the region (Song et al. 2015). In many cases the Chinese government makes decisions about which climate and environmental conditions are not conducive to human habitation. Since the mid-1980s the Chinese government (from the county to national level) has endeavoured to move some rural residents out of extremely vulnerable environments to reduce population pressure on the environment, alleviate poverty, rehabilitate ecosystems, and combat natural disasters (State Council of China 2001). As a result, by 2010, over 7.7 million farmers and herdsmen, mostly in western China and absolutely poor, had been relocated and resettled by the government to (mainly) other rural communities in China deemed less hazardous by the government (SCC 2011). Usually conceptualised as *ecological migration* by Chinese scholars (e.g. Bao 2006; Zheng 2013), since 2001 this strategy has evolved into a national policy to direct environment-related human resettlement (SCC 2001). Ecological migration in China is largely permanent, and sometimes involuntary. Most resettlement has been within western China and within a household's home county boundary, and is thus termed *near resettlement*. Where enormous environmental stressors and constraints in natural resources (especially land and water) are experienced, resettling the relocated people outside their counties via *distant resettlement* schemes becomes important, but this still occurs within the boundary of their original province.

The consequences of ecological migration in China are contested. Many studies (e.g. Liu 2002; Jiao and Wang 2008) have documented some beneficial outcomes of livelihood reconstruction after displacement and environmental rehabilitation in migrant-sending areas. In addition, Tan (forthcoming) has identified significant negative social and cultural impacts on both resettled people and their resettlement communities (e.g. cultural discontinuity and loss of sense of identity). Other recent research shows that resettlement adversely impacts the household asset base (particularly financial and natural capital), and concludes that resettlement has the potential to act as a driver of vulnerability, rather than lessening the vulnerability of rural households to climate change relative to non-resettled households (Rogers and Xue 2015). A number of studies also show that ecological migration to some arid and semi-arid areas is environmentally and economically unsustainable (Jiao and Wang 2008; Fan, Li, and Li 2015; Sun et al. 2016). They suggest that resettlement may trigger new environmental risks (e.g. deteriorating water quality) and exacerbate local social and ecological problems (especially regional water shortages) in the destination area. However, to date, no studies have evaluated how resettled households adapt to current and future climate (environmental) impacts in ecologically vulnerable resettlement areas, nor analysed the factors that influence their intentions to relocate again or adapt *in situ*.

The social-ecological system (SES) approach (outlined in the Methodology section of this paper) provides a useful perspective for unravelling the nexus between climate change and population dynamics, including migration (Hummel et al. 2008). Although some research (e.g. Wrathall 2012) has attempted to explore the influence of climate and other environmental changes on migration from a SES perspective, to date the application of the SES approach to empirical studies remains very limited. This is due primarily to two methodological and data-related challenges: (1) measuring dynamic and complex natural and social systems and quantifying the complex determinants of actual migration or migration intentions; and (2) a lack of, and therefore the need for collecting, suitable data for analysis. This paper seeks to bridge the knowledge gap by applying the SES approach to examine how climatic hazards impact on migration intentions of resettled rural households. Migration intentions are defined as an anticipated relocation of the whole household (decided by the household or the government) within 2 years from when the household was surveyed. The focus is on a major rural resettlement district of Ningxia Hui Autonomous Region[1] in western China where primary data were collected through a household survey in September 2012 and used for analysis. In bridging this knowledge gap, the paper furthers theoretical understandings of the intersections between population mobility and climate vulnerability. This study specifically addresses two issues: first, what aspects of livelihoods are perceived by resettled people to be significantly affected by climatic hazards in the resettlement area? Second, how do such climate impacts, alongside local/contextual and household socio-economic and demographic factors influence intentions of resettled people to relocate their households again within 2 years?

The paper begins with an overview of the limited literature on migration intention in the context of climate (environmental) change. This is followed by a discussion of methodology that involves three parts: an overview of SES and a conceptual framework that this study constructed for disentangling the nexus between climatic hazards and migration intention of resettled people in Hongsibu district of Ningxia, China; a description of the research setting in Hongsibu; and the collection of survey data in the study area. An analysis of those data, employing the SES framework and using the two-step Probit regression models, follows. The paper concludes with policy implications for ecological migration in China.

Literature review: migration intention and environmental change

Intention, as encapsulated in goal-pursuit theories used in psychology research, can be classified into two forms: 'goal intentions' and 'implementation intentions' (Sheeran, Webb, and Gollwitzer 2005; Gollwitzer and Sheeran 2006). The former refers to what one 'intends to achieve', while the latter 'specifies when, where, and how one intends to achieve it (Gollwitzer and Sheeran 2006, 82). Goal intentions are construed as the 'most immediate and important predictor[s] of attainment' (Gollwitzer and Sheeran 2006, 71), although holding a strong goal intention does not ensure goal realisation. The concept of implementation intentions helps us to understand the processes of goal attainment (e.g. migration, or relocation) and provides a self-regulatory strategy, at the micro-level (individual, household), to help people reach their goals.

Research into *migration intention* in the context of climate change is not as systematic and comprehensive as research into *actual migration* as an outcome of climate change. To conceptualise the non-linear relationship between climate change and migration, Perch-Nielsen, Bättig, and Imboden (2008) used 'direct effects' and 'indirect effects' to link 'climate change' and 'adaptation options' in their models explaining the climate change–migration nexus. Internationally, a few studies have examined the factors that influence migration intention. In a case study in Funafuti (Tuvalu), Mortreux and Barnett (2009) noted that migration intention is not significantly influenced by climate impacts but is more associated with people's age, income, social connections, and cultural and spiritual ties. Van Dalen, Groenewold, and Fokkema (2005) stated that the remittances migrant workers send to families can either trigger or inhibit other members' intention to migrate. They also showed that fundamental demographic and socio-economic factors (e.g. gender, marital status, education, occupation and social networks) are important predictors of migration intention, as did the study by Abu, Codjoe, and Sward (2014) in the forest–savannah transition zone of Ghana. Kuruppu and Liverman (2011) suggested that people's perceptions of both climate impacts and their adaptive capacity shape their intentions to migrate or stay. Further, a recent study by Otrachshenko and Popova (2014) showed that migration intention is influenced by individual life satisfaction, in addition to socio-economic characteristics (e.g. financial situation), experience of migration and macro-economic factors (e.g. employment rate, gross domestic product (GDP) per capita and income inequality).

In the Chinese context, a number of studies have examined how environmental factors combine with demographic, social, economic and political factors to influence migration behaviour of people, mostly in migrant-sending, rather than migrant-receiving, communities. At the household level, Feng and Nie (2013) analysed key factors influencing the willingness to undertake ecological migration and the actual migration behaviours of farmers in Ningxia Hui Autonomous Region. They found that a small household size, poor quality of farmland, lack of irrigation water, poor environmental conditions, and poor accessibility to a well-established mosque in the original villages are significant 'push' factors stimulating both the willingness and the actual out-migration of people. Additional variables—experience of previous out-migration and household income—are significantly and positively related to both the willingness of farmers to consider pursuing migration as well as to actually migrate. A case study in the Sangong River watershed of Xinjiang Uyghur Autonomous Region showed that a smaller household size, low net income per capita, main source of income being non-agriculture based (especially from financial remittances that migrant workers sent/brought to their households in the hometown), and farmers' participation in the 'Grain for Green' project[2] significantly affect people's willingness to undertake ecological migration (Tang, Zhang, and Yang 2011). The study by Li, López-Carr, and Chen (2014) in Minqin county, an extremely arid area of Gansu province, found that environmental factors such as water quantity, groundwater quality, land quantity and changing trends of these factors significantly shape the intention of whether or not farmers wish to migrate. Moreover, their study showed that governance policies on environmental management of inland river basins had mixed and complex effects on household livelihoods and environmental integrity, which then influenced migration intentions. For example, they found that, increasing water availability and quality, improved soil quality and vegetation coverage in the arid region could impede migration intentions. Notably, a study by Dong, Liu, and Klein (2012) in Inner Mongolia examined the factors that affect the willingness of displaced households to stay in their new resettlement villages. They found that significant factors include the age of the household head, years post-relocation, the ratio of governmental subsidies against the household's total income, and the household's level of fixed, durable and current assets. This study differs from those summarised above by using an SES approach and building a conceptual econometric framework which differentiates two steps that drive migration intention at the household level.

Methodology

Conceptual framework

The late Professor Graeme Hugo (1996) was one of a small group of pioneers in the 1990s who charted a new field of research on the many potential impacts of environmental factors (including climate change) on human societies, especially their influence on human migration. He recognised that the effects of the environment and environmental change were complex and mediated not only by the severity and nature of environmental change but also by the vulnerability, resilience, resources and situation of impacted communities (Hugo 1996). Inspired by Hugo's insightful understanding of the population–environment nexus, adaptation studies have now moved towards a holistic conceptualisation of SESs by integrating both ecological and social spheres of the complex human–environment system (Berkes and Folke 1998). A SES can be defined as a system involving ecological and social components at all scales from the household to the globe (Gallopín et al. 2001). From the SES perspective, social-ecological problems are viewed as being caused by the resilience of the supply (or environmental) system, which is shaped by social-ecological dynamics (Hummel et al. 2013). A system's resilience is the ability to tolerate shock and stress without significant changes (Folke 2006). When a system is unable to tolerate the stress on its fundamental relationships, the system may undergo a major transformation such as massive human movements (Walker et al. 2004). Migrants can be regarded as either passive respondents to social-ecological problems or proactive stability-seeking components of a shifted regime (Wrathall 2012). In a resilient SES, migration induced by environmental change has the potential to become an opportunity for development.

The SES framework, constructed by Hummel et al. (2008), conceptualises the relationship between natural and societal spheres as a supply system within which the two spheres are linked tightly and their linkage is mediated by economic, technical, political and ecological factors. This framework views migration as an outcome of the self-organisation of a supply system which is shaped and influenced by the interactions between individuals, societies and nature. There are two principal components in the SES: natural resources (e.g. food, water and energy) and their users (e.g. producers and consumers). Hummel et al. (2013) stated that the SES approach depicts all factors explaining the relationship between resources in the natural sphere and users in the social sphere, but 'specific factors that might actually matter must be identified in each case' (Hummel et al. 2013, 499). They called 'for the identification of pathways to more sustainable provisioning structures and corresponding governance initiatives, a mixed methodology of surveys, participatory research and more formalized modelling' (Hummel et al. 2013, 500). This paper contributes to the emerging SES studies in climate (environmental) change-migration-displacement research by investigating specific factors that influence the nexus between climatic hazards and migration intention of resettled households living in a major climate change hotspot of western China.

Applying the SES framework to the case study area, 'resources' that are particularly relevant include irrigation water and arable land, and 'users' are resettled rural households

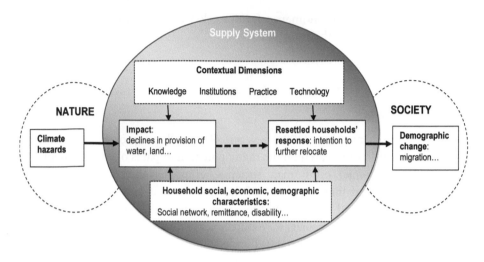

Figure 1. Conceptual framework: intention of further relocation of resettled households. *Source:* adapted from Hummel et al. (2008, 48).

Note: that the solid arrows represent direct influence and the dashed arrow represents indirect influence.

whose livelihoods are essentially reliant upon such resources. Informed by SES, this study builds a conceptual framework for unravelling the relationship between climatic hazards and migration intention of resettled people in Hongsibu district of Ningxia. The relationship between climatic hazards and migration, as depicted in Figure 1, is influenced and mediated by contextual factors (knowledge, practice, institutions and technology) and specific demographic and socio-economic factors at the household level in a community.

As Figure 1 shows, '*impact*' refers to a household's experience of climate impact on their agriculture-based livelihood and resources (e.g. water scarcity and land degradation) in an earlier time (e.g. in 2011). '*Response*' refers to the household's intention to relocate the whole household within the next 2 years (e.g. 2013–14). A household's prior experience of climate impacts interacts with local contextual factors, and household socio-economic and demographic factors, to stimulate their intention to migrate (and later on, their actual migration). As this paper shows, these contextual and household factors impose *direct* effects on a household's migration intention, and *indirectly* influence its migration intention through shaping the severity of climate impacts on livelihood domains.

Research setting: environmental resettlement in Hongsibu district of Ningxia, China

Hongsibu district in central Ningxia was selected as the case study area for the present analysis (Figure 2). Hongsibu is the largest single environmental resettlement area in China; it has received about 177 000 migrants relocated from eight poverty-stricken counties in the mountainous region of southern Ningxia between 2001 and 2010 (Hongsibu Resettlement Office 2012). Conventional farming has been practised chiefly on a rain-fed basis in this mountainous region. Hongsibu is an arid region where evident climate change, severe desertification, land salinisation and water shortage overlap. Harsh climatic conditions and a warming trend in host areas are considered when the government plans

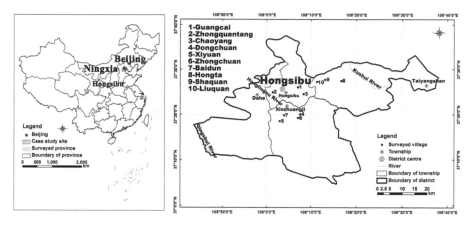

Figure 2. Surveyed towns and villages in Hongsibu district of Ningxia, China. *Source*: author.

large-scale ecological resettlements in this arid region (e.g. Hongsibu Poverty-Alleviation Office 2010). However, there is abundant uncultivated land in Hongsibu, which is perceived by the government to be a vital resource that the relocated farmers can depend on to develop self-sufficient agricultural production and eliminate poverty.

Hongsibu district's development was initially driven by the Irrigation Project for Poverty Alleviation by Lifting Water from the Yellow River in Ningxia—henceforth referred to as the Irrigation Project (which started in March 1998 and completed its first stage in June 2016). The Irrigation Project was a major hydro project under the country's '80/7 Plan to Combat Poverty (1994–2000)', a program that aimed to help some 80 million rural people living under the national poverty line to escape poverty in 7 years (State Council Leading Group Office of Poverty Alleviation and Development 1994). The Irrigation Project has enabled vast desertified barren land and grassland in the arid region of central Ningxia to be developed into farmland by utilising pumped water diverted from the Yellow River for irrigation. The total volume of water designated to Hongsibu is 188 million m^3 in normal years, but the available amount of water varies over time depending on runoff to the Yellow River. On average, each eligible resettled person has been allocated 2.5 mu (1 mu = 1/15 ha) of irrigated farmland by the government. Facilitated by rich land resources and the large irrigation project, and stimulated by the success of resettling thousands of poor farmer households relocated from southern Ningxia at an earlier time (in 1998–2000), this district was designated by the State Council in 2001 as a major experimental area for the ecological migration scheme in Ningxia. Since then, it has developed to be China's largest single environmental resettlement community, receiving massive numbers of migrants relocated from the aforementioned mountainous areas of southern Ningxia (Li et al. 2013). The district administers four townships, with a total land area of 2767 km^2. These four townships encompass 64 administrative villages, 91 per cent of which are resettlement villages.

Hongsibu is characterised by high climate vulnerability, irrigation-based agriculture, and a mix of two ethnic groups (Hui and Han Chinese) amongst the resettled population. This arid zone has experienced a rise in mean surface air temperature of 0.3–0.4°C every decade since 1987, which is greater than the national average rate in China (0.25°C) or the global average rate (0.13°C) over the same time period (Chen, Ma, and Zhou 2012). The total precipitation has exhibited a declining trend, dropping by 4.77 mm per decade since 1961. Annual mean precipitation has been 277 mm in the last 50 years, and the annual rate of evaporation remains as much as 2388 mm. As a result, 98 per cent of the rainfall evaporates, yielding an amount of groundwater (4.75 million m^3) which is of little use because of poor quality caused by high levels of salinity and fluoride (Hongsibu Water Bureau 2010). Yearly rainfall is mostly (61.8 per cent) concentrated in the summer months (from June to September). Seasonal variation in precipitation over the last 50 years is apparent, decreasing at the rates of 2.3 mm per decade in autumn, 2.1 mm in summer and 0.71 mm in spring (Chen, Ma, and Zhou 2012). Corresponding to these changes has been an increase in the frequency (at a high incidence rate of 9 in 10 years) and severity of drought and strong wind (at a daily average speed of 21.7 m/s for more than 30 days in a year) since 2000 (HWB 2010). Frequent and severe climatic hazards (e.g. drought, dry and hot winds, sandstorms, and hail) have severely threatened agricultural production and economic development in the resettlement areas. Hence, the Chinese government's key ecological resettlement location is itself being confronted by major environmental challenges.

An overwhelming proportion (95.5 per cent) of the total population (168 707 in 2011) in this district are resettled people and some in-migrants (Hongsibu Resettlement Office 2012). There is a slight discrepancy between the number of people planned to be resettled (177 000) and the actual number of people (176 864) who transferred their household registration status (*hukou*) from the counties of origin to their resettlement destination, between 2001 and 2010 (HRO 2012). According to the same source, however, the actual number of resettled people who transferred their *hukou* to the resettlement communities and physically resided in Hongsibu was 120 004 in the decade to 2010. Of these, 86.2 per cent were resettled in country areas, while the other 13.8 per cent were resettled in the centre of Hongsibu district and four town centres within the district. Strikingly, a significant number of people (60 097) who have their *hukou* registered in the resettlement area did not reside there, leading to a low resettlement rate (less than 66 per cent of the 177 000 targeted for resettlement) over the 2001–10 period (HRO 2012). Therefore, by 2010 nearly one in three resettled people had become emigrants living and working beyond Hongsibu district.

Ethnicity is a factor in this area, in resettlement, poverty alleviation and adaptation to climate change (Li et al. 2013). Ethnic minority *Hui* people, the overwhelming majority of whom are Chinese Muslim, constituted the majority (62.5 per cent) of the total resident population in 2011, while the 'majority' *Han* Chinese in China's overall population made up a smaller proportion (37.5 per cent) in this district. Nearly half of the people (47 per cent) resettled by the government from the eight counties in southern Ningxia to Hongsibu district still lived under the national poverty line in 2010 (HPAO 2010). Despite this, the prevalence rate of poverty has declined dramatically when compared with the poverty rate (85 per cent) of these resettled people before they were relocated (HPAO 2010). Clearly, relocation has improved their living circumstances. In addition, 14 600 resettled people, many of whom are *Hui* people, have various physical and/or intellectual disabilities (HPAO 2010). According to the 2010 census data of Ningxia, the illiteracy rate in Hongsibu district (12.6 per cent in 2010) was higher than the average rates in Ningxia (7.8 per cent) and China (4.1 per cent). These factors add a further challenge to livelihood reconstruction and climate adaptation in this resettlement area.

Agriculture has been the largest sector in the economic structure of the district, making up 39.2 per cent of its GDP, while the secondary and tertiary industries accounted for 31 per cent and 29.8 per cent, respectively, in 2011 (Hongsibu Census and Statistics Office 2012). The average net per capita income of farmers in Hongsibu district was 4027.7 yuan in 2011 (HCSO 2012). According to data from our survey in 2012, remittances that migrant workers from resettled households in Hongsibu district working elsewhere in China sent or brought back to their households was the largest income source in the district, contributing 51.4 per cent of the total mean income of the resettled households. This was followed by agricultural production, accounting for 35.5 per cent of the total average income of the resettled households. Non-agricultural production (e.g. family business and industry) was rather weak, making up a negligible share (3.3 per cent) of the resettled households' income.

Many resettled households have confronted climate-related constraints on water and land. The major constraint lies in water shortage. In 2011, about 5.5 per cent of the total arable land in Hongsibu district (29 740 ha) could not be irrigated due to water shortage (Ninxia Institute of Water Resources Research 2012). Declining rainfall and increased rainfall variations, inadequate quantity of irrigation water mainly supplied from the Yellow River, low efficiency of water use, and water-intensive cropping (e.g. corn and wheat) are key reasons for severe water shortages in the district. Per capita annual water supply in Hongsibu was 821 m^3 in 2010, which falls into the category of 'water scarcity' (annual 1000 m^3 per person) as defined by the United Nations (2014). The agricultural sector consumes the largest proportion (94.9 per cent in 2010) of the overall water resources in the district (HWB 2010). The efficiency rate of irrigation water was very low (0.52). Inefficiency of water use is attributed to various factors, including high evaporation of water, and the high rate of loss due to leaking (by 15 per cent) caused by some poorly constructed segments of the canal and waterway networks in the study area, severe land desertification and soil salinisation, and the widespread use of traditional methods for irrigation (HRO 2012).

Furrow or flood irrigation has continued to be used by resettled people as the dominant irrigation method in the resettlement area. As observed in some experimental sites during my fieldwork in September 2012, some advanced water-saving irrigation techniques and methods (e.g. sprinkler and micro-irrigation) had come into trial on a small scale. Corn, wheat and potato were the dominant subsistence crops, while cash crop farming of grapes, *gouji*, liquorice, and vegetables was practised essentially on a household basis and thus on a small scale. Scaling-up agriculture in the form of dragon-head enterprises,[3] cooperatives, or family farms is at the early stages of development in the study area. At the time of data collection, agricultural restructuring involved two aspects: (1) reducing the land area of high-water-consuming but low-yielding crops (e.g. corn and wheat); and (2) increasing the acreage of low-water-consuming and value-added crops and fruit groves (e.g. *gouji*, liquorice, red dates, and grapes for wine making).

Data collection

Primary data collection was undertaken in Hongsibu district in September 2012.[4] The surveyed area involves three townships where the majority of relocated people were resettled between 2001 and 2010, namely Taiyangshan, Xinzhuangji and Hongsibu (see Figure 2).

The total sample size ($N = 289$ resettled households) was proportionately distributed to each selected township in terms of population size in 2012. Then, within these township areas, 10 administrative villages in total were selected by using a probability proportionate to size (PPS) sampling method in each selected township. This PPS method ensures that each household in the sampling population has the same probability of being sampled regardless of the size of clusters involved in the sampling. Finally, a random sampling method was used to select potential households in each selected village for survey. A structured questionnaire survey was conducted through face-to-face interviews.[5] The interviewers asked the head of the household questions about their experience of climatic events, key aspects of livelihood perceived to be affected by climatic extremes, household demographic, social and economic characteristics, and their household's intentions to relocate or not relocate within next 2 years (2013–14). The interviewers wrote the respondent's answers on the questionnaire paper directly. Where the household head was unavailable, their spouse or the household member who best understood the household's situation answered the questions. The average age of the respondents was 42.4 years. The primary working age group (25–54 years) accounted for 72.3 per cent of the sample, while the young age group (15–24 years), old working age group (55–64 years), and the elderly group (65 years or above) made up 8.7 per cent, 9.3 per cent and 9.7 per cent, respectively. In addition, in-depth interviews with village leaders and some resettled households were also conducted during the household survey process. These interviews covered the main issues of livelihood reconstruction of the resettled people (e.g. water quantity and quality, living expenses, re-establishing local Muslim groups).

Analysis: a modelling approach to assess the relationship between climate impact and migration intention

Based on the conceptual framework constructed in Figure 1 which draws on the SES framework, a two-step regression procedure was adopted to analyse the questionnaire survey data. As per Figure 1, a core focus for understanding the questionnaire survey data is in relation to impact and response, which in the case of this study is climate *impact* and resettled household *response*, specifically migration intention. The two-step regression procedure was used to address the unobserved variables such as contextual dimensions as well as household socio-economic and demographic characteristics (see Figure 1) that influence, directly and indirectly, both climate impacts and migration intentions. In particular, the regression procedure was used to address some of the problems of sample-induced endogeneity, that is, resettled households that intend to relocate would be more likely to claim adverse climate impacts. For the purposes of carrying out the two-step regression procedure, it was necessary to create and distinguish between dependent variables and independent variables, as explained below.

Dependent variables

Impacts of climatic hazards on livelihoods
The questionnaire asked resettled people about their perceived frequency of climatic hazards (e.g. drought, dry and hot winds, sandstorms, and hail) in 2011 compared to

the prior 5–10 years since resettling in Hongsibu district. Their answers were coded according to 11 levels on a Likert scale from 0 ('very rare') to 10 ('very frequent').

The questionnaire further asked, first: 'to what extent did climatic hazards in 2011 affect any aspect of your household's livelihood in 2011?', and second: 'what particular aspects of livelihood were mainly affected by climatic hazards?' Answers to the first question were coded as 0 (never affected) and 1 (much affected). The survey data show that climatic hazards imposed relatively large effects on three aspects of livelihood: water supply (decreased), land loss (less land available for farming as farmland is abandoned due to drought, and water and wind erosion), and living cost (increased). The mean scores for corresponding aspects of climate impact are 0.11, 0.18, and 0.33, suggesting that the resettled people perceived that climatic hazards had the biggest influence on their increased living costs. Analysing how climatic hazards affected living costs in the resettlement locations is especially important for resettlement and adaptation policy making and policy adjustments. This is because a fundamental resettlement objective (or principle), as committed by the Chinese government, is to facilitate displaced and resettled people to rehabilitate their livelihoods and maintain their basic living standards at a level that is not lower than the level before resettlement, or not lower than the average level of the host population in the resettlement region. It is important to note that these three domains (decreased water supply, land loss, and increased living costs) could also be influenced by other factors, according to the in-depth interviews with village leaders and resettled people. For example, loss of arable land could be a result of salinisation, poor farming practice and land acquisition by the government for infrastructure projects. Increased living costs could relate to rising capital investment in agricultural production (e.g. irrigation water, seeds, fertiliser), daily living expenses (e.g. tap water), and education fees for school students. The total living cost in the resettlement area was much higher than in the homelands where the resettled people had resided (Development and Reform Commission of Ningxia 2004). Nevertheless, this study focuses on the respondents' perceptions of the livelihood impacts of climatic hazards. In my analysis, to predict each household's migration intention, the probabilities of these three aspects of climate impacts are considered as dependent variables for the impact outcome, whilst the predicted probability of each of the three aspects of impact is considered as an independent variable for the subsequent response—migration intention outcome. Through the two-step regression procedure, the indirect effect of climatic hazards on people's intention to further relocate within 2 years could be analysed.

Migration intention

The questionnaire further asked a direct question: 'to what extent does your whole household intend to relocate in the next two years (2013–14) because of experienced impact of climatic hazards in 2011?' Answers to the question, which represent the strength of migration intention, were coded according to 11 levels on a Likert scale ranging from 0 ('no intention') to 10 ('very strong intention'). The survey data show that responses to the question were essentially distributed at both ends of the spectrum. Therefore, responses were recoded as 1 'planning to relocate' (for responses coded 6–10) and 0 'no migration intention' (for responses coded 0–5). Accordingly, the households in the sample were categorised into two groups: one group having an intention to relocate, and the other having no migration intention. It is noted that migration intention

Table 1. Summary of the dependent variable in the migration intention model.

Households	Observations	%
Households did not intend to relocate again (= 0)	237	82.0
Households intended to relocate again (= 1)	52	18.0
Total	289	100.0

Source: author's own survey, 2012.

should not be treated as a binary choice, but rather as existing on a sliding scale along a continuum of options. Therefore, recoding migration intention as a dichotomous variable is quite a strong assumption, but reasonable as a first approximation. Treating migration intention as a dichotomous variable, and accordingly deploying binary logistic or logit regression models, which are similar to the Probit model used in the present paper, has been common practice in empirical studies examining the determinants of migration intentions. Studies on rural-to-urban migration intentions in Hubei province of central China (Yang 2000), farmers' intentions to move out of the ecologically vulnerable areas of arid western China (Li, López-Carr, and Chen 2014), and rural people's intentions to migrate in Thailand (De Jong 2000) and Russia (Bednaříková, Bavorová, and Ponkina 2016) provide some examples.

Table 1 summarises the observations in the response model. Nearly one in five of sampled households ($N = 289$) expressed an intention to undertake further relocation, while the majority (82 per cent) of the households in the sample did not plan to relocate their households again. As a reminder, *migration intention* is defined in this paper as a further relocation of the whole household within 2 years (2013–14) from the date of being surveyed because of experiencing an impact of a climatic hazard.

Independent variables

Except for the 'frequency of climate-related events' variable, four dimensions of contextual factors (knowledge, institutions, practice, and technology) were incorporated into both the *impact* model and *intention* model to examine how they impact on the key domains of a household's livelihood and subsequently on their migration intention. *Knowledge* was measured by the number of information sources that people can access before, during, and after climatic events take place. Information on 'knowledge' was collected in the questionnaire survey by directly asking the respondents: 'what information sources did you use to get information on climatic events?' *Institutional factors*, which measure people's access to political power ('political affiliation') and to 'government support' when natural disasters occur, are particularly important in rural China (Nee and Su 1998). For example, we asked respondents 'to what extent did your household receive support from local governments when your household experienced climatic impacts or economic difficulties?' Governmental assistance includes the provision of living subsidies; access to small bank loans; provision of improved varieties of crops, fruits and livestock; and on-site training in agricultural techniques and skills. Survey responses were coded according to 11 levels on a Likert scale ranging from 0 ('no support') to 10 ('very much'). On average, the sample had a mean score of 2.55, indicating that the level of governmental assistance was low. The questionnaire asked the respondents about two facets of *technology* that are essential for livelihood reconstruction after resettlement. One facet is the household's accessibility to irrigation facilities, and the other is whether or not the household used water-saving techniques in farming. As per agricultural *practice*, the questionnaire asked the respondents about two aspects fundamental to agricultural restructuring: whether or not the household cultivated drought-resistant crops, and whether or not the household developed high-value-added agriculture. This study also included important social, economic and demographic characteristics of the households, such as spatial distribution of a household's social network, remittance ratios, amount of debt, level of education, and family demographic composition (as measured by 'working-age ratio', 'disability ratio', and ethnicity).

Table 2 presents the names and definitions of the dependent and independent variables of both the impact and response (migration intention) models, as discussed in the conceptual framework in the Methodology section and depicted in Figure 1. A major difference between the models is the inclusion, in the response model, of the predicted probability of climate impact on the three key domains of household livelihood (land loss, water supply, and living cost). The advantage of using the predicted impact approach is that it considers a range of influences on climate impact, and accordingly solves the potential problem of endogeneity.

Implementing the two-step regression models

A two-step regression procedure was employed to quantify how climatic hazards, households' demographic and socio-economic characteristics and contextual factors in the study area influenced some key aspects of resettled households' livelihoods and their subsequent response to climate impacts—migration intention. In Step 1, a Probit model was estimated to examine how the frequency of climatic hazards, contextual, and household socio-economic and demographic variables in 2011 influenced the three major domains of household livelihood in that year, as each domain of livelihood is measured as a dichotomous variable. Greene (2008, 772–775) provided the detailed econometric specification for Probit models. Note that one may choose the Probit model instead of the logistic or logit model for the binary choices modelling. However, as shown in the literature (e.g. Amemiya 1981), a similarity exists between the two approaches in terms of comparing the marginal effects of two regressors. One purpose of the first step model is to obtain the predicted probability (severity) of each impact, and the Probit model best meets this purpose. The three probabilities of climate impact are predicated probabilities from Step 1 regression rather than subjective belief, and thus should not be endogenous. Using the predicted probabilities is similar to the technique in two-stage least squares (2SLS) to address endogeneity. These predicted probabilities of climate impact are used as the independent variables in Step 2 (response) model. Step 2 also uses a Probit model to investigate how these specific impacts (i.e. predictions of the Probit model in Step 1) influence households' migration intention in the next 2 years.

Results

Tables 3 and 4 present the estimated coefficients of the independent variables and the overall fit statistics in Step 1 and Step 2, respectively. Table 4 (in the third column) also

Table 2. Definitions, means, and standard deviation (SD) of variables.

Variables	Definition	Impact	Response	Mean	SD
Outcomes					
Domains of climate impact in 2011					
Land loss	1 if arable land decreased due to climatic hazards; 0 otherwise	x		0.18	0.24
Water supply	1 if water supply decreased due to climatic hazards; 0 otherwise	x		0.11	0.13
Living cost	1 if living cost increased due to climatic hazards; 0 otherwise	x		0.33	0.22
Response to climate impact (intention to migration in next 2 years)					
Migration intention	1 if the household plans to relocate in next 2 years; 0 otherwise		x	0.18	0.38
Predictors					
Climatic events and impact in 2011					
Frequency of climatic hazard	Frequency of climatic hazards: [0, 10] = [very rare, very frequent]	x		3.09	2.76
Predicted land loss	Predicted probability, calculated from the impact model: [0, 1]		x	0.18	0.24
Predicted water supply	Predicted probability, calculated from the impact model: [0, 1]		x	0.11	0.13
Predicted living cost	Predicted probability, calculated from the impact model: [0, 1]		x	0.33	0.21
Knowledge					
Information source	Number of sources to receive information on climatic conditions: [0, 6]	x	x	3.07	1.30
Institutions					
Governmental support	Governmental assistance received by the family when climatic or natural disasters occur or when it faces economic difficulties: [0, 10] = [no support, very much]	x	x	2.55	2.69
Political affiliation	1 if the household has any member of the Communist Party of China (CPC) or minority political parties; 0 otherwise	x	x	0.12	0.33
Technology					
Irrigation facility	1 if household can access water storage or irrigation facilities; 0 otherwise	x	x	0.69	0.46
Water-saving technique	1 if household used water-saving techniques; 0 otherwise	x	x	0.39	0.49
Practice					
Drought-resistant crop	1 if household adopted drought-resistant cropping; 0 otherwise	x	x	0.48	0.50
Value-added agriculture	1 if household developed income-generating crops, fruits and vegetables; 0 otherwise	x	x	0.51	0.50
Social factors in 2011					
Connection with local relatives	1 if a family has any relative or friend who lives in other towns within the same district (i.e. Hongsibu); 0 otherwise	x	x	0.69	0.46
Connection with distant relatives	1 if a family has any relative or friend living outside Ningxia; 0 otherwise	x	x	0.19	0.39
Economic factors					
Remittance ratio	Proportion of household income from financial remittances: [0, 1]	x	x	0.51	0.34
Debt	Debt of household (000s yuan)	x	x	12.10	24.21
Demographic characteristics					
Working-age ratio	Proportion of the number of members of working age against total number of household members: [0, 1]	x	x	0.72	0.23
Disability ratio	Proportion of those disabled or with chronic medical conditions against total number of household members: [0, 1]. Disability includes any intellectual problem, physical disability, mental problem, and psychological disorder	x	x	0.10	0.20
Schooling	Average number of years of education of household members	x	x	5.30	2.49
Ethnicity	1 if all members of the household are Hui-ethnic people; 0 otherwise	x	x	0.82	0.38

Table 3. Probit regression results: Step 1 'impact model'.

Predictors	Impact on land loss Coefficient	Impact on water supply Coefficient	Impact on living cost Coefficient
Climatic events in 2011			
Frequency of climatic hazard	0.038	0.019	0.113***
Knowledge			
Information source	−0.226*	−0.104	0.129
Institutions			
Governmental support	−0.332***	−0.096**	−0.052
Political affiliation	−0.655	−0.167	0.211
Technology			
Irrigation facility	0.504*	−0.020	0.239
Water-saving technique	−0.770**	0.696**	0.653***
Practice			
Drought-resistant crop	−1.157***	0.905**	−0.443
Value-added agriculture	−0.384	−0.030	0.351
Social factors in 2011			
Connection with local relatives	1.295***	0.344	−0.016
Connection with distant relatives	−0.681*	−0.374	0.163
Economic factors			
Remittance ratio	0.160	−0.901***	−0.265
Debt	−0.007	−0.005	0.001
Demographic characteristics			
Working-age ratio	−0.409	0.619	0.271
Disability ratio	0.217	−2.724*	−0.560
Schooling	0.076	−0.022	−0.050
Ethnicity	0.555	−0.205	0.317
Constant	−0.908	−1.247**	−1.548***
Observations	289	289	289
Count R^2	0.858	0.896	0.723
Wald χ^2	95.43	34.12	59.28

*$p < 0.10$; **$p < 0.05$; ***$p < 0.01$.

presents the marginal effects of the climate impact and explanatory variables in the Step 2 model. The changes in probability are compared to the baseline. None of the models suffered from serious multicollinearity, and robust standard errors were used to minimise heteroskedasticity. Values of correct prediction in all models range between 72 per cent and 89 per cent, suggesting above-average prediction accuracy of these models.

Climate impact

The frequency of climatic hazards exhibited significant influence, at the 1 per cent significance level, on the living costs of the resettled households, as shown in Table 3. The results show that the higher the frequency of climatic hazards experienced by households, the greater the likelihood of those households attributing climatic hazards to be a significant factor causing them to pay higher living costs.

Among the contextual factors, those households having received government support were less likely to suffer loss of farmland or reduction of water supply due to climatic hazards. This is because such support particularly improved the household's accessibility to, and affordability of, irrigation water. Adopting water-saving techniques and planting aridity-resistant crops significantly (at a statistical significance level of 1–5 per cent) expanded the scale of the household's agricultural land use, thereby reducing land loss when experiencing droughts. However, adopting these water-saving techniques and new crop species brought about increasing demand for water (as a result of the increased

Table 4. Probit regression results: Step 2 'response model' and predicted marginal effects of factors influencing migration intention of whole households.

Predictors	Coefficient	dy/dx (change in probability of household's migration intention)[a]
Climatic impact in 2011		
Predicted land loss	−0.029	−0.006
Predicted water supply	3.571**	0.734**
Predicted living cost	2.753***	0.566***
Knowledge		
Information source	0.076	0.566
Institutions		
Governmental support	0.199***	0.041***
Political affiliation	−0.686**	−0.141**
Technology		
Irrigation facility	−0.041	−0.009
Water-saving technique	−1.072***	−0.220***
Practice		
Drought-resistant crop	−0.662*	−0.136*
Value-added agriculture	−0.135	−0.028
Social factors in 2011		
Connection with local relatives	−0.771**	−0.158**
Connection with distant relatives	0.217	0.045
Economic factors		
Remittance ratio	1.126***	0.231***
Debt	0.007*	0.001*
Demographic characteristics		
Working-age ratio	−0.159	−0.033
Disability ratio	−1.635*	−0.336*
Schooling	−0.036	−0.008
Ethnicity	−0.493*	−0.101*
Constant	−1.627***	
Observations	289	
Count R^2	0.837	
Wald χ^2	60.02	

Notes: [a]In the baseline model, the dummy variables are set as 0; and the continuous variables are set at their respective means (which is the usual practice for continuous variables in generating predicted probabilities). The marginal effect measures the change in probability for every one unit change in corresponding independent variables (i.e. change from 0 to 1 for a dummy variable, or increase by one unit above the mean for a continuous variable).
*$p < 0.10$; **$p < 0.05$; ***$p < 0.01$.

amount of farmland in use) and higher living costs (because of increased capital investment to purchase the necessary facilities, equipment and seeds).

Among a number of household socio-economic and demographic factors, local kinship (friendship) and proportion of financial remittances that migrant workers located outside Hongsibu district sent to their households located in Hongsibu exhibited a significant influence on the effect of climatic hazards (at the 1 per cent significance level). The group of households that have relatives or friends residing only in Hongsibu had a higher risk of losing farmland (i.e. abandonment of arable land) than their counterparts who do not have local kinship or friendship connections in Hongsibu. The households with local kinship networks reported difficulties in obtaining assistance from local relatives or friends to prevent cultivated land from being abandoned when experiencing climatic hazards (especially drought) as most local people experienced the same climatic hazards and thus similar impacts. The resettled households that kept social ties with relatives or friends residing in places beyond Hongsibu usually received outside assistance which helped to prevent some land loss when drought and other climatic events occurred. The households with a higher remittance rate in the household's income composition were less likely to experience reduction in water supply than those households with no or a lower ratio of remittances when facing climatic hazards.

Migration intention

As a key finding, those households which experienced a decline in water supply and/or a rise in living costs significantly (at the 1–5 per cent significance level) and dramatically increased their probability of migration intention (by 0.73 and 0.57, respectively), compared to those that did not suffer such climatic impacts (see the third column of Table 4). This finding is in line with other empirical evidence which demonstrates how climatic impacts can stimulate migration in developing countries or regions, such as rural Sahel (Mertz et al. 2009) and Bangladesh (Martin et al. 2014).

Looking into the 'institutional' factors, households that had any family member who was a member of the Communist Party of China (CPC or minority party) were less likely to plan to relocate their households again than those without such political affiliation. It is not an unusual practice in rural China that those holding CPC membership play administrative leadership roles at the grassroots (village) level. Thus they have convenient and frequent connections with local government officials (at township and district levels), so that these households can access additional resources (e.g. governmental subsidies offered to low-income families) that are not available beyond the community. This privilege might constrain their motivation to relocate again. Local governments did provide assistance to cope with climatic hazards and to facilitate a second relocation of some resettled households which still lived in harsh environments with poor access to irrigation (and drinking) water and poorly constructed houses. Those households which received more support from formal sources of government had a greater propensity to plan further relocation within the 2-year time period. The probability of further relocation would increase by 0.04 for an increase of one unit on a Likert scale from 0 ('no support') to 10 ('very much support') above the mean (2.55) if receiving government assistance, everything else being equal and at the 1 per cent significance level.

Providing access to agricultural 'techniques' can significantly lower the propensity of resettled households to relocate a second time. Those having applied water-saving techniques in farming had less probability (at 0.22, at the 1 per cent significance level) to plan for a second movement than their counterparts who could not access, or did not adopt, such water-saving techniques (the third column in Table 4). Similarly, those who cultivated drought-tolerant cropping had a lower probability (at 0.14, at a lesser statistical significance level of 10 per cent) of relocating their whole households compared to their counterparts who did not engage in such cropping practice.

Finally, a number of household social, economic and demographic factors can significantly motivate, or constrain, migration intention. Notably, for households who were in receipt of financial remittances, this performed as a significant 'push' factor for migration intention. The greater the proportion of remittances in the household's income structure, the higher the tendency for the household to plan for relocation within the 2-year time period, increasing the probability by 0.23 for every one unit of increase in the remittance ratio above the mean (at the 1 per cent significance level). Remittances are a major income source for many resettled households, making up 51 per cent of the household's total income on average. Our in-depth interviews with the remittance-receiving households in the study area indicate that the group of households that received a higher proportion of remittances usually had more household members who were migrant labourers outside Hongsibu to send remittances and also had a longer duration for which remittances were received (reflecting that their migrant household members experienced more years of employment predominantly as wage earners) than the group of households that received a lower proportion of remittances. Compared to the latter group, the former group had greater affordability, and intention, of further relocation through either the government-assisted initiative or self-motivated migration by which they would live with household members who were migrant workers. In contrast, local kinship/friendships played a strong 'pull' force, impeding intentions to relocate whole households. Households having relatives or friends living within the resettlement area, Hongsibu district, were less likely (at a probability of 0.16, at the 5 per cent significance level) to plan entire household migration within 2 years. Importantly, those households that had a higher proportion of disabled members had a much lower probability (by 0.34, at a lesser statistical significance level of 10 per cent) of relocation again within 2 years compared to their counterparts who did not have (or only had a lower proportion of) members suffering disabilities. Hui-ethnic households had a lower probability (by 0.10) of planning to relocate in the next 2 years than Han-Chinese families (at a lesser statistical significance level of 10 per cent). According to my in-depth interviews with village leaders, a possible reason is that further migration means leaving the existing local Muslim groups in which *Hui* people are strongly anchored and from which Hui-ethnic households easily seek support and reciprocity. For example, in the study area, Islamic religious leaders (imams), acting as guarantors, assist the *Hui* villagers in obtaining labour support and loans from each other and from local credit organisations. This enhances the resilience of *Hui* households to external shocks, thus reducing the intention of further relocation.

Conclusion and policy implications for ecological migration in China

A publication by De Sherbinin et al. (2011), in which Hugo was an important contributor, asserted that climate change related relocation is already underway in many countries, and that both planned resettlement and self-motivated resettlement are likely in the future in response to climate risks. Hugo acknowledged that relocation can be seen either as a failure (because *in situ* adaptation has been unsuccessful) or as part of a portfolio of effective adaptation measures (Bardsley and Hugo 2010; De Sherbinin et al. 2011). His insights did much to deepen the field of research on climate adaptation and population mobility.

Understanding social-ecological transformation is crucial if we are to unravel the nexus between climate impact and population mobility, but there is limited empirical evidence addressing it, especially in a vulnerable rural resettlement setting. This paper addresses this gap in knowledge and considers not only household factors but also contextual factors influencing complex natural and social systems by measuring four major domains— *knowledge, institutions, practice,* and *technology*. The study provides evidence that these four domains have significant associations with the three major aspects of adverse climate impacts (decreased water supply, land loss, and increased living costs) experienced by resettled households in Hongsibu district of central Ningxia, and also with their subsequent response to these impacts—*intention to relocate their whole households* within 2 years. Vulnerable households with little access to government support, limited use of water-saving techniques and little practice of cultivating aridity-resistant crops were shown to experience more severe climate impacts in terms of land loss. Households that adopted water-saving techniques and engaged in cultivating aridity-tolerant cropping experienced more severe climate impacts in terms of decreased water supply and increased living costs. This was because their water-saving techniques and crops had actually led them to increase the extent of farmland in use with a resultant effect that they had a need for higher volumes of water than their counterparts who did not adopt such water-saving techniques or crops.

Households' subsequent responses to climate impacts were greatly determined by the severity of adverse impacts in the two distinct aspects of decreased water supply and increased living costs. Several contextual factors (particularly limited access to water-saving techniques in farming and lack of political affiliation to local government) consistently motivated their intention to plan for a second relocation, leading to unstable resettlement. The group of households with low socio-economic status (characterised as having a high rate of disabled members, having a low remittance rate in the household's income structure, and having all members belonging to the Chinese Muslim minority ethnic group) and the group maintaining mainly local kinship ties exhibited a lower tendency to plan for relocation again within 2 years. The findings about remittances align with the view of Findley (1994) that remittances can help families and local communities to make it through climatic extremes but disagree with her view that remittances reduce the risk of household pursuing out-migration.

Ecological migration, as practised in China, is an important approach to overcoming the limits of *in situ* adaptation where vulnerable groups are unable to adjust adequately to stressors associated with climate change and other environmental changes. However, the risk of unstable resettlement in some resettlement communities, as shown in this case study, is immense, and thus planning for ecological migration is essential. For adaptation and resettlement policy to be effective, decision makers must recognise the particular needs of people affected by climate (environmental) stressors in their new resettlement areas. In Hongsibu district, which is environmentally vulnerable and economically underdeveloped, as in many other rural resettlement communities in China, capacity-building policies and programs are especially needed to stabilise resettlement mainly through increasing the capability of the resettled to adapt to climate/environmental changes *in situ*. Key areas of capacity building should focus on dealing with water shortages and increasing water-use efficiency, controlling rising living costs, and assisting households to generate more income from non-agricultural production and activities. Some examples include shifting from canal-based to pipeline-equipped water transmission and incorporating advanced irrigation methods like sprinkler, drip, and micro-irrigation to enhance the water-transferring and use efficiency; and facilitating agricultural restructuring by shifting from currently water-consuming cropping to water-saving and value-added farming and livestock husbandry. Migration is another important strategy in climate adaptation. For those households that have sent some members to be migrant workers, and where those members have stable employment and long-term residency in cities, local government authorities should facilitate other household members, who have expressed strong intentions to migrate, to relocate voluntarily by helping them to deal effectively with any regulatory hurdles (e.g. the transfer of *hukou* and associated social security status) encountered during migration, especially enabling them to settle in urban areas and live with their migrant household members.

Notes

1. This is an Autonomous Region in which there is a high concentration of the *Hui* ethnic group, who are mainly Muslim. The region is an administrative sub-division equivalent to a province-level unit of China.
2. A program aiming to return cropping land with a gradient of 25 degrees or greater to forest or grassland.
3. Dragon-head enterprises are large-scale enterprises that bring many enterprises and farmers together by involving them in their supply chains and providing them with guidance on agricultural production practices that improve food safety and quality.
4. Thanks to the help of Professor Wenbao Mi at the College of Resources and Environment, Ningxia University, I obtained official permission from the Department of Development and Reform, Government of Hongsibu District, to conduct the household survey in selected towns and villages.
5. I led the survey and the team of interviewers included my two PhD students, eight postgraduate students majoring in geography, and one scholar working in the College of Earth Environmental Sciences at Lanzhou University. Before going to the study area I provided an intensive skills training course to the interviewers about how to carry out household surveys in rural settings.

Acknowledgements

The author would like to express her great gratitude to the anonymous reviewers, and the editors of this special issue, Dr Natascha Klocker (University of Wollongong) and Dr Olivia Dun (University of Melbourne), for their insightful suggestions and thought-provoking comments on the earlier manuscript.

Disclosure statement

No potential conflict of interest was reported by the author.

Funding

This work was supported by the Australian Research Council Discovery project [DP110105522].

References

Abu, M., S. Codjoe, and J. Sward. 2014. "Climate Change and Internal Migration Intentions in the Forest-savannah Transition Zone of Ghana." *Population and Environment* 35: 341–364.

Amemiya, T. 1981. "Qualitative Response Models: A Survey." *Journal of Economic Literature* 19 (4): 1483–1536.

Bao, Z. M. 2006. "Definition, Categorisation and Some Other Issues of Ecological Migration (in Chinese)." *Journal of the Central University for Nationalities* 33 (1): 27–31.

Bardsley, D. K., and G. J. Hugo. 2010. "Migration and Climate Change: Examining Thresholds of Change to Guide Effective Adaptation Decision-making." *Population and Environment* 32 (2): 238–262. doi:10.1007/s11111-010-0126-9.

Bednaříková, Z., M. Bavorová, and E. V. Ponkina. 2016. "Migration Motivation of Agriculturally Educated Rural Youth: The Case of Russian Siberia." *Journal of Rural Studies* 45: 99–111. doi:10.1016/j.jrurstud.2016.03.006.

Berkes, F., and C. Folke, eds. 1998. *Linking Social and Ecological Systems: Management Practices and Social Mechanisms for Building Resilience.* Cambridge: Cambridge University Press.

Chen, Y., Q. Ma, and B. Zhou, eds. 2012. *Qihou bianhuaxia Ningxia nongye zhonghe kaifa de tansuo yu shiqian (Exploration and practice of comprehensive agricultural development in Ningxia Hui Autonomous Region in the context of climate change).* Yinchuan: Yangguang Publishing.

De Jong, G. F. 2000. "Expectations, Gender, and Norms in Migration Decision-making." *Population Studies* 54 (3): 307–319.

De Sherbinin, A., M. Castro, F. Gemenne, M. M. Cernea, S. Adamo, P. M. Fearnside, G. Krieger, et al. 2011. "Preparing for Resettlement Associated with Climate Change." *Science* 334 (6055): 456–457. doi:10.1126/science.1208821.

Dong, C., X. Liu, and K. K. Klein. 2012. "Land Degradation and Population Relocation in Northern China." *Asia Pacific Viewpoint* 53 (2): 163–177. doi:10.1111/j.1467-8373.2012.01488.x.

DRCN (Development and Reform Commission of Ningxia). 2004. *Crucial Issues of, and Countermeasures for, Ecological Migration in Ningxia.* Government document provided by the Development and Reform Commission of Ningxia, Yinchuan, China.

Fan, M., Y. Li, and W. Li. 2015. "Solving One Problem by Creating a Bigger One: The Consequences of Ecological Resettlement for Grassland Restoration and Poverty Alleviation in Northwestern China." *Land Use Policy* 42: 124–130. doi:10.1016/j.landusepol.2014.07.011.

Feng, X., and J. Nie. 2013. "Migration Intention and Behavior of the Ecological Migrants among the Hui Ethnic Groups of Ningxia (in Chinese)." *Journal of Lanzhou University(Social Sciences)* 06: 53–59.

Findley, S. E. 1994. "Does Drought Increase Migration? A Study of Migration from Rural Mali During the 1983-1985 Drought." *International Migration Review* 28: 539–553.

Folke, C. 2006. "Resilience: The Emergence of a Perspective for Social-ecological Systems Analyses." *Global Environmental Change* 16 (3): 253–267. doi:10.1016/j.gloenvcha.2006.04.002.

Gallopín, G. C., S. Funtowicz, M. O'Connor, and J. Ravetz. 2001. "Science for the Twenty-first Century: From Social Contract to the Scientific Core." *International Social Science Journal* 53 (168): 219–229.

Gollwitzer, P. M., and P. Sheeran. 2006. "Implementation Intentions and Goal Achievement: A Meta-analysis of Effects and Processes." *Advances in Experimental Social Psychology* 38: 69–119. doi:10.1016/S0065-2601(06)38002-1.

Greene, W. H. 2008. *Econometric analysis pper.* Saddle River, NJ: Pearson/Prentice Hall.

HCSO (Hongsibu Census and Statistics Office). 2012. *2011 Statistical Report on Social and Economic Development in Hongsibu.* Government document provided by Hongsibu Census and Statistics Office, Hongsibu, China.

HPAO (Hongsibu Poverty-Alleviation Office). 2010. *Rural Poverty-Alleviating Plan in Hongsibu.* Government document provided by the Hongsibu Poverty-Alleviation Office, Hongsibu, China.

HRO (Hongsibu Resettlement Office). 2012. *Report on the Survey into 'Four Unclear Situations of Resettlers' in Hongsibu.* Government document No. 36. Ningxia, China.

Hugo, G. 1996. "Environmental Concerns and International Migration." *International Migration Review* 30 (1): 105–131.

Hugo, G. 2013. *Migration and Climate Change.* Cheltenham, UK: Edward Elgar Publishing.

Hummel, D., S. Adamo, A. de Sherbinin, L. Murphy, R. Aggarwal, L. Zulu, Liu Jianguo, et al. 2013. "Inter- and Transdisciplinary Approaches to Population-environment Research for Sustainability Aims: A Review and Appraisal." *Population and Environment* 34 (4): 481–509. doi:10.1007/s11111-012-0176-2.

Hummel, D., C. Hertler, S. Niemann, A. Lux, and C. Janowicz. 2008. "The Analytical Framework." In *Population Dynamics and Supply Systems: A Transdisciplinary Approach*, edited by D. Hummel, 11–69. Frankfurt, New York: Campus Verlag.

HWB (Hongsibu Water Bureau). 2010. *The 12th Five-Year Water Conservation Plan of Hongsibu District, Wuzhong Prefecture.* Government document provided by the Hongsibu Water Bureau, Hongsibu, China.

Jiao, K., and R. Wang. 2008. "Investigating the Effectiveness of Eco-migration in Ethnic Areas: A Case Study in Luanjingzitan of Inner Mongolia (in Chinese)." *Inner Mongolia Social Sciences* 5: 84–88.

Kuruppu, N., and D. Liverman. 2011. "Mental Preparation for Climate Adaptation: The Role of Cognition and Culture in Enhancing Adaptive Capacity of Water Management in Kiribati." *Global Environmental Change* 21 (2): 657–669. doi:10.1016/j.gloenvcha.2010.12.002.

Li, Y., D. Conway, Y. Wu, Q. Gao, S. Rothausen, W. Xiong, Ju Hui, et al. 2013. "Rural Livelihoods and Climate Variability in Ningxia, Northwest China." *Climatic Change* 119 (3-4): 891–904.

Li, Y., D. López-Carr, and W. Chen. 2014. "Factors Affecting Migration Intentions in Ecological Restoration Areas and their Implications for the Sustainability of Ecological Migration Policy in arid Northwest China." *Sustainability* 6 (12): 8639–8660. doi:10.3390/su6128639.

Li, S. Y., P. H. Verburg, S. H. Lv, J. L. Wu, and X. B. Li. 2012. "Spatial Analysis of the Driving Factors of Grassland Degradation Under Conditions of Climate Change and Intensive Use in Inner Mongolia, China." *Regional Environmental Change* 12: 461–474.

Liu, X. 2002. "Effects and Problems of Ecological Migration in Northwest China." *Chinese Rural Economy* 4: 47–52.

Martin, M., M. Billah, T. Siddiqui, C. Abrar, R. Black, and D. Kniveton. 2014. "Climate-related Migration in Rural Bangladesh: A Behavioural Model." *Population and Environment* 36: 85–110.

MEA (Millennium Ecosystem Assessment). 2005. *Ecosystems and Human Well-being*. Washington, DC: Island Press.

Meng, J., Y. Zhang, and P. Zhou. 2010. "Ecological Vulnerability Assessment of the Farming-pastoral Transitional Zone in Northern China: A Case Study of Ordos City." *Journal of Desert Research* 30: 840–856.

MEP (Ministry of Environmental Protection). 2008. Planning Outline of Protection of the Ecologically Vulnerable Areas in China. http://www.gov.cn/gongbao/content/2009/content_1250928.htm.

Mertz, O., C. Mbow, A. Reenberg, and A. Diouf. 2009. "Farmers' Perceptions of Climate Change and Agricultural Adaptation Strategies in Rural Sahel." *Environmental Management* 43 (5): 804–816. doi:10.1007/s00267-008-9197-0.

Mortreux, C., and J. Barnett. 2009. "Climate Change, Migration and Adaptation in Funafuti, Tuvalu." *Global Environmental Change* 19 (1): 105–112.

NDRC (National Development and Reform Commission). 2013. China's National Strategies for Adapting to Climate Change. < http://www.gov.cn/gzdt/att/att/site1/20131209/001e3741a2cc140f6a8701.pdf>.

Nee, V., and S. Su. 1998. "Institutional Foundations of Robust Economic Performance: Public Sector Industrial Growth in China." In *Industrial transformation in Eastern Europe in the light of the East Asian experience*, edited by J. Henderson, 167–187. New York: St. Martin's.

NIWRR (Ningxia Institute of Water Resources Research). 2012. *Report on Optimising Water Resource in Hongsibu District of Ningxia*. Government document provided by Ningxia Institute of Water Resources Research, Yinchuan, China.

Otrachshenko, V., and O. Popova. 2014. "Life (Dis)satisfaction and the Intention to Migrate: Evidence from Central and Eastern Europe." *The Journal of Socio-Economics* 48: 40–49.

Perch-Nielsen, S. L., M. B. Bättig, and D. Imboden. 2008. "Exploring the Link Between Climate Change and Migration." *Climatic Change* 91 (3-4): 375–393.

Rogers, S., and T. Xue. 2015. "Resettlement and Climate Change Vulnerability: Evidence from rural China." *Global Environmental Change* 35: 62–69. doi:10.1016/j.gloenvcha.2015.08.005.

SCC (State Council of China). 2001. Outline of Development-Oriented Poverty Reduction Program for Rural China (2001-2010). http://news.xinhuanet.com/zhengfu/2005-07/19/content_3239424.htm.

SCC (State Council of China). 2011. New Progress in Development-Oriented Poverty Reduction Program for Rural China. http://usa.chinadaily.com.cn/china/2011-11/16/content_14106364.htm.

SCLGOPAD (State Council Leading Group Office of Poverty Alleviation and Development). 1994. *80/7 Plan to Combat Poverty (1994-2000)*. Government document No. 30. Beijing, China.

Sheeran, P., T. L. Webb, and P. M. Gollwitzer. 2005. "The Interplay Between Goal Intentions and Implementation Intentions." *Personality and Social Psychology Bulletin* 31 (1): 87–98. doi:10.1177/0146167204271308.

Song, Y., C. Li, L. Jiang, and L. Lu. 2015. "Ecological Indicators for Immigrant Relocation Areas: A Case in Luanjingtan, Alxa, Inner Mongolia." *International Journal of Sustainable Development & World Ecology* 22 (5): 445–451.

Sun, D., X. Yu, X. Liu, and B. Li. 2016. "A New Artificial Oasis Landscape Dynamics in Semi-arid Hongsipu Region with Decadal Agricultural Irrigation Development in Ning Xia, China." *Earth Science Informatics* 9 (1): 21–33. doi:10.1007/s12145-015-0228-0.

Tang, H., X. Zhang, and D. Yang. 2011. "Ecological Migration Willingness and its Affecting Factors: A Case of Sangonghe River Watershed, Xinjiang (in Chinese)." *Journal of Natural Resources* 10: 1658–1669.

United Nations. 2014. International Decade for Action 'Water for Life' 2005-2015. Accessed June 1, 2016. http://www.un.org/waterforlifedecade/scarcity.shtml.

Van Dalen, H. P., G. Groenewold, and T. Fokkema. 2005. "The Effect of Remittances on Emigration Intentions in Egypt, Morocco, and Turkey." *Population Studies* 59 (3): 375–392. doi:10.1080/00324720500249448.

Walker, B., C. S. Holling, S. R. Carpenter, and A. Kinzig. 2004. "Resilience, Adaptability and Transformability in Social-ecological Systems." *Ecology and Society* 9 (2): 5.

Wang, X., F. Chen, and Z. Dong. 2006. "The Relative Role of Climatic and Human Factors in Desertification in Semiarid China." *Global Environmental Change* 16 (1): 48–57.

Wrathall, D. J. 2012. "Migration Amidst Social-Ecological Regime Shift: The Search for Stability in Garífuna Villages of Northern Honduras." *Human Ecology* 40 (4): 583–596. doi:10.1007/s10745-012-9501-8.

Yang, X. 2000. "Determinants of Migration Intentions in Hubei Province, China: Individual versus Family Migration." *Environment and Planning A* 32 (5): 769–787.

Zheng, Y. 2013. "Environmental Migration: Concepts, Theoretical Ground and Policy Implications (in Chinese)." *China Population, Resources and Environment* 23 (4): 96–103.

OBITUARY

Graeme John Hugo, 1946–2015

Graeme Hugo, Professor of Geography at the University of Adelaide, died in Adelaide on 20 January 2015, barely a few months after being diagnosed with cancer. He was only 68 years into a valuable and productive life that substantially influenced Australian geography, demography and wider society. Graeme was one of the most distinguished and dedicated geographers that Australia has produced and he was certainly the most productive. His main research areas covered a smorgasbord of activities centred on migration, its changing patterns and causes and the implications for social and economic change, especially in Asia and Australia.

Graeme grew up in the western Adelaide suburb of Findon where he went to Flinders Park Primary School and then Findon High School. Findon was an unpretentious suburb and Graeme is quite probably the only person from Findon High School to have ever become a professor. He was the first person from his family to go to university. His fascination with geography was much influenced by his high school geography teacher, Vic Mashford, and when he received the John Lewis medal for topping the State in geography in 1963 his future direction was becoming clear. By then he was already appreciating the scope of geography as a subject with a range that could encompass the subtleties and intrigue of social geography with the 'science' of census data and statistics.

Graeme began academic life with a BA at the University of Adelaide, where he majored in geography and history, and concurrently attended Adelaide Teachers' College, training as a secondary school teacher. In 1967 he completed his Honours year with a thesis on 'Service provision in a sector of western Metropolitan Adelaide'. Three months of backpacking in Southeast Asia (officially a study-tour) ensued and ensured that another direction in life, away from western Adelaide, would follow. In 1968 he began his academic career as a Tutor at Flinders University, while completing an MA thesis there on 'Internal migration in South Australia' (1971). That was followed by a PhD at ANU, after which he returned permanently to Adelaide, where he eventually became the ARC Australian Professorial Fellow in the Discipline of Geography, Environment and Population, and Director of the Australian Population and Migration Research Centre at the University of Adelaide.

That might seem to be a trajectory within a localised small-scale geographical world, but nothing could be further from the truth. Surely no Australian geographer has been more mobile than Graeme, and that is entirely in keeping with an academic lifetime spent studying migration, mobility and development in Australia and Southeast Asia. Indeed, Graeme long took that mobility into personal realms, as his forays into distance running demonstrated, and in his earlier days he had himself been a 'boat-person' directly experiencing the illegal movement of Indonesians to neighbouring Malaysia.

At ANU Graeme temporarily moved away from geography into the Department of Demography, where his thesis was on 'Population mobility in West Java' (1975), supervised primarily by Jack Caldwell. At that time ANU had developed a strong focus on the demography and economics of Indonesia. That ultimately gave him a breadth of interest and purpose that stemmed from blending the disciplines of geography, demography and other social sciences, and from developing a theoretical and practical understanding of the challenges of development in rural and urban Indonesia. Thirty years later he was to return as part of a review of the department. Ever meticulous, Graeme carefully recorded how, from Honours to MA and PhD, the number of thesis pages had gone from 108 to 287 and then 699.

After the thesis Graeme returned to Flinders in 1975 as a Lecturer, subsequently rising to Reader in Geography. In 1991 he crossed the city to rejoin the University of Adelaide, as Professor and a year later as Head of the Department. Between 2002 and 2007 he held an ARC Federation Fellowship. Flinders did not forget him; in 2006 he was an inaugural recipient of the University's Distinguished Alumnus Award for his already exceptional contribution to research, teaching and public service.

It is simply impossible to do justice to the full extent of Graeme's extraordinary academic life—and digest a CV that extends to over 100 single-spaced pages. Graeme produced 32 books (and more are in press), more than 200 refereed articles (and more of those too are on the way) and over 260 book chapters. Then there are the three theses, 43 monographs and 89 working papers. He frequently participated in conferences, both academic and with such agencies as UNFPA, the World Bank and the International Organisation of Migration. There were therefore a phenomenal 1066 conference papers (some 25 per year between 1975 and 2013). In his last year, until early December 2014, he had given 62 conference presentations in such places as Ottawa, Maastricht, Bogotá, Nairobi, Washington and, of course, Adelaide. And that total excludes the 20 plenary addresses, 120 assorted reports and 34 book reviews. I wrote in 2014 that 'it is impossible to imagine the flow suddenly drying up, when Graeme has so many insights still to impart, and so many requests for him to impart them'. Sadly that is now no longer true. Some of the key publications are listed below.

Quantity never erased quality. That was demonstrably apparent in the citation counts that Graeme accumulated. He is the most cited geographer in Australia. His work with Douglas Massey and others on international migration and migration theory—a seminal book and article—have received 2250 and 3250 citations respectively, and the numbers grow daily. Numerous other publications have passed the century mark. More to the point, because of Graeme's work we now have a much more sophisticated understanding of the theory and practice of migration in the Asia-Pacific region, and Australian geography is itself more cosmopolitan in outlook. He was not just an expert on population migration in the Asia-Pacific region: he was *the* expert. Beyond that his expertise in fertility and mortality and more arcane aspects of demography was formidable. He was revered in Australian and Asian population studies. It has long been impossible to go to a meeting in Asia on a related issue without being confronted by someone disappointed that Graeme was not there (in itself unusual!) and then remarking on what a great scholar he was and how his inputs and insights were invaluable. He was highly respected in Asia and far beyond.

That respect came from his intellectual contributions, but also from beyond them, because they were underpinned by a quiet but persuasive concern for social justice. Beyond the multiple intellectual outputs that grace some of the finest global journals, and have been cited many times, Graeme was an activist, concerned with the development of equitable migration policies and the welfare and rights of migrants and refugees.

He was no mere scholar—quietly piling up papers and books and attending obscure conferences—but was committed to developing appropriate population and development policies that linked in different ways the destinies of Australia and Asian nations, and might contribute to building positive relationships between them. One of his last papers, only just published, was entitled 'The economic contribution of humanitarian settlers' and focused on the implications of migration into Australia and the significance, role and contribution of refugees in a multicultural Australia. That paper is a condensed version of a much larger report for the former Department of Immigration and Citizenship (DIAC). Typically many of Graeme's most influential papers followed a substantial piece of work commissioned by a government policy agency.

He could work as easily in a small Australian country town as speaking Indonesian in the villages and cities of Indonesia. A holism of spirit and purpose meant that he was as comfortable working in small societies in remote parts of Indonesia as addressing elite international United Nations conferences—they were all part of a whole. It was so typical of Graeme that the first song at his memorial service was 'Imagine' ('Imagine all the people living life in peace ... No need for greed or hunger, a brotherhood of man, imagine all the people sharing all the world'). But Graeme did not merely dream and imagine, he actively sought to make the world a better place.

That quiet and so often self-effacing activism was evident in Graeme's involvement with an enormous number of bodies in population studies and geography, both academic (such as the IUSSP and APA) and professional/practical (such as the Ministerial Advisory Board for the Ageing, of the South Australian government, and the Irregular Migration Advisory Group of the Department of

Immigration and Border Protection). At the time of his death he had served for 8 years on the Australian Statistical Advisory Council. He was a regular television and radio commentator. He gave very valuable service to geography through being Chair of the ARC's Expert Advisory Committee on the Social, Behavioural and Economic Sciences between 2000 and 2004, and in a host of other ways. He was a member of International Geographical Union's Committee on Famine and Food Crisis Management, 1988–92. He has reviewed Geography Departments (and related organisations), been seconded to others in Asia and America, and at different times worked for such UN organisations as ESCAP, UNFPA and ILO, and in Indonesia for the Ministry of Population and Environment and the Central Planning Agency.

It was no surprise, then, that the Governor of South Australia, himself a refugee from Vietnam, expressed his apologies for being unable to attend the memorial service, or that the Premier of South Australia, Jay Weatherill, whose constituency embraces Findon, remarked 'This is an enormous loss to South Australia and the nation. Mr Hugo was an international thinker of the highest calibre and was greatly respected. I regarded him as a friend and I am deeply saddened at his passing.' At the same time the NTEU observed that Graeme

> epitomised the tradition of serving the public good. His analysis of the changing face of higher education, for instance, in terms of casualisation, the lost generation of permanent academics, the increasing mobility of academics and the failure of governments to fund adequately postgraduate training were all path breaking and in turn shaped the public debate in Australia and overseas. If there is a word that encapsulates Professor Hugo it is generosity—generosity of spirit, shared knowledge, public duty and collegiality. He will be greatly missed.

The Parliamentary Secretary to the Minister for Social Services stated:

> The highly respected geographer and academic contributed enormously to the body of evidence and analysis of migrant settlement and outcomes in Australia. In recent years he worked on ground-breaking research into the economic, social and civic contributions of first and second generation humanitarian entrants. He contributed to the understanding of the experiences of youth of culturally and linguistically diverse (CALD) backgrounds in Australia and I was privileged to launch his landmark research in this area CALD Youth Census Report 2014 at the Multi-cultural Youth Education and Development Centre last August in Adelaide. His particular area of interest was humanitarian migration to Australia and the economic contribution of those Australians. I am particularly grateful for his excellent research in *A Significant Contribution: The Economic, Social and Civic Contributions of First and Second Generation Humanitarian Entrants*, 2011.

From across a range of organisations and political divides he was acknowledged, respected and appreciated.

Beyond more formal activities, Graeme was renowned for his willingness to pitch in, when others of similar rank would decline, for example in refereeing papers for

Australian (and other) geography journals (he had refereed for no fewer than 53 different journals), for not just teaching first years but marking their exam papers and assignments, for refereeing grant applications and so on. He was a diligent, conscientious and committed teacher and inspired more than one generation of students, many of whom themselves went on to positions of eminence. Graeme supervised 32 Honours theses, more than 50 Masters theses and 62 PhD theses—some 20 Masters and PhDs are ongoing. He still found time to be a valuable and judicious marker for 30 Masters and PhD theses from elsewhere. All of this was done quietly—fitted in dutifully and carefully between the more stellar activities. One of my PhD students was probably the last whose thesis he examined late in 2014, when he was already ill—and should have had better, more personal things to do—but he had promised. The thesis was passed with his customary generosity of spirit—while providing invaluable comments and thoughtful directions. Duty was diligently done.

It is impossible to think of an Australian geographer who is better known in the real world beyond the ivory towers, and who has made such a contribution to it. Indeed he is one of the very few geographers to have been formally recognised outside the discipline. In 2012 he became an Officer of the Order of Australia (AO) 'for distinguished service to population research, particularly the study of international migration, population geography and mobility, and through leadership roles with national and international organisations'. Belatedly in 2014 the Institute of Australian Geographers awarded him the Australia-International Medal in recognition of his outstanding contribution, as an Australian geographer, to the advancement of geography worldwide. He has been an exemplary and thoroughly dedicated geographical citizen of Australia and the Asian world and his entire academic and practical career has been dedicated to making, and succeeding in making, outstanding intellectual and practical contributions to the advancement of geography.

Through all those years and those massive contributions Graeme was decidedly part of the real world outside the towers. He was a Port Adelaide Australian rules football tragic (long before that adjective became a noun, and long before Port Adelaide reached the national league) usually to be found carefully and dutifully recording the goals and behinds as they mounted up. The demographer was always there. Beyond that he was a cricket fanatic, with one of his last public outings being to the First Test between Australia and India at Adelaide early in December 2014. No one would have been more delighted at Australia's victory. And no one would have more enjoyed disputing whether India should have adopted the DRS review system, rather than poring over life tables. Fittingly his memorial service was held at Adelaide Oval where the scoreboard provided a vivid tribute to a remarkable and much loved man whose rich and full life enriched so many. Graeme leaves a partner, Sharon, two step-daughters, Melissa and Emily, and a daughter, Justine, all of whom emphasised Graeme's concern for a just society and generosity and warmth of spirit, at a memorial service attended by more than 500 people.

Selected books and other publications

(1975) 'Postwar settlement of Southern Europeans in Australian rural areas: the case of Renmark', *Australian Geographical Studies* 13, pp. 169–81.

(1977) 'Circular migration', *Bulletin of Indonesian Economic Studies* 13, pp. 57–66.

(1978) *Population mobility in West Java*, Gadjah Mada University Press, Yogyakarta.

(1980) (with MENZIES, B.J.) 'Greek immigrants in the South Australian Upper Murray', in Burnley, I.H., Pryor, R.J. & Rowland, D.T. (eds) *Mobility and community change in Australia*, University of Queensland Press, Brisbane, pp. 170–92.

(1981) 'Community and village ties, village norms and village networks in migration decision making', in DeJong, G.F. & Gardner, R.W. (eds) *Migration decision making: multidisciplinary approaches to microlevel studies in developed and developing countries*, Pergamon, New York, pp. 186–224.

(1984) (with CURREY, B.) (eds) *Famine as a geographical phenomenon*, Reidel, Dordrecht.

(1984) (with RUDD, D.M. & DOWNIE, M.C.) 'Adelaide's aged population: changing spatial patterns and their policy implications', *Urban Policy and Research* 2, pp. 17–25.

(1985) 'Circulation in West Java, Indonesia', in Prothero, R.M. & Chapman, M. (eds) *Circulation in Third World countries*, Routledge & Kegan Paul, London, pp. 75–99.

(1985) 'Structural change and labour mobility in rural Java', in Standing, G. (ed.) *Labour circulation and the labour process*, Croom Helm, London, pp. 46–88.

(1985) (with SMAILES, P.J.) 'Urban–rural migration in Australia: process view of the turnaround', *Journal of Rural Studies* 1, pp. 11–30.

(1986) *Australia's changing population: trends and implications*, Oxford University Press, Melbourne.

(1987) 'Demographic and welfare implications of urbanization: direct and indirect effects on sending and receiving areas', in Fuchs, R.J., Jones, G.W. & Pernia, E. (eds) *Urbanization and urban policies in Pacific Asia*, Westview Press, Boulder, pp. 136–65.

(1987) 'Postwar refugee migration in Southeast Asia: patterns, problems and policy consequences', in Rogge, J.R. (ed.) *Refugees: a Third World dilemma*, Rowman & Littlefield, Totowa, NJ, pp. 237–53.

(1987) (with JONES, G.W., HULL, T.H. & HULL, V.J.) *The demographic dimension in Indonesian development*, Oxford University Press, Kuala Lumpur.

(1989) *Atlas of the Australian people* (Volume V): *South Australia*, Australian Government Publishing Service, Canberra.

(1989) 'Australia: the spatial concentration of the turnaround', in Champion, A.G. (ed.) *Counterurbanization: the changing pace and nature of population deconcentration*, Edward Arnold, London, pp. 62–82.

(1989) (with BEED, T.W., Stimson, R.J. & PARIS, C.T.) *Stability and change in Australian housing*, Canberra Series in Administrative Studies No. 11, University of Canberra, Canberra.

(1990) (with WOODEN, M., HOLTON, R. & SLOAN, J.) *Australian immigration: a survey of the issues*, Australian Government Publishing Service, Canberra.

(1991) (with BOHLE, H.G., CANNON, T. & IBRAHIM, F.N.) (eds) *Famine and food security in Africa and Asia: Indigenous response and external intervention to avoid hunger*, Bayreuther Geowissenschaftliche Arbeiten, Bayreuth.

(1992) 'Women on the move: changing patterns of population movement of women in Indonesia', in Chant, S. (ed.) *Gender and migration in developing countries*, Belhaven Press, London, pp. 174–96.

(1992) (with MAUDE, A.) 'Mining settlements in Australia', in Neil, C., Tykkylainen, M. & Bradbury, J. (eds) *Coping with closure*, Routledge, London, pp. 66–94.

(1993) (with MASSEY, D.S., ARANGO, J., KOUAOUCI, A., PELLEGRINO, A. & TAYLOR, J.E.) 'Theories of international migration: an integration and appraisal', *Population and Development Review* 19, pp. 431–66.

(1994) *The economic implications of emigration from Australia*, Australian Government Publishing Service, Canberra.

(1994) (with WOODEN, M., HOLTON, R. & SLOAN, J.) *Australian immigration: a survey of the issues* (2nd edition), Australian Government Publishing Service, Canberra.

(1995) (with MAHER, C.) *Atlas of the Australian people 1991, national overview*, Australian Government Publishing Service, Canberra.

(1996) 'Brain drain and student movements', in Lloyd, P.J. & Williams, L.S. (eds) *International trade and migration in the APEC region*, Oxford University Press, Melbourne, pp. 210–28.

(1996) 'Environmental concerns and international migration', *International Migration Review* 30, pp. 105–31.

(1996) (with HASSELL, D.) *Immigrants and public housing*, Australian Government Publishing Service, Canberra.

(1997) 'Intergenerational wealth flows and the elderly in Indonesia', in Jones, G.W., Douglas, R.M., Caldwell, J.C. & D'Souza, R.M. (eds) *The continuing demographic transition*, Clarendon Press, Oxford, pp. 111–34.

(1997) (with OBERAI, A.S., ZLOTNIK, H. & BILSBORROW, R.) *International migration statistics: guidelines for improving data collection systems*, International Labour Office, Geneva.

(1998) (with MASSEY, D.S., ARANGO, J., KOUAOUCI, A., PELLEGRINO, A. & TAYLOR, J.E.) *Worlds in motion: understanding international migration at the end of the millennium*, Clarendon Press, Oxford.

(1999) (with HABERKORN, G., FISHER, M. & AYLWARD, R.) *Country matters: social atlas of rural and regional Australia*, Bureau of Rural Sciences, Canberra.

(2000) 'The impact of the crisis on internal population movement in Indonesia', *Bulletin of Indonesian Economic Studies* 36, pp. 115–38.

(2002) 'Effects of international migration on the family in Indonesia', *Asian and Pacific Migration Journal* 11, pp. 13–46.

(2003) 'Information, exploitation and empowerment: the case of Indonesian contract workers overseas', *Asian and Pacific Migration Journal* 12, pp. 439–66.

(2003) (with CHAMPION, T.) (eds) *New forms of urbanisation: beyond the urban–rural dichotomy*, Ashgate, Aldershot.

(2003) (with TAN, Y. & POTTER, L.) 'Government-organised distant resettlement and the Three Gorges Project, China', *Asia-Pacific Population Journal* 18, pp. 5–26.

(2005) 'Implications of demographic change for future housing demand in Australia', *Australian Planner* 42, pp. 33–41.

(2005) 'The new international migration in Asia: challenges for population research', *Asian Population Studies* 1, pp. 93–120.

(2005) 'Population movement in Indonesia: implications for the potential spread of HIV/AIDS', in Jatrana, S., Toyota, M. & Yeoh, B.S.A. (eds) *Migration and health in Asia*, Routledge, London, pp. 17–40.

(2005) (with MASSEY, D.S., ARANGO, J., KOUAOUCI, A., PELLEGRINO, A. & TAYLOR, J.E.) *Worlds in motion: understanding international migration at the end of the millennium*, Clarendon Press, Oxford.

(2005) (with TAN, Y. & BRYAN, B.) 'Development, land-use change and rural resettlement capacity: a case study of the Three Gorges Project, China', *Australian Geographer* 36, pp. 201–20.

(2006) 'Population geography', *Progress in Human Geography* 30, pp. 513–23.

(2006) 'Temporary migration and the labour market in Australia', *Australian Geographer* 37, pp. 211–31.

(2006) 'Women, work and international migration in Southeast Asia', in Kaur, A. & Metcalfe, I. (eds) *Mobility, labour migration and border controls in Asia*, Palgrave Macmillan, Basingstoke, pp. 73–113.

(2006) (with KHOO, S. & MCDONALD, P.) 'Attracting skilled migrants to regional areas: what does it take?', *People and Place* 14, pp. 26–47.

(2006) (with LIENERT, J.) 'Child obesity in South Australia: some initial findings', *Food, Culture and Society* 9, pp. 299–316.

(2008) 'In and out of Australia: rethinking Indian and Chinese skilled migration to Australia', *Asian Population Studies* 3, pp. 267–91.

(2008) (with YOUNG, S.) (eds) *Labour mobility in the Asia-Pacific region*, Institute of Southeast Asian Studies, Singapore.

(2009) 'Care worker migration, Australia and development', *Population, Space and Place* 15, pp. 189–203.

(2009) 'Migration, Labour Mobility and Poverty', in Stoler, A., Redden, J. & Jackson, L. (eds) *Migration, labour mobility and poverty reduction in the Asia-Pacific region*, Cambridge University Press, Cambridge, pp. 465–512.

(2010) 'Climate change induced mobility and the existing migration regime in Asia and the Pacific', in McAdam, J. (ed) *Climate change and displacement: multidisciplinary perspectives*, Hart, Oxford, pp. 9–36.

(2011) 'Geography and population in Australia: a historical perspective', *Geographical Research* 49, pp. 242–60.

(2012) 'Migration and development in low income countries: a role for destination country policy?', *Migration and Development* 1, pp. 24–49.

(2012) (with PINCUS, J.) (eds) *A Greater Australia: population, policies and governance*, Committee for Economic Development of Australia, Melbourne.

(2013) (ed.) *Migration and climate change*, Edward Elgar, Cheltenham.

(2014) 'The economic contribution of humanitarian settlers in Australia', *International Migration* 52, pp. 31–52.

(2014) 'Immigrant settlement in regional Australia: patterns and processes', in Dufty-Jones, R. & Connell, J. (eds) *Rural change in Australia: population, economy, environment*, Ashgate, Farnham, pp. 57–82.

JOHN CONNELL

Index

T - #0180 - 111024 - C328 - 246/174/15 - PB - 9780367891947 - Gloss Lamination